ACS SYMPOSIUM SERIES **395**

Multiphase Polymers: Blends and Ionomers

L. A. Utracki, EDITOR
National Research Council of Canada

R. A. Weiss, EDITOR
University of Connecticut

Developed from a symposium sponsored
by the Division of Polymeric Materials: Science
and Engineering of the American Chemical Society
and by the Division of Macromolecular Science
and Engineering of the Chemical Institute of Canada
at the Third Chemical Congress of North America
(195th National Meeting of the American Chemical Society),
Toronto, Ontario, Canada,
June 5–11, 1988

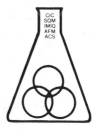

American Chemical Society, Washington, DC 1989

Library of Congress Cataloging-in-Publication Data

Multiphase polymers: blends and ionomers/ L. A. Utracki, editor;
R. A. Weiss, editor.
p.   cm.—(ACS Symposium Series, ISSN 0097–6156; 395).

"Developed from a symposium sponsored by the
Division of Polymeric Materials: Science and Engineering
of the American Chemical Society and by the Division of
Macromolecular Science and Engineering of the Chemical
Institute of Canada at the Third Chemical Congress of
North America (195th Meeting of the American Chemical
Society), Toronto, Ontario, Canada, June 5–11, 1988."

Bibliography: p.

Includes indexes.

ISBN 0–8412–1629–0
1. Polymers—Congresses. 2. Ionomers—Congresses.

I. Utracki, L. A., 1931– . II. Weiss, R. A., 1950– .
III. American Chemical Society. Division of Polymeric
Materials: Science and Engineering. IV. Chemical Institute
of Canada. Macromolecular Science Division.
V. American Chemical Society. Meeting (195th: 1988:
Toronto, Ont.). VI. Chemical Congress of North America
(3rd: 1988: Toronto, Ont.). VII. Series.

QD380.M85   1989
547.7—dc20   89–6987
                                            CIP

Foreword

The ACS SYMPOSIUM SERIES was founded in 1974 to provide a medium for publishing symposia quickly in book form. The format of the Series parallels that of the continuing ADVANCES IN CHEMISTRY SERIES except that, in order to save time, the papers are not typeset but are reproduced as they are submitted by the authors in camera-ready form. Papers are reviewed under the supervision of the Editors with the assistance of the Series Advisory Board and are selected to maintain the integrity of the symposia; however, verbatim reproductions of previously published papers are not accepted. Both reviews and reports of research are acceptable, because symposia may embrace both types of presentation.

Contents

Ionomers

NEW MATERIALS

MORPHOLOGY

SOLUTIONS AND PLASTICIZED SYSTEMS

INDEXES

Preface

THE APPEARANCE OF MANY NEW PRODUCTS AND APPLICATIONS based on multiphase polymer systems has spawned considerable research activity in recent years on this class of materials in both academic and industrial laboratories. The properties of multiphase polymers, such as their phase behavior, morphology, and mechanical behavior, are complicated by the problems associated with achieving equilibrium in highly viscous media. These problems are exacerbated by the poorly understood relationships between thermal and stress histories and the attainment of multiphase morphologies in transient stress, non-isothermal processing operations. The solution of these problems involves a multidisciplinary effort involving chemistry, physics, and engineering. The development of the underlying science remains in its infancy, despite major advances in theory, materials, and experimental instrumentation over the last decade. Yet, the development and commercialization of multiphase polymers has proceeded at a rapid pace.

The field of multiphase polymers is too broad for any single volume. Two of the more important topics within the field from the perspectives of both applications and scientific challenges are polymer blends and ionomers. The high level of interest in these areas is evidenced by the explosive growth of the literature and patents devoted to these subjects. With this in mind, we felt that a book devoted to recent advances in these fields was justified.

The chapters in this volume represent the current trends in the fields of polymer blends and ionomers, including materials development, characterization, theory, and processing. They are grouped into six sections: the first three are concerned with polymer blends and interpenetrating networks and the latter three with ionomers.

Although immiscible polymer blends and ionomers share a common feature in that both exhibit more than a single phase, a major difference between the two systems involves the dispersed phase size. For blends, this is generally of the order of micrometers and may be detected optically. Ionomers, however, are microphase-separated with domain sizes of the order of nanometers. Thus, blends and ionomers represent two extremes of the subject of multiphase polymers. In this book, the reader will observe similarities as well as differences in the problems associated with these materials and the approaches used to study them. In addition, we expect that the reader will find many parallels with multiphase systems that are not discussed here, such as block copolymers and liquid crystalline polymers.

Although this book provides only a small sampling of the kinds of materials and the activities in the field of multiphase polymers, the range of topics covered here clearly reflects the breadth of the field. For example, the subjects discussed include synthetic chemistry, theory, solution behavior, morphological characterization, and rheology. We expect that this material will be stimulating to academic and industrial scientists alike, whether their primary interests are in fundamental science or the development of the next generation of commercial polymer systems.

Finally, the editors thank the authors for their diligence and cooperation in ensuring timely publication of this volume. We especially thank Younghee Chudy for her invaluable assistance to the editors in the preparation of this book.

L. A. UTRACKI
Industrial Materials Research Institute
National Research Council of Canada
Boucherville, Quebec J4B 6Y4, Canada

R. A. WEISS
Polymer Science Program and
Department of Chemical Engineering
University of Connecticut
Storrs, CT 06260–3136

March 8, 1989

Chapter 1

Polymer Alloys, Blends, and Ionomers

An Overview

L. A. Utracki[1], D. J. Walsh[2], and R. A. Weiss[3]

[1]**Industrial Materials Research Institute, National Research Council of Canada, Boucherville, Quebec J4B 6Y4, Canada**
[2]**E. I. du Pont de Nemours and Company, Experimental Station, Wilmington, DE 19898**
[3]**Polymer Science Program and Department of Chemical Engineering, University of Connecticut, Storrs, CT 06269–3136**

This chapter provides a broad overview of the subjects of polymer blends and ionomers. Specific topics concerning polymer blends include the thermodynamics of mixing of polymer-polymer pairs, polymer interfaces, rheology, and mechanical properties. For ionomers, the chemistry, structure, rheology and solution properties are discussed.

Multiphase polymer systems are becoming an increasingly important technical area of polymer science. By definition, a multiphase polymer is one that has two or more distinct phases. The phases may differ in chemical composition and/or texture. Thus, in its broadest sense, the term includes not only multi-component systems, such as immiscible polymer blends and filled-polymers, but also semi-crystalline polymers, block copolymers, segmented polymers, and ionomers. The latter four systems are characterized by a microphase-separated morphology wherein a single polymer chain participates in more than one phase. In addition, even homopolymers that have experienced complex thermal and mechanical histories, such as encountered in most common polymer processing operations, may possess morphologies containing more than one crystalline texture. These may also be considered multiphase materials.

Because of the great diversity of multiphase polymers, coverage of the entire field in a single volume is neither possible nor practical. Instead, this book concentrates on two specific subjects: polymer blends, including interpenetrating polymer networks, and ionomers. Even with this specialization, a comprehensive treatise on both subjects is not possible, and this book focusses on selected contemporary topics from the two fields.

The purpose of this overview chapter is to provide a cursory

0097–6156/89/0395–0001$09.75/0
© 1989 American Chemical Society

introduction to these subjects and to outline the organization of the book. Those requiring a more detailed review of polymer blends and ionomers are directed to other monographs and review articles (1-22).

POLYMER BLENDS

There is some confusion in the literature regarding polymer blend nomenclature. Here the following definitions are assigned to the commonly used terms:

> POLYMER BLEND (PB) - the all-encompassing term for any mixture of homopolymers or copolymers;

> HOMOLOGOUS POLYMER BLENDS - a sub-class of PB limited to mixtures of chemically identical polymers differing in molar mass;

> POLYMER ALLOYS - a sub-class of PB reserved for polymeric mixtures with stabilized morphologies;

> MISCIBLE POLYMER BLENDS - a sub-class of PB encompassing those blends which exhibit single phase behavior;

> IMMISCIBLE POLYMER BLENDS - A sub-class of PB referring to those blends that exhibit two or more phases at all compositions and temperatures;

> PARTIALLY MISCIBLE POLYMER BLENDS - a sub-class of PB including those blends that exhibit a "window" of miscibility, i.e., are miscible only at some concentrations and temperatures;

> COMPATIBLE POLYMER BLENDS - a utilitarian term, indicating commercially useful materials, a mixture of polymers without strong repulsive forces that is homogeneous to the eye;

> INTERPENETRATING POLYMER NETWORK (IPN) - a sub-class of PB reserved for mixtures of two polymers where both components form continuous phases and at least one is synthesized or crosslinked in the presence of the other.

From the standpoint of commercial applications and developments, polymer blending represents one of the fastest growing segments of polymer technology. Both the open and the patent literature have become voluminous. In principle, blending two materials together in order to achieve a balance of properties not obtainable with a single one is an obvious and well-founded practice, one that has been successfully exploited in metallurgical science. With polymers, however, the thermodynamics of mixing do not usually favor mutual solubility and most binary polymer mixtures form two distinct phases. This is a direct consequence of their high molecular mass. Still, many immiscible systems form useful products and are commercial. Key examples include rubber-toughened plastics such as high impact polystyrene (HIPS) and ABS resins and blends of synthetic rubber with natural rubber.

The problems and challenges inherent to developing useful materials with optimal morphologies and properties from an immiscible or partially miscible polymer blend are not trivial and have spawned considerable industrial and academic research. Work on polymer miscibility, compatibilizing agents, reactive systems, and the influence of flow on the structure and properties of blends is described in later chapters.

The major technological problem in the use of polymer blends concerns determining correlations between composition, processing, structure and properties. Each variable has inherent characterization problems, e.g., of the preparation process, of the chemistry and morphology, and of what are meaningful properties. None of these correlations or characterizations are easy to make or particularly well understood.

Because polymer science is by nature interdisciplinary, the solution of the above problem involves contributions from many fields, including chemistry, physics, and engineering. The discussion that follows will highlight a number of areas where progress has recently been made in understanding the subject. Considerably more detail will be found in the subsequent chapters of this book.

Mechanical Mixing of Polymer Blends

Most commercial polymer alloys and blends are prepared by mechanical mixing, largely because of its simplicity and low cost. The preferred industrial method of mechanical mixing is to use a screw compounder or extruder that can be run continuously and generate a product in a convenient form for further processing. Not surprisingly, much effort has gone into trying to understand the flow of polymer blends.

Mixing from Ternary Systems and by Reaction

Other methods for forming blends such as by evaporation of a solvent or by polymerization of a monomer in the presence of a polymer involve at least three components in the preparation process. Mixing in a common solvent followed by its removal is a convenient way of making blends on a laboratory scale, but has obvious commercial disadvantages due to the cost and difficulty of solvent recovery as well as the potential environmental hazards associated with handling large volumes of often toxic chemicals. In specific applications, however, such as membrane formation or paints and coatings where thin films are required, the use of solvents is unavoidable.

The third component of such a blend, i.e., the solvent, and the kinetics of its removal can influence the resulting morphology. For example, if two miscible polymers are cast from a common solvent, one does not necessarily obtain a homogeneous mixture. A two-phase region can exist in the ternary phase diagram as shown in Fig. 1a, and as the solvent evaporates the composition may enter the two- phase region as shown by progressing from point A to point

B. As the evaporation of solvent continues, the composition may leave the two-phase region, but at that point the viscosity may be too high and the phase sizes may be too large for homogenization to occur.

The more common situation, illustrated in Fig. 1b, is where the two polymers are immiscible but form a homogenous solution in a common solvent. In this case, film casting along the line C to D generates a variety of structures depending on the selected solvent (and its interaction parameters χ_{12} and χ_{13}), the chemical nature of the two polymers (χ_{23}) as well as on the kinetics of the process. Three phase-separated types of morphologies can result: co-continuous, dispersed, and layered.

The co-continuous morphology with the polymers forming interpenetrating networks is the most interesting. This structure, which is known to exist even at concentrations as low as 10 to 15 vol%, can be created by judiciously selecting the casting conditions to assure dominance of the spinodal decomposition (SD) mechanism of phase separation. The correlation lengths of the generated structures vary with time from a few nanometers to about a micron (23, 24). This morphology allows for coexistence of the best characteristics of each polymer in the blend (25). For example, the combination of good mechanical properties with permeability, accomplished with a blend composition above the percolation threshold, has yielded a highly successful membrane technology (26).

The phase-separated droplet/matrix morphology is an outcome of the nucleation and growth mechanism (NG) of phase separation. The phase dimensions are similar to those observed for SD, but in this case the properties are dominated by the matrix polymer with the dispersed phase playing the role of a compatibilized filler. A similar dispersed morphology, but with large drops, can be obtained by allowing the SD or NG system to ripen. The coarsening usually leads to a non-uniformity of properties.

The layered structure of a cast film is controlled by the surface properties during evaporation. Significant compositional gradients can be generated by making use of the natural tendencies of one polymer to migrate toward the air-polymer interface and the other toward the substrate. Hydrophobicity/hydrophilicity of macromolecules is often cited as the driving force (27, 28).

Reactive mixing finds application in many commercial blends such as HIPS and rubber modified thermosets. Many IPN's can also be included here. In the case of the polymerization of monomer in the presence of a polymer, the monomer-1/polymer-1/polymer-2 ternary phase diagram also plays a role in determining the final morphology. Where the two polymers are immiscible, such as polystyrene and polybutadiene, a two-phase mixture will result. However, in cases where the polymers are miscible, single phase morphologies are not always achieved. For example, in the polymerization of vinyl chloride in the presence of poly(butyl acrylate) a two-phase region is present in the phase diagram, Fig. 2. Polymerization pathways that pass through this region, such as line A-B in Fig. 2, may yield a two-phase system for the same reasons as described above for solvent evaporation from a blend.

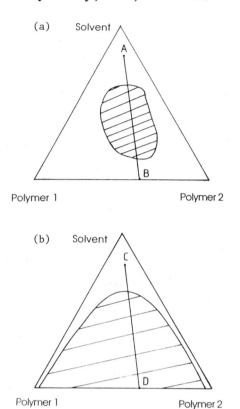

Figure 1. Schematic plots of polymer/polymer/solvent phase diagrams for (a) two miscible polymers plus a solvent inducing phase separation, (b) two immiscible polymers. The lines AB and CD show evaporation pathways.

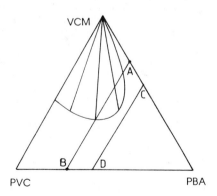

Figure 2. The phase diagram for vinyl chloride/PVC/poly (butyl acrylate) mixtures showing polymerization pathways AB and CD.

By avoiding the two-phase region of the phase diagram, such as choosing pathway C-D, one ensures a homogeneous mixture (29).

Many commerical blends are prepared by reactive processing. The technology is quite diverse, ranging from *in-situ* polymerization of one monomer to co-reaction of two polymers. In the latter case one may use the natural reactivity of the polymer groups (e.g., in transesterification of polyester with polycarbonate or the co-reaction between polyester and polyamide), one may chemically alter one polymer or both (e.g., blending maleated polypropylene or SEBS with polyamide or mixing polytetrachloroethylene-containing ionic groups with another similarly modified polymer), or one may add a third ingredient, monomeric or polymeric, to co-react both macromolecular components of the blend (e.g., addition of peroxide to polyolefin blends or quinone to polyamide/polyphenylene ether blends).

Polymer Miscibility

The most basic question when considering a polymer blend concerns the thermodynamic miscibility. Many polymer pairs are now known to be miscible or partially miscible, and many have become commercially important. Considerable attention has been focussed on the origins of miscibility and binary polymer/polymer phase diagrams. In the latter case, it has usually been observed that high molar mass polymer pairs showing partial miscibility usually exhibit phase diagrams with lower critical solution temperatures (LCST).

The criteria for polymer/polymer miscibility are embodied by the equation for the free energy of mixing,

$$\Delta G_{mix} = \Delta H_{mix} - T\Delta S_{mix} \qquad (1)$$

where G is the Gibbs free energy, H is enthalpy, S is entropy, and T is the absolute temperature.

The necessary conditions for miscibility are that $G < 0$ and that $d^2G/d^2\phi < 0$, where ϕ is the mole fraction of the ith component. In equation (1), the combinatorial entropy of mixing depends on the number of molecules present according to

$$\Delta S_m/RT = n_1 ln\phi_1 + n_2 ln\phi_2 \qquad (2)$$

Therefore, as the molar mass gets large the number of molecules becomes small, and the combinatorial entropy of mixing becomes negligibly small.

In order to explain phase separation on heating, i.e., LCST behavior, the effect of volume changes on mixing must be considered. This effect is described by equation-of-state theories such as that developed by Flory and co-workers (30). The free volume contributions to the free energy are unfavorable and increase with temperature.

This still leaves the question of why so many polymer pairs are miscible. Unless the polymers are very similar physically and chemically, it appears that the major driving force for miscibility must be enthalpic contributions to the free energy. There are two different but not mutually exclusive ways of explaining favorable enthalpic contributions in current thinking. One concerns favorable interactions between the two polymers, the other considers unfavorable interactions between groups on the same polymer to be the cause of an overall favorable interaction with another polymer.

In many cases a favorable interaction between the chains is known to exist. For example, in the case of mixtures of chlorine containing polymers with polyesters, hydrogen bonds of the type

$$O-C=O - - - H-C-Cl$$

are known from shifts in the carbonyl absorption frequency of the IR spectrum. In this case, the phase separation upon heating may be due to dissociation of the hydrogen bond. Fig. 3 shows how the carbonyl absorption frequency changes on heating for one such polymer mixture (31). The phase separation point coincides with a point where the shift of the carbonyl absorption has almost disappeared.

Unfavorable interactions between groups on the same chain can be the origin of miscibility between polymers, as shown schematically in Fig. 4. If groups one and two have a large enough unfavorable interaction, then they will prefer to mix with another polymer, group three, in order to minimize the number of one-two contacts. This has been used to explain the windows of miscibility which occur for copolymer blends. A copolymer is often found to be miscible with another polymer over some range of monomer composition and the use of a cross term in the free energy expression allows one to describe this phenomenon.

It should, however, be pointed out that in the above case, if one simply ascribes a single solubility parameter to each monomer, it is impossible to predict an overall negative enthalpy of mixing. It has also been noted that a window of miscibility can be explained by a favorable specific interaction without recourse to a cross term. If one separates the normal dispersive forces from the specific interaction, then as a first approximation, when the solubility parameters of the two polymers are similar the unfavorable dispersive interactions are small and specific interactions yield miscibility. For a copolymer/polymer mixture the solubility parameters might be expected to match at some specific copolymer composition (32). A method of combining the features of both the specific interaction and the cross term is to use something similar to the UNIFAC group contribution system and model all the interactions, both favorable and unfavorable within the system.

Polymer Interfaces

The interface in a polymer blend is important both in the preparation of the blend and in determining its ultimate properties. During mechanical melt mixing, the break up of the drops of one polymer within a matrix of the other is determined in part by the interfacial tension. For a Newtonian liquid drop in a Newtonian matrix the process can be simply described. In a shear

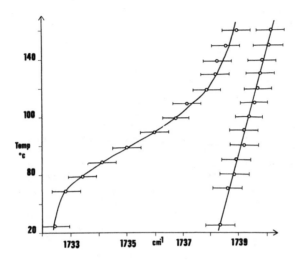

Figure 3. Plot of temperature versus the carbonyl peak position for EVA and an 80:20 wt. % PVC-EVA blend.

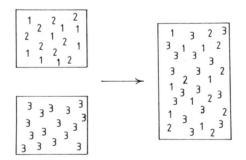

Figure 4. Schematic diagram for mixing copolymer 1,2 with homopolymer 3 segments.

field the drop deforms into an ellipsoid due to the viscous forces upon it and this deformation is resisted by the interfacial tension. The interfacial force depends on v_0/d, i.e., the interfacial tension divided by the drop diameter, and the viscous force depends on $\eta_m \dot{\gamma}$ the matrix viscosity times the shear rate. The ratio of these is sometimes called the Taylor number,

$$\kappa = v_0/d\eta_m\dot{\gamma} = v_0/d\sigma_{12} = 1/W_b \qquad (3)$$

Where W_b is the Weber number and σ_{12} is the shear stress, the same on both sides of the interface. The critical condition for drop breakup in Newtonian liquid mixtures is given by:

$$(19\lambda + 16)/(16\lambda + 16) > \kappa \qquad (4)$$

where λ is the viscosity ratio of dispersed and matrix liquids, $\lambda = \eta_d/\eta_m$. Note that in the full range of $0 \le \lambda \le \infty$ left hand side of inequality (4) changes from 1.0 to 1.19. This theoretically predicted insensitivity of the dispersing process to the viscosity ratio is valid only within the range of first order perturbations, i.e. at small deformations. Experimentally, drops with equilibrium diameter varying by more than two decades have been generated by shearing mixtures with different values of λ (33).

Interfacial agents, such as block copolymers, are known to reduce the interfacial tension and hence are expected to increase the degree of dispersion in blends. The measurement of interfacial tension for polymer systems is not easy. Most measurements have been made by the pendant drop technique. Measurements of interfacial thickness are also difficult. They have been made using electron microscopy and, mostly in the case of block copolymers, by x-ray and neutron scattering. Recent results using neutron reflection suggest that this will be a useful technique in the future.

Theories of the polymer interface have been presented by, among others, Helfand using lattice calculations or diffusion calculations. They yield the following scaling relations,

$$v \propto \chi^{1/2} \quad a_I \propto \chi^{-1/2} \quad v \propto a_I \qquad (5)$$

where a_I is the interface thickness and χ is an interaction parameter. Recent attempts have been made to develop theories of copolymers at interfaces.

According to van Oene (34) the interfacial tension coefficient depends on the difference between the first normal stress differences, $N_{1,i}$, of the blend components:

$$v = v_0 [1 + (d/12) (N_{1,d} - N_{1,m})] \qquad (6)$$

The dependence, derived with several limiting assumptions, was never experimentally verified. However, there is no doubt that the dynamic interfacial coefficient v diverges from its static value, v_0, as the magnitude of the term $|d(N_{1,d}-N_{1,m})|$ increases.

Of the three popular methods of ν-determination the sessile drop is the slowest, the pendant drop faster and the spinning drop the most rapid (35). For commercial resin pairs the first two may require days before the drop reaches its equilibrium shape. During this time there is diffusion of the low molar mass fractions toward the interface gradually decreasing the value of the interfacial tension coefficient (36). These two factors, the normal stress and the time scale, are generally responsible for the poor correlation between the predicted and measured droplet diameters in commercial blends.

Microrheology

The microrheology of Newtonian liquids has been summarized by Goldsmith and Mason and by Grace (33). That of polymeric melts is in its infancy (37).

When discussing drop deformability it is convenient to consider first shear deformation and then uniaxial extension. As shown in Fig. 5, in a shear field the equilibrium droplet diameter, d, depends mainly on the viscosity ratio, λ. Here, the diameter scaling factor is ν/σ_{12}. The most efficient drop breakup has been observed for $1/3 < \lambda < 1$. At higher values, as λ increases so does the drop diameter d; for $\lambda \geq 3.8$ only formation into an ellipsoid is possible. Also for $\lambda \leq 0.05$ the overall efficiency of dispersing decreases. However, on the one hand the decrease is less dramatic than that for high values of λ and on the other hand there are two mechanisms present: the satellite breaking and so called tip spinning, schematically illustrated in Figs. 6a and 6b, respectively.

Droplet breakup in uniaxial extensional flow is more efficient (10). The theoretical calculations estimate that (21):

$$d_\varepsilon/d_\sigma \cong 3.3 \ [1 + (1.9\lambda\kappa)^2]^{1/2}/[1 + 1.5\kappa] \qquad (7)$$

where subscripts ε and σ identify the equilibrium drop diameter in extension and in shear, respectively. Moreover, in uniaxial extension, the process is insensitive to elongational viscosity ratio, $\lambda_E = \eta_{E,d}/\eta_{E,m}$ and fine dispersion is always obtained. Only a slight loss of dispersing efficiency is noted at the limiting values of λ_E.

The most efficient mechanism of drop breakup involves its deformation into a fiber followed by the thread disintegration under the influence of capillary forces. Fibrillation occurs in both steady state shear and uniaxial extension. In shear (= rotation + extension) the process is less efficient and limited to low-λ region, e.g. $\lambda < 2$. In irrotational uniaxial extension (in absence of the interphase slip) the phases codeform into thread-like structures.

The method of morpholgy control should involve the fibrillation step. When fine droplet dispersion is desirable, the resins should be selected to assure low λ, the uniaxial extension should be maximized and imposition of a fiber disturbing dynamic field should be used at the end of the process. On the other hand, if preservation of extended morphology is desired, (i.e., the

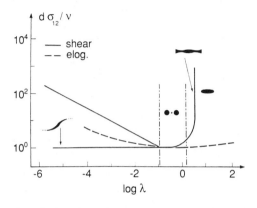

Figure 5.   Reduced droplet diameter vs. viscosity ratio, λ, in shear and extensional flows.   The type of shear drop deformation within each of the four zones of λ is indicated.

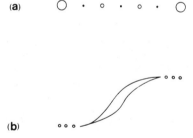

Figure 6.   Schematic representation of drop breakup in shear field at two different values of the viscosity ratio λ: (a) 0.1 < λ ≤ 1, (b) λ < 0.01.

fibrillas in uniaxial extension or lamellae in biaxial elongation)
then a system with large λ and large initial drop diameters should
be used. Disintegration of the extended structures will be
prevented if the material forming the dispersed phase shows the
apparent yield stress behavior.

It was recently reported that extended structures are also
created in low amplitude oscillatory shear field (38). However,
judging by the geometry of the generated threads, coalescence
rather than drop elongation was the responsible mechanism. The
coalescence also takes place in convergent (elongational) as well
as in the shear flow fields. Elmendorp stressed the need for
taking the coalescence into account while evaluating
microrheological models (37). Thus, the dispersing process must be
seen as a dynamic event, tending to different "equilibrium" states
under different sets of imposed conditions.

In summary, microrheology, combining the thermodynamic and
rheological principles, provides a powerful tool for understanding
and controlling the mixing process. However, while microrheology
looks at the influence of each factor in isolation, in commercial
compounding all of these take place at once; there is an interplay
between various deformation fields, their intensities, and diverse
process variables.

Blend Rheology

The most frequently used rheological function is the shear
viscosity, η. An old argument states that in multiphase systems
the shear stress, σ_{12}, is a continuous function at the interface
whereas the shear rate, $\dot{\gamma}$ is not (10). Of course, the argument
ignores the existence of systems with interlayer slip where
both σ_{12} and $\dot{\gamma}$ are discontinuous, but indeed these are less
frequent. Thus, conditionally accepting the argument, the blend
rheological functions should be examined at constant stress; i.e.,
$\eta = \eta\ (\sigma_{12})$ not $\eta = \eta\ (\dot{\gamma})$ and $G' = G'(G'')$ not $G' = G'(\omega)$.
Consequently, let us set $P = \eta = \eta(\sigma_{12})$ in Eq (8). Note that at
low enough stress η equals the zero-shear viscosity, $\eta_o = \eta(\sigma_{12} \to 0)$.

It is convenient to define the log-additivity rule as:

$$\ln P = \Sigma_i w_i \ln P_i \qquad (8)$$

where P is a rheological function (e.g., viscosity, η, dynamic
storage and loss moduli, G' and G'', stress, σ_{12}, etc.) and w_i is
the weight fraction of ingredient "i" in the mixture.

Equation (8) can now be used to separate all blends into four
categories: 1. additive blends where η follows the relation,
i.e., there is agreement between the experimental value of
viscosity, η, measured for the blend and that calculated from
viscosities of neat resins by means of Eq. (8): $\eta = \eta_{calc}$; 2.
positively deviating blends, PDB, with $\eta > \eta_{calc}$; 3. negatively
deviating blends, NDB, with $\eta < \eta_{calc}$; and 4. mixed, positively and
negatively deviating blends, PNDB. Hopefully, different flow
mechanisms can be assigned to each of these categories. A general
picture which emerges from analysis of many alloys and blends is
that all four categories are found in miscible and immiscible

systems. Different mechanisms are responsible for the deviation in miscible systems than in immiscible ones and the intensity of deviation from the log-additivity rule in miscible blends is never very large. Unfortunately, however, one cannot say that all blends of a given phase behavior belong to a single rheological category (21).

In miscible systems the category of ln η vs. w_i relation depends on the free volume variation with composition, $f = f(w)$. For non-interacting, homologous liquids (for example: fractions of a given polymer), this relation can be computed from statistical thermodynamics and variations of viscosity with composition have been predicted with good precision (39). In that case only a small PDB was observed.

For most miscible blends the miscibility stems from specific interactions which usually lead to a reduction of volume which in turn causes an increase in η; thus PDB in miscible blends predominates. On the other hand, in systems where miscibility originates from reduction of unfavorable interactions between groups of the same polymer an expansion of volume and NDB are expected (21).

In immiscible blends the changes in rheological properties may originate either from the volume change in the interphase region or from the presence of the second phase. The effect of the dispersed phase can be modeled by either comparing the blend to an emulsion ($\lambda<1$) or to a suspension ($\lambda>>1$); in both cases η rapidly increases with composition leading to PDB. About 60% of blends belong to this type.

NDB immiscible systems are about half as numerous. The only mechanism which explains this behavior is interlayer slip, and indirect evidence of this has been shown with micrographs of extruded non-compatibilized blends. A more direct affirmation was recently provided by Lyngaae-Jorgensen (40), who prepared a multi-layered sandwich consisting of hundreds of layers of selected immiscible polymers per millimeter of specimen thickness. The viscosity showed strong negative deviation and was analysed by considering the blend as a three component system, including an interphase regime. The viscosity of this region was found to be about one-tenth of the viscosity of the less viscous component. Interlayer slip is associated with the repulsive forces between immiscible polymers which lead to a reduction of density at the interface.

In most cases, the viscosity follows the theoretical relation (41):

$$1/\eta = [1 + (\beta/\sigma_{12}) \ (w_1 w_2)^{1/2}] \cdot [w_1/\eta_1 + w_2/\eta_2] \qquad (9)$$

where β is a characteristic interlayer slip factor for the system. Eq (9) predicts a reduction of the importance of slip with increasing σ_{12}. The relation is symmetrical forecasting a maximum slip effect for a 50:50 composition, though due to asymmetry of the thermodynamic interactions, this is not always observed.

At low concentration of the dispersed phase discrete droplets are expected. As the concentration increases the shear field extends the drops, coalesces them and finally at the critical volume fraction, ϕ_c, forms a co-continuous structure and inverts. Paul and Barlow proposed the simple formula for the phase inversion (42):

$$\phi_i/(1-\phi_i) = \lambda \qquad (10)$$

More elaborate formulas are also available in the literature (21). The concentration at phase inversion, .ϕ_i, represents the mid-value of two ϕ_c's determined for both polymers as the matrix, though there are no theories which can predict ϕ_c. Experimentally the breadth of the co-continuous region, $\phi_{c1} - \phi_{c2}$, increases with lowering the component viscosities, η (43).

From the morphological point of view, the existence of these critical concentrations, has a more profound influence on some rheological functions than on others, ie., strain recovery, S_R. Thornton et al. (44) reported step-increases in S_R vs. w_i function at concentrations ϕ_c and ϕ_i. The steps were not visibile in η vs. w_i plots.

In emulsions and suspensions the interacting particles may form a three-dimensional network that results in a yield stress, σ_y. In blends, σ_y has been reported for both dispersed and co-continuous morphologies, i.e., below and above ϕ_c. Presence of σ_y increases PDB behavior in ln η vs. w_i plots.

PNDB occurs for blends that are miscible only within a small range of concentration. To understand these systems, the interaction between the flow and the phase separation must be considered. This has become an area of considerable scientific activity. One needs to consider both how the phase separation influences the rheological behavior and how the flow (or stress) affects the thermodynamics of phase separation.

Theoretical calculations for liquid/liquid systems predict that the viscosity goes through a maximum at the spinodal. Depending on the type of system and its regularity, the increase may be quite large; for example, Larson and Frederickson (45) predicted that for block copolymers. These authors concluded that in the spinodal region a three-dimensional network is formed and that the system exhibits non-linear viscoelastic behavior. Experimentally, sharp increases of η near the phase separation have been reported for low molar mass solutions as well as for oligomeric and polymeric mixtures (21).

Melt flow, however, also affects the phase separation, usually enhancing the miscibility for partially miscible blends that show LCST behavior. From Lyngaae-Jorgensen work (46) one may derive the following relation between the shear stress and the change in the spinodal temperature, T_s, (21),

$$\sigma^2_{12} = \alpha(T-T_s)T \qquad (11)$$

where α is a characteristic material parameter. Several groups have observed reduction of opacity of sheared immiscible polymer blends by measuring light scattering during the flow within the two-phase region (47-49). An important question that arises from these results is whether they represent true phase mixing or are due to a reduction of the phase size below the limit needed to

scatter light. Winter (50) has recently shown by fluorescence
measurements that in the case of polystyrene/polyvinylmethylether
blends that high strain ($\varepsilon \cong 44$) flows lead to molecular mixing.
    The structure of immiscible blends is seldom at equilibrium.
In principle, the coarser the dispersion the less stable it is.
There are two aspects of stability involved: the coalescence in a
static system and deformability due to flow. As discussed above
the critical parameter for blend deformability is the total strain
in shear $\gamma = t\dot{\gamma}$, or in extension, $\varepsilon = t\dot{\varepsilon}$. Provided t is large
enough in steady state the strains and deformations can be quite
substantial; one starts a test with one material and ends with
another. This means that neither the steady state shearing nor
elongational flow can be used for characterization of materials
with deformable structure. For these systems the only suitable
method is a low strain dynamic oscillatory test. The test is
simple and rapid, and a method of data evaluation leading to
unambiguous determination of the state of miscibility is discussed
in a later chapter.
    There are four popular measures of liquid elasticity: the
first normal stress difference, the extrudate swell, B, the Bagley
entrance-exit pressure drop correction, P_E, and the storage shear
modulus, G'. For multiphase systems there is no simple correlation
between N_1 and G', although the Sprigg's theoretical relation (51):

$$\sigma_{12}\dot{\gamma} = G''(\omega)/C; \quad N_1(\dot{\gamma}) = 2G'(\omega)/C^2 \quad \text{Where } C = \omega/\dot{\gamma}, \qquad (12)$$

has been found to work when $1 \leq C \leq 2$. Qualitatively C increased
with heterogeneity of the melt (21). Both N_1 and G' (at constant
stress) show additivity, PDB, NDB or PNDB behavior paralleling the
η vs. w_1 plot.
    The extrudate swell, B, dependence on composition is
determined by λ. For deformable drops in systems where $\lambda < 4$,
strong PDB behavior is observed, independent of the η vs. w_i
relation. This behavior originates in shrinking of fibrils created
in the convergent flow region at the capillary entrance. The
tendency for the droplet to regain sphericity causes the extrudate
to swell considerably. This has little to do with the mechanism
responsible for extrudate swell in homogeneous melts. On the other
hand, in blends where $\lambda > 4$ the swell is small, frequently NDB, and
independent of the η vs. w_i behavior. The highly viscous drop can
be equated with a solid^{-1} particle; rheologically these blends
resemble suspensions.
    The fourth measure of liquid elasticity, P_E, reflects the
extensional properties rather than elasticity. For polymer blends
it is difficult to determine P_E with a sufficient degree of
accuracy. The data indicate that this parameter is most sensitive
to morphological changes. For example, the degree of droplet
coalescence in the instrument reservoir drastically changes the
values of P_E.
    Polymer processes such as the film blowing, blow molding or
wire coating involve extensional flows and are, therefore, most
appropriately studied using extensional rheometry. Extensional
flow data provide unique information on stress hardening, SH, which
are important for good bubble stability during film blowing. Some

polymers, such as low density polyethylene, LDPE, have strong SH, whereas some like polypropylene, PP, do not show it at all. Blending offers an easy way to introduce SH into materials lacking this ability. SH can be incorporated into both miscible and immiscible blends, although in the latter case the phenomenon seems to be restricted to systems with co-continuous structures.

Compounding and Processing Polymer Blends

Commercial blends are designed either to optimize their morphological stability or their ability to attain a specific morphology. In the first case the manufacturer's aim is generation of materials that are relatively insensitive to the method of processing. Most of these blends contain $w_i \le 30\%$ of well dispersed and compatibilized minor phase ingredient. Stabilization of morphology at higher concentration is more difficult, but possible, as evidenced by new, commercially successful PA/ABS = 50/50 blends. The second category includes blends in which the minor phase must deform into lamellae or fibers during the final processing step.

There is no single recipe for preparation of blends. Before the ingredients are selected and the compounding/processing equipment is chosen, one should have a clear idea of the use of the blend. A program of blend development was recently proposed (21).

Past experience teaches that synergism is rare and selection of blend components is usually based on compensation of natural properties. For example, the rationale for blending poly (phenylene ether), PPE, with high impact polystyrene, HIPS, or with polyamide, PA, is shown in Table I. The compensation of properties in PPE/HIPS and PA/PPE blends is obvious. What is less certain is the method of compatibilization. Since PPE/Polystyrene blends are miscible there is no need for a compatibilizer in PPE/HIPS blends. However PA and PPE are antagonistically immiscible and compatibilization is required; low or high molar mass additives with acid (or anhydride) groups are often used in a reactive compounding operation. The advantage of blends with fine droplet dispersion is their homopolymer-like behavior. These materials are easy to process in a reproducible manner, weld lines do not cause serious problems and scrap recycling is possible. On the other hand, the ability of anisotropic reinforcements by oriented inclusions is lost.

Fibrillar or lamellar morphologies are desirable in some products either as reinforcements or barriers against gas or liquid permeation. Here the compounding step must generate relatively large domains of a dispersed, compatiblized component, and the desired morphology is developed in the subsequent processing step.

Processing of blends usually involves two steps: compounding, and forming (52). In about 80% of cases these two steps are carried out at different locations, e.g. the material is precompounded either by the resin manufacturer or a compounder then transformed into the final article by a processor. Most compounding is currently done in a twin-screw extruder, twin-shaft continuous intensive mixer or other type of extruding machine (single-screw, disk, speciality coupounder, etc.). The twin-screw extruders are expensive but they provide flexibility for changes, uniformity of mixing, short residence times and narrow residence time distribution. With single screw extruders, control is not as good.

During the compounding step the ingredients are often compatibilized by addition of a third component (e.g., copolymer or co-solvent) or by one of several methods of reactive processing. Sometimes the generated morphology must be further stabilized by enhanced crystallization or partial crosslinking of one phase. As evidenced by the extensive patent literature, reactive compounding dominates the field. These processes require fine control of process variables, narrow distribution of residence times and uniform stress fields, that can be provided only by the more sophisticated, and therefore, expensive machines. Compounding can add from 50 to 500 US$ per metric ton to the cost of a blend.

The forming operations for polymer alloys and blends use the same methods and equipment as for homopolymers, but the process variables usually need be more precisely controlled. The key to a successful product is optimization of the morphology; exceeding the recommended temperature or shear stress can readily annihilate the fragile balance of forces responsible for the blend structure. For example, in injection molding, layering and delamination is always possible. In spite of compatibilization there is invariably a skin/core morphology which may lead to high notch-sensitivity. With the growing popularity of blends compounded by the reactive processing method there is also a new danger. Frequently two batches of the same blend, reacted to a different degree, behave like two immiscible polymers. Adding too much recycle or mixing different batches may result in sudden dramatic weakening of weld lines or, in extreme conditions, delamination of the processed article.

Mechanical Properties

Predicting the modulus of a two phase structure is not a trivial problem, but it can be measured reasonably well. Failure properties are more difficult to measure in a reproducible and meaningful way to give a measure of material properties and not just a function of the test method and sample geometry. The best solution lies in fracture mechanics (53). The basis of fracture mechanics is the analysis of the stress around a pre-existing crack of known size as shown in Fig. 7. The toughness, given by K_c the stress concentration factor is:

$$K_c = Y\sigma a_i^{1/2} \tag{13}$$

where σ is the stress at break, a_i is the crack length and Y is a factor which depends on the specific geometry.

The problem in the case of blends is that the plastic zone must be smaller than the other dimensions as shown in Fig. 7. In particular, the thickness, B, must be large enough to maintain a plane strain condition, otherwise excess yielding along the width of the crack, i.e., changing B, occurs. For tough plastics, which

Table I. Advantages and Disadvantages of Selected Polymers

No.	Polymer	Advantage	Disadvantage
1	PPE	High Heat Distortion T, Rigidity, Flammability, Moisture Absorption	Processability, Impact Strength, Solvent Resistance
2	HIPS	Processability, Impact Strength	Heat Distortion T, Rigidity
3	PA	Processability, Impact Strength, Solvent Resistance	Heat Distortion T, Rigidity, Moisture Absorption

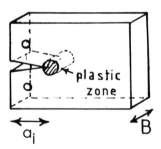

Figure 7. Compact tension specimen showing the plastic zone around the crack tip.

include many plastic/rubber blends, B needs to be several centimeters, which is not very practical.

Various ways have been tried to overcome this problem, an example of which is the J method. A three point bending method is used and the energy per unit area for initiation of crack growth measured as the area under a load-deflection curve up to the point of initiation. In order to determine the point of initiation, several samples are loaded to different post-initiation stages, the amount of crack growth measured after freezing and breaking, and this is extrapolated to $\Delta a_i = 0$ with a correction being made for crack blunting. Measurements of J have been reported for a variety of polymers and comparison with equivalent K_c values found to be satisfactory (54).

In practice, much of the literature still reports notched Izod numbers, but these can be deceptive, especially for materials close to a ductile-brittle transition. Better indicators of real toughness are obtained when samples of different thickness are considered. As one adds more rubber to a plastic, a critical concentration is often found where the toughness increases dramatically. In some cases, e.g., polymers that are capable of shear yielding, this correlates better with the interparticle distance or "ligament thickness" than with any other variable.

Fatigue is probably the commonest cause of failure in plastics, but it is less studied than other forms of failure. It has been reviewed by Bucknall (6). Data is often shown in terms of a plot of the stress amplitude against the log of the number of cycles to failure. It is notable that for unnotched fatigue tests, high impact polystyrene actually preforms more poorly than pure polystyrene when studied this way. It is possible that the ease of initiation of multiple crazes, which gives rise to improved impact, actually leads to failure in a fatigue test. With notched samples studied by the methods of fracture mechanics one plots a crack growth rate against the stress intensity factor.

IONOMERS

The introduction of a small amount of bonded salt groups into a relatively nonpolar polymer has a profound effect on its structure and properties. These materials, termed ionomers, have been used in a variety of applications, including permselective membranes, thermoplastic elastomers, packaging and films, viscosifiers and compatibilizing agents for polymer blends. For all these applications, the interactions of the ionic groups and the morphologies that result are critical for establishing the unique properties. Despite this, however, the spatial arrangement of the ions in ionomers remains an open question, and considerable research has been and continues to be devoted towards understanding the microstructure of these materials. This, however, does not appear to have inhibited progress in the technological exploitation of ionomers as novel materials.

Comprehensive reviews and discussions of ionomers can be found in a number of monographs and review articles (55-64). Four general topics are discussed here: (1) new materials, (2) structure, (3) solutions, and (4) membranes.

Synthesis of Ionomers

Ionomers have been prepared by two general routes: (1) copolymerization of a low level of functionalized monomer with an
olefinically unsaturated comonomer or (2) direct functionalization
of a preformed polymer. Almost all ionomers of practical interest
have contained either carboxylate or sulfonate groups as the ionic
species. Other salts, such as phosphonates, sulfates,
thioglycolates, ammonium, and pyridinium salts have been studied,
but nowhere to the extent of the carboxylate and sulfonate
anionomers. (An aniomer is defined as an ionomer in which the
anion is bonded to the polymer. Conversely, ionomers that have the
cation bonded to the polymer are termed cationomers). Relatively
little information is available on the structure and properties of
these types of ionomers.

Typically, carboxylate ionomers are prepared by direct copolymerization of acrylic or methacrylic acid with ethylene,
styrene or similar comonomers by free radical copolymerization
(65). More recently, a number of copolymerizations involving
sulfonated monomers have been described. For example, Weiss et al.
(66-69) prepared ionomers by a free-radical, emulsion copolymerization of sodium sulfonated styrene with butadiene or styrene.
Similarly, Allen et al. (70) copolymerized n-butyl acrylate with
salts of sulfonated styrene. The ionomers prepared by this route,
however, were reported to be "blocky" with regard to the incorporation of the sulfonated styrene monomer. Salamone et al. (71-76)
prepared ionomers based on the copolymerization of a neutral
monomer, such as styrene, methyl methacrylate, or n-butyl acrylate,
with a cationic-anionic monomer pair, 3-methacrylamidopropyl-
trimethylammonium 2-acrylamide-2-methylpropane sulfonate.

The second method used to prepare ionomers involves
functionalizing a preformed polymer. This has been the more common
strategy for obtaining sulfonate-ionomers. Makowski et al. (77, 78)
prepared lightly sulfonated polystyrene (SPS) and ethylene-
propylene-diene terpolymers (SEPDM) by reaction of the polymers
with acetyl sulfate in homogeneous solution. This chemistry yields
a controlled concentration and a random distribution of sulfonate
groups. Bishop et al. (79, 80) described the sulfonation of poly
(ether ether ketone) with chlorosulfonic acid or sulfuric acid.
Zhou and Eisenberg (81) prepared sulfonated poly-cis-1,4-isoprene
using acetyl sulfate, but this reaction also yielded a considerable
amount of cyclized polymer. Rahrig et al. (82) sulfonated polypentenamer using a 1:1 complex of SO_3 with triethylphosphate. Polyam-
pholytes were prepared by Peiffer et al. (83) who sulfonated a
copolymer of styrene and 4-vinylpyridine with acetyl sulfate.
Huang et al. (84) made zwitterionomers by reacting a polyurethane
segmented copolymer with g-propane sultone. This was converted to
an anionomer (85) by reaction with a metal acetate.

A special class of ionomer in which the salt groups are only
at the chain ends, i.e., telechelic ionomers, have recently
received considerable attention as model ionomer systems. Kennedy
and coworkers (86-88) prepared linear and tri-arm star telechelic
sulfonated polyisobutylene by heterogeneous sulfonation of an·
olefin-terminated polyisobutylene with acetyl sulfate. Omeis et

al. (89) synthesized telechelic sulfonated polystyrene and polybutadiene by terminating an anionic polymerization with 1,3-propane sultone. In a later chapter, Storey and George describe the synthesis of star-branched telechelic block copolymer ionomers.

Blends

As discussed earlier in this chapter, physical blending of two polymers generally results in a two-phase material as a consequence of a positive enthalpy of mixing. Various approaches have been taken to introduce specific interacting functionalities on polymer pairs in order to improve their miscibility. The most common method, though by no means the only method, is to employ hydrogen bonding (90). Recent work has shown that it is possible to improve miscibility of two immiscible polymers by using interactions involving one or more ionized species.

Eisenberg and coworkers have employed acid-base interactions to improve the miscibility of a number of polymer-polymer pairs. Miscible blends were prepared using acid-base interactions, e.g., with SPS (acid derivative) and poly (ethylacrylate-co-4-vinylpyridine) (91), sulfonated polyisoprene and poly (styrene-co-4-vinylpyridine) (92), and using ion-dipole interactions, e.g., poly (styrene-co-lithium methacrylate) and poly (ethylene oxide) (93). Similarly, Weiss et al. (94) prepared miscible blends of SPS(acid) and amino-terminated poly (alkylene oxide). In addition to miscibility improvements, the interactions between two functionalized polymers offers the possibility for achieving unique molecular architecture with a polymer blend. Sen and Weiss describe the preparation of graft-copolymers by transition metal complexation of two functionalized polymers in another chapter.

STRUCTURE

There is a considerable body of experimental and theoretical evidence for two types of ionic aggregates termed multiplets and clusters (95). The multiplets are considered to consist of small numbers of associated contact ion-pairs that are dispersed in the matrix of low dielectric constant, but do not themselves constitute a second phase. The number of ion-pairs in a multiplet is sterically limited by the fact that the salt groups are bound to the polymer chain. On the other hand, clusters are considered to be small microphase separated regions (<5 nm) of aggregated multiplets. Thus, the clusters are rich in ion-pairs, but they also contain an amount of the organic polymer.

The proportion of the salt groups that reside in either of the two environments in a particular ionomer is determined by the polarity (i.e., dielectric constant) of the backbone, the total concentration of the salt groups, the nature of the anion and cation, and to some extent by non-equilibrium effects that arise from the fact that equilibration of the structure requires the diffusion of the ionic species through a viscous and low dielectric medium.

Theory

There have been relatively few attempts to formulate a theoretical
basis for explaining and predicting ionic aggregation in ionomers.
These have been recently reviewed by Mauritz (41). Some theories
(96-99) are semi-empirical in that they include experimentally
determined parameters. Several others (95, 100, 101) are derived
from first principles. Various parameters, including the size and
the ionic density of the cluster, are predicted by these models.
Confirmation and, indeed, further theoretical developments, have
been hindered by the lack of quantitative experimental data for the
cluster structure and energetics. Furthermore, except for some
recent work (102), there have been no experimental observations of
the actual clustering or its kinetics.
     The first theory of ionic clustering (95), though simplistic
in its approach, still contains many of the important features of
the more elegant, and mathematically more complex, subsequent
theories. This theory states that the driving force for ionic
aggregation involves a basic thermodynamic immiscibility between
the hydrophobic polymer matrix and the salt groups. Thus, for
example, in Eisenberg's theory, the ion-dipole interactions between
the salt groups favor aggregation, which is opposed by the entropic
penalty associated with the expansion of the chain conformation.
Although the concept of chain expansion in ionomers is not
universally accepted, the only experimental measurements to date
(103) show that the radius of gyration of an ionomer increases with
increasing fuctionalization.
     Although the theories cited above predict many aspects of
ionomers that are in agreement with the available experimental
observations, an impasse has unfortunately been reached with regard
to further developments. The reason is the lack of sufficient and
adequate experimental data concerning the detailed features of the
microstructure and the current lack of a technique for providing
this detailed information.

Experimental Studies

A variety of different experimental techniques have been used to
study and characterize the microstructure of ionomers. These have
included small-angle x-ray scattering (SAXS), small-angle neutron
scattering (SANS), extended x-ray absorption fine structure
(EXAFS), Fourier transform infrared spectroscopy (FTIR), electron
spin resonance (ESR), Mossbauer spectroscopy, nuclear magnetic
resonance (NMR), fluorescence spectroscopy, and electron
microscopy. Although these techniques have provided evidence for
strong molecular association of the ionic groups that lead to
microphase separated ion-rich domains, the size, shape and
distribution of these domains have not yet been determined.
     SAXS results have been of central importance in the
interpretation of the structure of ionomers. A peak is typically
observed in the scattering pattern of most ionomers at a scattering
vector, k, of 1 to 3 nm. Examples of the SAXS patterns for a
series of SPS ionomers with varying sulfonate concentration and
different cations are shown in Fig. 8. Although the "ionic peak"

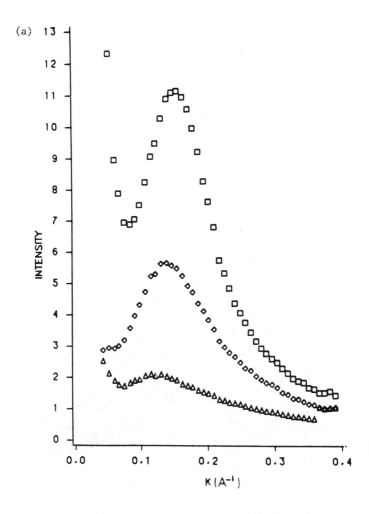

Figure 8a. Effect of sulfonation level on SAXS of a sulfonated polystyrene, Cs-SPS: Δ, 0.90 mol %; ◇, 2.65 mol %; □, 4.65 mol %. (Reproduced with permission from ref. 113.)

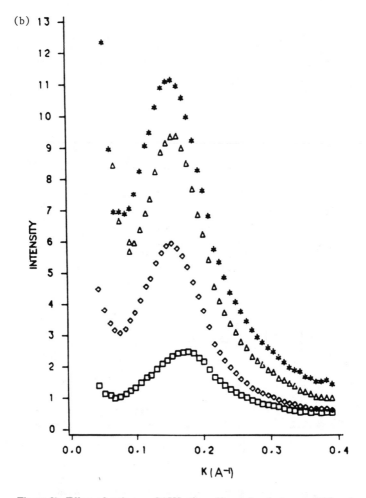

Figure 8b. Effect of cation on SAXS of a sulfonated polystyrene, 4.85 mol % SPS: *, Cs; △, Rb; ◇, k; □, Li. (Reproduced with permission from ref. 113.)

represents the most incontrovertible evidence for microphase separation in ionomers, no model has yet been able to unequivocally describe the size, shape and distribution of the ionic domains. One of the unexplained features of the scattering patterns of ionomers is the very large upturn in intensity at very small scattering angles, see Fig. 8. Over the years, it has been debated as to whether this upturn is an artifact due to impurities or voids in the samples or whether it is real and related to large correlation distances arising from the presence of the salt groups. Recent SAXS results (102, 104) have indicated that the upturn is a consequence of the presence of the ions. Using real-time SAXS studies of cluster growth in SPS ionomers, Galambos et al. (48) observed that the zero-angle scattering decreased in intensity as the ionic peak developed. They suggested that the two features of the SAXS scattering pattern, the peak and the zero-angle scattering, may represent two distinct environments of the ions, namely clusters and multiplets.

In a later chapter, Ding et al. describe a new SAXS technique, anomalous SAXS, for characterizing the zero-angle upturn in intensity. Their results demonstrate that the zero-angle scattering is due to the presence of the ions, and they suggest that the origin of the zero-angle scattering is an inhomogeneous distribution of isolated ionic groups. They propose a model for the SAXS scattering that combines the liquidlike interference model of Yarusso and Cooper (105) and a Debye-Bueche random-two-phase model.

The only characterization technique that can directly view heterogeneities of the order of 1 nm, i.e., the size of the ionic clusters is electron microscopy. There have been very few electron microscopy studies of ionomer microstructure and these have failed to provide a satisfactory picture of ionic clusters (106). With the advancements that have been made in recent years in the optics and theories of electron microscopy, microscopists are in a better position to reexamine these materials (See the chapter by Williams).

Viscoelastic measurements of ionomers have been used to indirectly characterize the microstructure and to establish property structure relationships. Forming an ionomer results in three important changes in the viscoelastic properties of a polymer. First, T_g usually increases with increasing ionization. This is a consequence of the reduced mobility of the polymer backbone as a result of the formation of physical, ionic crosslinks. Second, an extended rubber plateau is observed in the modulus above T_g, again as a result of the ionic network. Third, a high temperature mechanical loss is observed above T_g, which is due to motion in the ion-rich phase. The dynamic mechanical curves for SPS ionomers shown in Fig. 9 clearly demonstrate these three characteristics.

Using melt viscosity measurements, Lundberg et al. (107) showed that one could selectively plasticize either the ion-rich phase or the nonpolar hydrocarbon phase of SPS ionomers by varying the chemistry of the diluent used. Fitzgerald et al. (108) studied the effects of dioctyl phthalate (DOP) and glycerol on the dynamic mechanical properties of SPS ionomers. The addition of DOP lowered

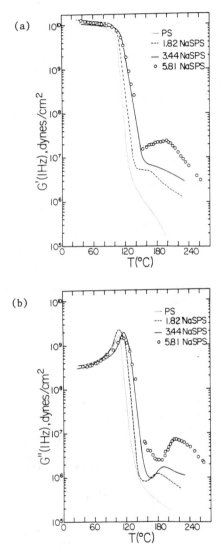

Figure 9.   Effect of sulfonation level on the (a) dynamic and
(b) loss moduli vs. temperature curves of Na-SPS (f = 1 Hz).

the T_g of the hydrocarbon phase and shifted the entire modulus-temperature curve to lower temperatures, see Fig. 10. When the change in T_g was accounted for, the viscoelastic response of the DOP-plasticized SPS was identical to that of the unplasticized ionomer. However, when glycerol was the diluent, little change in T_g resulted, but the high temperature transition associated with the ion-rich phase was shifted to lower temperatures. Thus, the relatively non-polar plasticizer, DOP, preferentially plasticized the hydrocarbon phase, while polar glycerol interacted with the ionic phase.

SOLUTIONS

The properties of ionomer solutions are sensitive to not only the degree of the ionic functionality and the polymer concentration, but perhaps even to a greater extent, the ability of the solvent to ionize the ion-pairs (64). Thus, non-ionizing solvents, usually those with relatively low dielectric constant, favor association of the ionic groups even in dilute solutions. In contrast, ionomer solutions may exhibit polyelectrolyte behavior in polar solvents due to solvation of the ion-pair that leaves the bound ions unshielded.

Most of the literature on ionomer solutions has concentrated on viscosity measurements. Fig. 11 contrasts the viscosity-concentration behavior of SPS dissolved in a tetrahydrofuran (THF), a non-ionizing solvent, and in dimethylformamide (DMF), an ionizing solvent. The viscosity in THF exhibits a monotonic increase as the polymer concentration is increased. The data, however, deviate significantly from that of the unmodified PS at both high and low polymer concentrations. At high polymer concentration, the viscosity of the SPS solution is much greater than that of the PS solution, and the difference increases dramatically as the polymer concentration increases. In fact, at relatively low polymer concentrations (e.g., 5% Mg-SEPDM in paraffinic oil), the ionomer solution may gel in non-polar solvents depending on the molecular mass of the polymer and its ionic functionality (109). The elevation of the viscosity compared with the PS solution and the gelation phenomenon are a consequence of intermolecular dipolar interactions of the ionic species, which increase the effective polymer molar mass and give rise to a transient crosslink network. At low polymer concentrations in a non-polar solvent, the ionomer viscosity curve crosses that of the PS such that the ionomer viscosity is lower. Recent scattering studies have shown that this is a consequence of chain aggregation, which presumably shields some of the frictional sites of the polymer. The chapters by Hara and Wu and Lantman et al. respectively discuss light scattering and SANS of SPS solutions in THF.

Polyelectrolyte behavior is exhibited by solutions of SPS in DMF, Fig. 12. At low polymer concentrations, the viscosity increases as a result of repulsion between the unshielded anions, which increases the hydrodynamic volume of the polymer. The structure of SPS in DMF solutions as determined by static and dynamic light scattering is also discussed in the chapter by Hara and Wu.

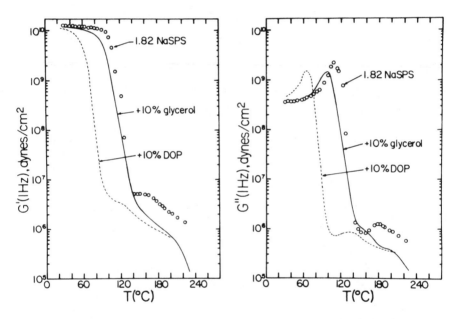

Figure 10. Dynamic (left) and loss moduli (right) vs. temperature for 1.82 mol % NaSPS with and without plasticizers. (Reproduced with permission from reference 108. Copyright 1986 Wiley.)

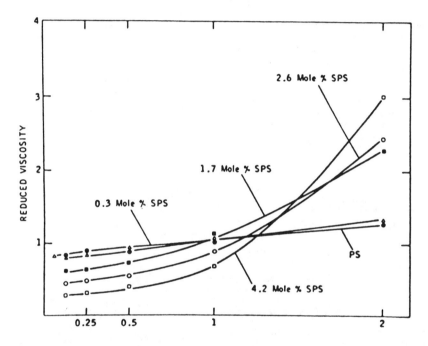

Figure 11. Reduced viscosity vs. concentration relationships of PS and sulfonated polystyrene with different sodium sulfonate levels in tetrahydrofuran. (Reproduced with permission from reference 114. Copyright 1982 Wiley.)

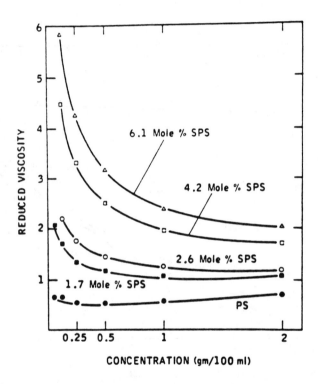

Figure 12.   Reduced viscosity vs. concentration for Na–SPS solutions in dimethylformamide: (●) polystyrene, (■) 1.7, (o) 2.6, (□) 4.2, (Δ) 6.1 mol % Na–SPS. (Reproduced with permission from reference 114. Copyright 1982 Wiley.)

MEMBRANES

Ionomer membranes based on perfluorocarbon polymers became available in the late 1960's. These materials have excellent chemical resistance, thermal stability, mechanical strength and strong acid strength. A number of functionalities have been studied, including carboxylate, sulfonate and sulfonamide, but only the first two are available as commercial materials. Perfluorinated ionomers have been evaluated as membranes in a variety of applications, such as water electrolysis, fuel cells, air driers, Donnan dialysis in waste metal recovery, and acid catalysts, but the primary interest in these materials is for the permselective membrane in electrochemical processes such as in the production of chlorine and caustic (58).

Until recently, nearly all studies of perfluorinated ionomers have been carried out with Nafion (110). Materials with shorter side chains were developed by Dow Chemical Co. in the early 1980's (111) and in a later chapter, Tant et. al. contrast the properties of the short side chain perfluorosulfonic acid ionomers with those of a comparable Nafion material.

Barnes et al. (112) considered the use of sulfonate-ionomers, SPS and perfluorosulfonate membranes, as catalysts for carrying out chemical reactions. In a variation on this theme, Mauritz et al. have prepared unique microcomposite membranes by the in situ growth of silicon oxide in solvent swollen Nafion membranes. This work is described later in this book.

Literature Cited

1. Polymer Modifications of Rubber and Plastics; Keskkula H., Ed.; Appl. Polym. Symposium Series No. 7; J. Wiley and Sons: New York, 1968.
2. Polyblends and Composites; Bruins, P. F., Ed.; Appl. Polym. Symposium Series No. 15; J. Wiley and Sons: New York, 1970.
3. Copolymers, Polyblends and Composites; Platzer, A. J., Ed.; ACS Symposium Series No. 142; Adv. Chem.: Washington, DC, 1975.
4. Manson, J. A.; Sperling, L. H. In Polymer Blends and Composites; Plenum: New York, 1976.
5. Polymer Alloys; Klempner, D.; Frisch, K. C., Eds.; Plenum: New York, 1977; Polymer Alloys II; Plenum: New York, 1980; Polymer Alloys III; Plenum: New York, 1983.
6. Bucknall, C. In Toughened Plastics; Appl. Sci. Publ.: London, 1977.
7. Polymer Blends; Paul, D. R.; Newman, S., Eds.; Academic: New York, 1978.
8. Multiphase Polymers; Cooper, S. J.; Estes, G. M., Eds.; ACS Adv. Chem. Series No. 176: Washington, DC, 1979.
9. Olabisi, O.; Robeson, L. M.; Shaw, M. T. In Polymer-Polymer Miscibility; Academic: New York, 1979.
10. Han, C. D. In Multiphase Flow in Polymer Processing; Academic: New York, 1981.
11. Sperling, L. H. In Interpenetrating Polymer Networks and Related Materials; Plenum: New York, 1981.

12. Polymer Compatibility and Incompatibility; Solc, K., Ed.; MMI: Chur, Switzerland, 1982.
13. Olabisi, O., "Polyblends"; In Kirk-Othmer Encyclopedia of Chemical Technology; J. Wiely and Sons: New York, 1982.
14. The Plastics ABC's: Polymer Alloys Blends and Composites; Shaw, M. T. Ed.; S.P.E. NATEC, Miami, FL., Oct. 1982, Polym. Eng. Sci. 1983, 23 (11,12).
15. Polymer blends and Composites in Multiphase Systems; Han, C. D. Ed.; ACS Adv. Chem. Series No. 206: Washington, DC, 1984.
16. Polymer Blends and Mixtures; Walsh, D. J.; Higgins, J. S.; Maconnachie, A., Eds.; NATO Series E. No. 89; Martinns Nijhoff: Dordrecht, 1985.
17. Integration of Fundamental Polymer Science and Technology; Kleintjens, L. A.; Lemstra, P. Eds.; Elsevier: New York, 1985.
18. Multicomponent Polymer Materials; Paul, D. R.; Sperling, L. H., Eds.; ACS Adv. Chem. Series 211: Washington, DC, 1986.
19. Current Topics in Polymer Science; Ottenbrite, R. M.; Utracki, L. A.; S. Inoue, Eds.; Hanser Ver.: Muchen, 1987.
20. Rheological Measurements; Collyer, A. A.; Clegg, D. W. Eds.; Elsevier; Appl. Sci. Publ.: London, 1988.
21. Utracki, L. A. In Industrial Polymer Alloys and Blends; Hanser Ver.: Muchen, 1989.
22. Polyblends; Annual NRCC/IMRI Series Symposia; Polym. Eng. Sci.: 1982, 22 (2); 1982, 22 (17); 1984, 24 (2);1984, 24 (17); 1986, 26, (1);1987, 27 (5);1987, 27 (6);1987, 27 (20);1987, 27 (21);1988, 28 (00).
23. Reich, S., Phys. Lett. 1986, 114A, 90.
24. Voigt-Martin, I. G.; Leister, K.-H.; Rosenau, R.; Koningsveld, R. J. Polym. Sci.: B. Polym. Phys. 1986, 24, 723.
25. Hashimoto, T. Chapter 6.1 in ref. 19.
26. Inoue, T.; Ougizawa, T.; Niyasaka, K., Chapter 6.2 in ref. 19.
27. Tsukahara, Y.; Kohno, K.; Yamashita, Y. Chapter 6.7 in ref. 19.
28. Blahovici,T. F.; Brown, G. R., NRCC/IMRI Symposium Polyblends-'88; paper No. 6; Bouchervill: Quebec, Canada; April 5 and 6, 1988; Polym. Eng. Sci. 1989, 29, 000.
29. Walsh, D. J.; Cheng, G. L. Polymer 1982, 23, 1965.
30. Flory, P. J.; Orwoll, R. A.; Vrij, A. J. Am. Chem. Soc. 1984, 86, 3507.
31. Coleman, M. M.; Moskala, E. J.; Painter, P. C.; Walsh, D. J.; Rostami, S. Polymer 1983, 24, 1401.
32. Walsh, D. J.; Cheng, G. L. Polymer 1984, 25, 499.
33. Goldsmith, H. L.; Mason, S. G. In Rheology, Theory and Applications; Eirich, F. R. Ed.; Academic: New York, 1969; Grace, H. P. Chem. Eng. Commun. 1982, 14, 225.
34. Van Oene, H. J. Colloid Interface Sci. 1972, 40, 448.
35. Wu, S. In Polymer Interfaces and Adhesion; M. Dekker Inc.: New York, 1979.
36. Miles, I., ICI report to VAMAS Working Party on Polymer Blends; 1985.
37. Elmendorp, S. Ph.D. Thesis, Delft Univ. Technolog., The Netherlands, 1986.
38. Utracki, L. A.; Sammut, P., NRCC/IMR Symposium Polyblends-'88; April 5 and 6, 1988; Polym. Eng. Sci. 1988, 28, 000.

39. Utracki, L. A., J. Rheol. 1986, 30, 829.
40. Lyngaae-Jorgensen, J., private communications, 1987.
41. Lin, C.-C. Polym. J. 1979, 11, 185.
42. Paul, D. R.; Barlow, J. W. J. Macromol. Sci.: Rev. 1980, C18, 109.
43. Elmendorp, J. J.; van der Vegt, A. K. 3rd Annual Meeting Polymer Processing Society, Stuttgart, FRG, April 7-10, 1987.
44. Thorton, B.; Villasenor, R. G.; Maxwell, B. J. Appl. Polym. Sci. 1980, 25, 653.
45. Larson, R. G.; Fredrickson, G. H. Macromolecules 1989, 20, 1897.
46. Lyngaae-Jorgensen, J. In Processing, Structure and Properties of Block Copolymers; Folkes, M. J. Ed.; Elsevier Appl. Sci. Publ.: London, 1985.
47. Katsaros, J. K.; Malone, M. F.; Winter, H. H. Polym. Bull. 1986, 16, 83.
48. Lyngaae-Jorgenson, J.; Sondergaard, K. Polym. Eng. Sci. 1987, 27, 344.
49. Weiss, R. A.; Higgins, J. S.; Galambos, A.; Hindawi, I. unpublished results.
50. Winter, H. H. private communication, 1988.
51. Spriggs, T. W. Chem. Eng. Sci. 1965, 20, 931.
52. Utracki, L. A.; Bata, G. L. Mater. Techn.; Paris, 1982, 70, 223, 282.
53. Williams, J. G.; Hashemi, S. Chapter 14 in ref. 16.
54. Huang, D. D.; Williams, J. G. J. Metr. Sci. 1987, 22, 2503.
55. Ionic Polymers; Holliday, L., Ed.; Applied Science Publ.: London, 1975.
56. Eisenberg, A.; King, M. In Ion-Containing Polymers; Academic: New York, 1977.
57. Ions in Polymers; Eisenberg, A. Ed.; Adv. Chem. Ser. 187; Amer. Chem. Soc.: Washington, DC, 1980.
58. Perfluorinated Ionomer Membranes; Eisenberg A.; Yaeger, H. J. Eds.; ACS Syposium Series No. 180; Amer. Chem. Soc.: Washington, DC, 1982.
59. a. Developments in Ionic Polymers-1; Wilson, A. D.; Prosser, H. J., Eds.; Applied Science Publ.: London, 1983; b. Developments in Ionic Polymers-2; Elsevier Applied Science Publ.: London, 1986.
60. MacKnight, W. J.; Earnest, T. R. Jr., J. Polym. Sci.: Macromol. Rev., 1981, 16, 41.
61. Bazuin, C. G.; Eisenberg, A. Ind. Eng. Chem. Prod. R & D 1981, 20, 271.
62. MacKnight, W. J.; Lundberg, R. D. Rub. Chem. Technol. 1984, 57, 652.
63. Coulombic Interactions in Macromolecular Systems; Eisenberg, A.; Bailey, F.E., Eds.; ACS Symposium Series No. 302; Amer. Chem. Soc.: 1986
64. Fitzgerald J. J.; Weiss, R. A. J. Macromol. Sci.: Rev. Macromol. Chem. Phys. 1988, C28, 99.
65. Rees, R. W. U.S. Patent 3 322 734, 1966, to E. I. du Pont de Nemours & Co.
66. Weiss, R. A.; Lundberg, R. D.; Werner, A. J. Polym. Sci.: Polym. Chem. Ed. 1980, 18, 3427.

67. Weiss, R. A.; Turner, S. R.; Lundberg, R. D. J. Polym. Sci.:
    Polym. Chem. Ed. 1985, 23, 525.
68. Turner, S. R.; Weiss, R. A.; Lundberg, R. D. J. Polym. Sci.:
    Polym. Chem. Ed. 1985, 23, 535.
69. Weiss, R. A.; Turner, S. R.; Lundberg, R. D. J. Polym. Sci.:
    Polym. Chem. Ed. 1985, 23, 549.
70. Allen, R. D.; Yilgor, I.; McGrath, J. E. p 79 in Ref. 63.
71. Salamone, J. C.; Watterson, A. C.; Hsu, T. D.; Tsai, C. C.;
    Mahmud, M. U. J. Polym. Sci.: Polym. Chem. Ed. 1977, 15, 487.
72. Salamone, J. C.; Watterson, A. C.; Hsu, T. D.; Tsai, C. C.;,
    Mahmud, M. U.; Wisniewski, A. W.; Israel, S. C. J. Polym.
    Sci.: Polym. Symposium, 1978, 64, 229.
73. Salamone, J. C.; Tsai, C. C.; Watterson, A. C. J. Macromol.
    Sci.: Chem. 1979, A13, 665.
74. Salamone, J. C.; Tsai, C. C.; Alson, A. P.; Watterson, A. C.
    J. Polym. Sci.: Polym. Chem. Ed. 1980, 18, 2983.
75. Salamone, J. C.; Tsai, C. C.; Watterson, A. C.; Olson, A. P.
    In Polymer Amines and Ammonium Salts; Goethals, E. J. Ed.;
    Pergamon: Oxford, 1980.
76. Salamone, J. C.; Muhmud, N. A,; Mahmud, M. U.; Nagabhushanam,
    T.; Watterson, A. C. Polymer 1982, 23, 843.
77. Makawski, H. S.; Lundberg, R. D.; Singhal, G. H. U.S. Patent
    3 870 841, 1975.
78. Makowski, H. S.; Lundberg, R. D.; Westerman, L.; Bock, J. P3
    in Ref. 57.
79. Bishop, M. T.; Karasz, F. E.; Russo, P. S.; Langley, K. H.
    Macromolecules 1985, 18, 86.
80. Jin, K.; Bishop, M. T.; Ellis, T. S.; Karasz, F. E. Br. Polym
    J. 1985, 17, 4.
81. Zhou, Z. L.; Eisenberg, A. J. Polym. Sci.: Polym. Phys. Ed.
    1982, 27, 657.
82. Rahrig, D.; MacKnight, W. J.; Lenz, R. W. Macromolecules
    1979, 12, 195.
83. Peiffer, D. G.; Lundberg, R. D.; Duvdevani, I. Polymer 1986,
    27, 1453.
84. Hwang, K. K. S.; Yang, C. Z.; Cooper, S. L. Polym. Eng.
    Sci. 1981, 21, 1027.
85. Miller, J. A.; Hwang, K. K. S.; Cooper, S. L. J. Macromol.
    Sci., Phys. 1983, B22, 321.
86. Kennedy, J. P.; Storey, R. F.; Mohajer, Y.; Wilkes, G. L.
    Proc. IUPAC, Macro '82 1982, p 905.
87. Kennedy, J. P.; Ross, L. R.; Lackey, J. E.; Nuyken, O.
    Polym. Bull. 1981, 4, 67.
88. Kennedy, J. P.; Storey, R. F. Am. Chem. Soc., Prepr. Div.
    Org. Coat. Appl. Polym. Sci. 1982, 46, 182.
89. Omeis, J.; Muhleisen, E.; Moller, M. Polym. Prepr. 1986,
    27(1), 213.
90. Cangelosi, F.; Shaw, M. T. Polym. Plast. Tech. Eng. 1983,;
    21, 13.
91. Eisenberg, A.; Smith, P.; Zhou, Z. L. Polym. Eng. Sci. 1982,;
    22, 455.
92. Zhou, Z. L.; Eisenberg, A. J. Polym. Sci.: Polym. Phys. Ed.
    1983, 21, 595.
93. Eisenberg, A.; Hara, M. Polym. Eng. Sci. 1984, 24, 17.

94. Weiss, R. A.; Beretta, C. A.; Garton, A.; manuscript submitted to J. Appl. Polym. Sci.
95. Eisenberg, A. Macromolecules 1970, 3, 147.
96. Mauritz, K. A. In Structure and Properties of Ionomers; Pineri, M.; Eisenberg, A., Eds.; Reidel Publ. Co.: Dordrecht, 1987, p 11.
97. Mauritz, K. A.; Hora, C. J.; Hopfinger, A. J. p 123 in ref. 57.
98. Hsu, W. Y.; Gierke, T. D. Macromolecules 1982, 15, 101.
99. Mauritz, K. A.; Rogers, C. E. Macromolecules 1985, 18, 483.
100. Forsman, W. C. Macromolecules 1982, 15, 1032.
101. Dreyfus, B. p 103 in ref. 63.
102. Galambos, A. F.; Stockton, W. B.; Koberstein, J. T.; Sen, A.; Weiss, R. A.; Russell, T. P. Macromolecules 1987, 20, 3091.
103. Earnest, Jr., T. R.; Higgins, J. S.; Handlin, D. L.; MacKnight, W. J. Macromolecules 1981, 14, 192.
104. Chu, B. Macromolecules 1988, 21, 523.
105. Yarusso, D. J.; Cooper, S. L. Macromolecules 1983, 26, 1871.
106. Handlin, D. L.; MacKnight, W. J.; Thomas, E. L. Macromolecules 1981, 14, 795.
107. Lundberg, R. D.; Makowski, H. S.; Westerman, L. p 67 in ref. 57.
108. Fitzgerald, J. J.; Kim, D.; Weiss, R. A. J. Polym. Sci. Polym. Let. Ed. 1986, 24, 263.
109. Agarwal, P. K.; Lundberg, R. D. Macromolecules 1984, 17, 1928.
110. Registered trademark of the E. I. Dupont de Nemours & Co.
111. Ezzell, B. R.; Carl, W. P.; Mod, W. A. U.S. Patent 4 358 412, 1982.
112. Barnes, D. M.; Chaudhuri, S. N.; Chryssikow, G. D.; Mattera, Jr., V. D.; Pelusso, S. L.; Shim, I. W.; Tsatsas, A. T; Risen, Jr., W. M. 1986, p 66. in ref. 63.
113. Fitzgerald, J. J. Ph.D. Thesis, Univ. of Connecticut, 1986.
114. Lundberg, R. D.; Phillips, J. J. Polym. Sci.: Polym. Phys. Ed. 1982, 20, 1143.

RECEIVED April 27, 1989

THERMODYNAMICS

Chapter 2

Interface Modification in Polymer Blends

R. Fayt, R. Jérôme[1], and P. Teyssié

Laboratory of Macromolecular Chemistry and Organic Catalysis, University of Liège, Sart Tilman B6, B-4000 Liège, Belgium

Two strategies are now available that allow a poor interfacial adhesion to be improved in any type of multiphase polymeric systems: use of a diblock copolymer designed as an effective interfacial agent, and specific interactions between the partners to be associated. Meaningful examples for the power of the interfacial activity of diblock copolymers are discussed: they cover a large spectrum of situations, such as commodity and engineering polymer blends and dispersions of fine solid particles in a liquid phase or a polymeric matrix. Blending mutually interacting telechelic polymers is a promising way to promote phase morphologies very similar to those seen in covalently bonded block copolymers. That approach based on mutual interactions is also useful to stabilize dispersions of solid particles in a liquid or a solid phase.

It may seem incredible that the molecular structure of a high molecular weight compound was completely unknown at the beginning of the century, although polymer based materials have burst since to a nearly 100 M ton/year total production. Actually, polymer science and engineering started in the 1920's when Herman Staudinger postulated a macromolecular structure for the few big molecules known at that time like cellulose and natural rubber. During the period of evolution that followed what one can call a "Big Bang" in polymer science (1), three main strategies have been used in order to develop new polymeric materials. The search for new monomers, and accordingly for new polymers, is quite a common approach, the issue of which is however questionable due to stringent economic and technical restrictions. This is particularly true for

[1]Author to whom correspondence should be directed

monomers polymerizing by the addition mechanism, while,
now and then, the step-growth polymerization gives rise
to a new polymer, as examplified by poly(ether-ether-
ketone) (PEEK). Copolymerizing the available monomers is
a rather old strategy that promoted the production of
more or less random copolymers, several decades ago. This
type of copolymers exhibit all the characteristic features of chemically homogeneous materials with properties
intermediate to those of the related homopolymers. As a
result, random copolymers do nothing but extend the range
of the available homopolymers, which is advantageous but
still not creative enough. It was not until 1955, when
Szwarc discovered the living anionic polymerization (2),
that the synthesis of block and graft copolymers could be
considered. These copolymers are essentially multiphase
materials because of the immiscibility of most of their
polymeric constituents. Accordingly, they have the great
advantage to sum up the properties of each of the partners
and possibly to exhibit a new and sometimes unexpected
behavior in relation to a particular phase morphology. So,
the way was open to technological innovations as illustrated by the discovery of the thermoplastic elastomers
(Solprene, Kraton, Hytrel,...) (3). Today, blending of the
available polymers and copolymers in existing processing
equipments is a still more general answer to challenges
raised by modern technology. It is worth mentioning that
the growth rate of polymer blends was approximately 15% a
year over the last decade leading to a 1 billion U.S.$
market (4,5).
    Focusing on polymer blending two extreme situations
have to be pointed out, depending on the polymer miscibility. Would the polymers be miscible, then their blends
are quite similar to the random copolymers comprising the
same monomer units used in the same molar ratio. As
already stressed, this approach cannot generate technological breakthroughs, but it can however extend the application range of some polymers as supported by the commercialization of poly(phenylene ether) (PPE) as homogeneous
blends with polystyrene (i.e. Noryl) (6). Blending of
completely immiscible polymers might be much more fruitful
since phase separation of multicomponent materials seems
to be a prerequesite for novel properties. Unfortunately,
a very poor interfacial adhesion is very often responsible
for disappointing results. That situation could however be
alleviated when two immiscible polymers in the melt are
compared to e.g. oil-in-water emulsions, the stability of
which is usually promoted by the well-known surfactants
(7). By analogy with the structure of surfactants that is
comprised of a polar head soluble in water and a hydrophobic tail soluble in oil, it might be suggested that a PA-
PB diblock copolymer is a surface-active agent for the PA
and PB homopolymers blends (Figure 1). This concept has
been proposed two decades ago and convincingly substantia-
ted in the meantime (8). The next sections will illustrate

how fruitful that strategy can be for providing immiscible
polymer blends with improved mechanical performances

Interfacial activity of block copolymers in commodity polymer blends

The effect that block copolymers could have on the melt
blending of immiscible polymers and the ultimate properties of the polyblends has been first considered for very
widespread nonpolar commodity polymers, like polyethylene
(PE) and polystyrene (PS) (9-26). It is obvious that the
melt blending of low-density polyethylene (LDPE) and PS
generates polyblends that combine over a broad composition
range the low strength of LDPE and the brittleness of PS
(Figure 2). The origin of these very poor performances has
to be found in a lack of adhesion between the phases as
highlighted by scanning electron microscopy of fracture
surfaces (11,14,19). 15 years ago, two pioneer research
groups published data supporting that the situation could
be improved using LDPE chains grafted with PS (9,11,14).
Beneficial effects on mechanical strength, especially
modulus and impact resistance, have been attributed to the
interfacial activity of the graft copolymers (9,12,14).
Later on, poly (ethylene-b-styrene) copolymers have been
purposely synthesized by the sequential living anionic
polymerization of styrene and butadiene, followed by the
selective hydrogenation of the polydiene blocks into polyolefin ones (12,16-18). When butadiene is anionically
polymerized in a nonpolar solvent, the microstructure of
the polydiene comprises ca. 10-15 mole % of randomly distributed vinyl units. The hydrogenation of the polydiene
block leads accordingly to a block (hPBD) comparable to
LDPE with approximately 30 ethyl branches per 1000 C
atoms.
Electron microscopy provides convincing evidences of the
interfacial activity of the block copolymer. SEM of fracture surfaces do not allow the two components to be easily distinguished from each other supporting an intimate
connection of the phases (19,20). Transmission electron
microscopy shows that a PS-hPBD copolymer modified by a
short OsO4-stained central block forms a continuous layer
around the dispersed phases of either PS or PE (21). As
previously reported for graft copolymers (9,22), the
addition of block copolymers significantly enhances both
the ultimate tensile strength and the elongation at break
resulting in a striking increase of the energy to break
(Figure 3). In this study, the content of diblock copolymer has been calculated with respect to the total weight
of the blended polymer, and the original cross-section
area of tensile bars has been considered in the calculation of stress. Furthermore previous papers have reported
how tensile bars were machined, stress-strain curves
performed (cross head speed of 2 cm min^{-1}) and specimens
for electron microscopy prepared (17,19). The increase in

Figure 1. Analogy of traditional surfactants and diblock copolymers as potential surface-active agents in immiscible polymer blends.

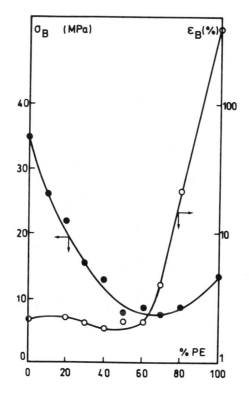

Figure 2. Ultimate tensile strength (σ_B) and elongation at break (ε_B) of LDPE ($\bar{M}n$: 40K) and PS ($\bar{M}n$: 100K) blends.

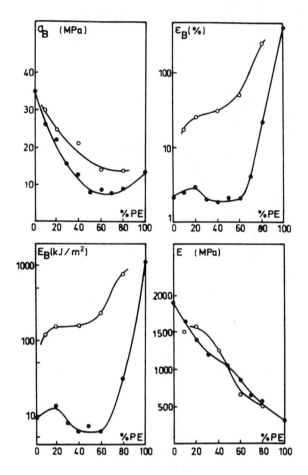

Figure 3. Ultimate tensile strength (σ_B), elongation at break (ε_B), energy to break (E_B) and Young modulus (E) of LDPE ($\bar{M}n : 40K$) and PS ($\bar{M}n : 100K$) blends; ($\bullet$) without copolymer ; (O) with 9% of a PS/hPBD diblock.

tensile strength promoted by the block copolymer (Figure
3) has been rationalized by a substantial increase in the
activation energy for craze initiation (from 40 to 180
KJ/mol) (23). It means that void formation is prevented
by the diblock as a result of an enhanced interfacial
adhesion. Interestingly enough, the modulus remains
essentially unmodified upon the addition of the PS-hPBD
copolymer; this effect has been explained elsewhere (20).
    When the question of the molecular architecture of
the block copolymer used as an interfacial agent is
addressed, the superiority of the diblock structure over
the graft copolymer and the linear and star-shaped triblock copolymers leaves no doubt (16). As an example, SEM
illustrates the poor interfacial activity of a triblock
copolymer compared to the related diblock one (Figure 4).
The same conclusion arises from the improvement in
strength and elongation at break that LDPE/PS blends
exhibit when modified by copolymers of various architec-
tures (16). It might be suggested that a diblock suffers
from less drastic conformational restraints at the interface, and that its constitutive blocks can penetrate more
deeply the corresponding homopolymer phases providing
them with a stronger mutual anchorage. An interesting
piece of comparison between diblock and triblock copolymers has been provided by Paul et al (24) who modified
blends of PS and high density polyethylene (HDPE) using a
poly(styrene-b-ethylene-co-butylene-b-styrene) (Kraton
G1652) triblock copolymer. Paul made comparisons and contrasts where possible (25) with a similar work from our
laboratory based on PS-hPBD diblock copolymers (17,26). In
contrast to graft copolymers, diblock copolymers are
known for their tendency to associate into micelles and
mesophases upon increasing concentrations (27). Therefore,
the interfacial activity that a block copolymer has in a
polyblend, could be limited by the tendency to phase
separate in one phase instead of being located at the
interphase (12,22).
    Let us consider exclusively now PS-hPBD diblocks,
with an approximately 50/50 composition. Blocks of the
same length should indeed favour the location of the copolymer at the interface, since hydrophobic interactions
and chain entanglements between each block and the related
homopolymer are expected to match closely. The question
arises as to whether the total molecular weight has an
effect on the interfacial activity. By comparison with
Figure 4a, Figure 5 shows that a 18K molecular weight is
too small to stabilize efficiently the mechanically induced phase dispersion. Expectedly, the immiscibility of
the very short blocks is not great enough to impart a
significant interfacial activity. On the other hand, when
the total molecular weight is too high (400K), the interfacial situation is also disappointing as evidenced by
the mean size of the dispersed phases and the apparent
lack of adhesion between the phases (Figure 5). This might

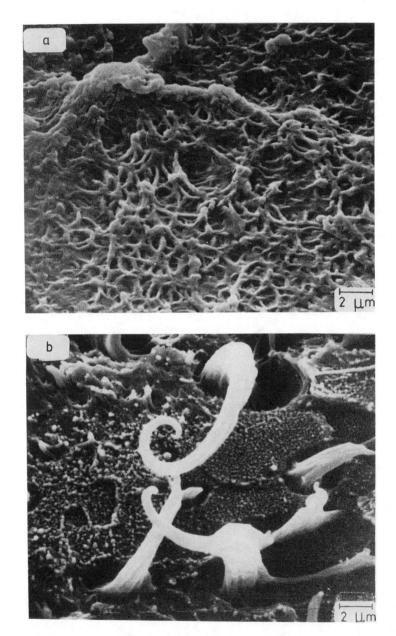

Figure 4. Scanning electron micrographs of a 20/80 blend of LDPE and PS, melt blended with 10% of a PS/hPBD diblock (a) and a PS/hPBD/PS triblock (b), respectively.

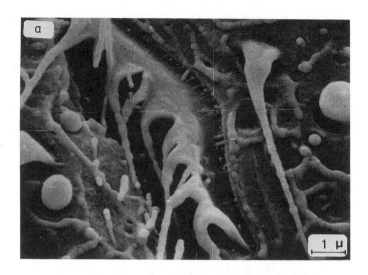

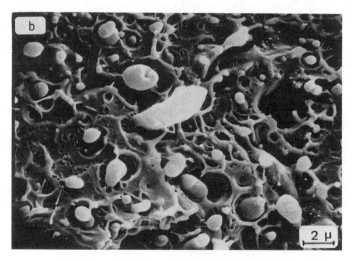

Figure 5. Scanning electron micrographs of a 20/80 blend of LDPE and PS, melt blended with 10% of a PS/hPBD diblock of 18K (a) and 400K (b) molecular weight, respectively.

be due to a prohibitively high melt viscosity of the diblock copolymer and a substantial decrease in its critical
micellar concentration. Would homopolymers have to diffuse
into phase separated domains of the copolymer, information on that process has been made available recently
using elastic recoil detection (28). It has been found
that homopolymers diffuse into symmetric diblock structures, above Tg's, at a rate inversely proportional to
the square of their length, when the latter is shorter
than the related block. In cases where homopolymer is
longer than the parent block, the dependence of the diffusion coefficient on the chain length is considerably
more severe than the reciprocal square of that length.
Diffusion coefficients have been found to be smaller by
an order of magnitude than the values reported for the
homopolymer chains diffusing into their respective homopolymer hosts. Undoubtedly, geometrical factors (tortuosi-
ty and orientation of the microphases) control diffusion
in addition to the molecular weight between entanglements
and the monomeric friction coefficient as in the case of
homopolymer-homopolymer diffusion.
According to these observations, it ensues that the optimal molecular weight should be between 18K and 400K.
Figure 6 compares two diblocks of intermediate molecular
weight, i.e. 80K and 270K, respectively. As a rule, a
shorter diblock copolymer gives rise to higher ultimate
tensile strength, while a higher total molecular weight
leads to a more ductile material with values of the elongation at break higher than additive values. In order to
account for these clear-cut effects, the phase morphology
and the relative contribution of crazing and shearing to
deformation should be established. Indeed, not only the
interface properties but also the phase morphology could
be operative in the deformation mechanism.
Quite interestingly, the mechanical properties of the PS-
rich blends strongly depends on the molecular weight of
the diblock copolymer used as interfacial agent. Upon the
addition of the 80K molecular weight diblock, a synergism
in strength is observed whatever the type of PE used, i.e.
low density or high density PE (Figure 6). In contrast,
using a higher molecular weight diblock prevents the synergism in strength to occur, but improves the elongation
at break (Figure 6) to such an extent that the mechanical
features of a toughened thermoplastic are reported
(Figure 7). The effect that the molecular weight of the
diblock copolymer has on the mechanical behavior of the
PS-rich blends should have something to do with the phase
morphology. Figure 8 illustrates the PE phase which is
left when the 80wt% of PS have been selectively extracted
from the related polyblends. Micrograph(a) shows that PE
forms a continuous phase when melt-blended with PS.
Accordingly, the parent related polyblend has to display
a co-continuous two-phase morphology. As shown elsewhere,
this morphology is however unstable, and PE coalesces

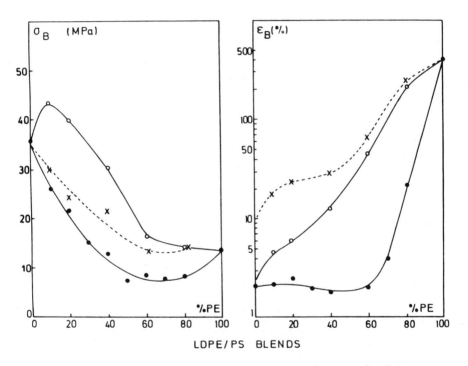

Figure 6. Ultimate tensile strength (σ_B) and elongation at break (ε_B) of LDPE (40K) and PS (100K) blends;
without copolymer (●); with 9% PS/hPBD diblock of 80K
(o) and 270K (X) molecular weight, respectively.

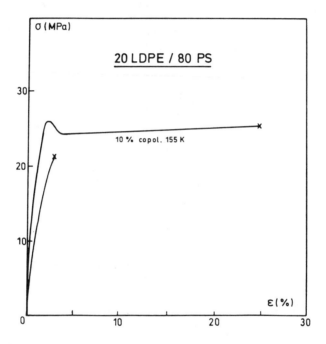

Figure 7. Effect of 10% of a PS/hPBD diblock on the stress-strain curve of a 20/80 blend of LDPE (40K) and PS (100K).

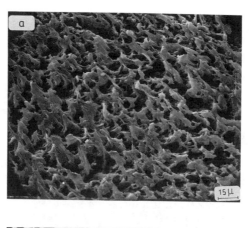

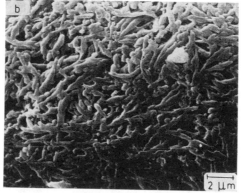

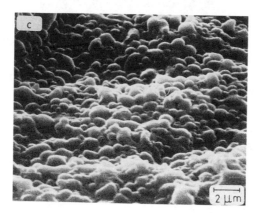

Figure 8. Scanning electron micrographs of a fractured 20/80 blend of LDPE (40K) and PS (100K) after selective extraction of PS by THF; no diblock (a); 10% PS/hPBD diblock of 80K (b) and 270K (c), respectively.

readily upon heating (16). Micrograph (b) illustrates
that a low M.W. diblock copolymer (80K) reduces the mean
size of the interlocked phases and stabilizes them against
any thermal or mechanical post-treatment. Finally, increasing the length of the diblock copolymer favours the
discontinuity of the PE microphases (micrograph(c)) which
are no longer intermingled within the PS matrix. Upon the
addition of a too viscous diblock copolymer, the melt
rheology of the polyblend seems to be modified to a point
where the phase morphology changes dramatically and, as a
result, the ultimate mechanical properties. Very clearly,
the molecular weight of the diblock copolymer has to be
optimized in relation to the property to be improved or
the performances to be bestowed on the polyblend. It is
not surprizing that the ideal molecular weight is unpre-
dictable when one refers to the complexity of phase
grams of polymer blends containing block copolymers (29-
32). For instance, when low amounts of a PS-PBD copolymer
is mixed within PS (or PBD), three main situations can
prevail in the phase diagram depending on temperature and
relative length of blocks and homopolymer: molecular dissolution of the copolymer in the homopolymer, micelliza-
tion of the diblock in the homopolymer-rich matrix, and
ordered arrays of densely packed micelles coexisting with
a dilute suspension of independent micelles. Would the
second homopolymer be added, the phase diagram increases
in complexity, resulting in a chancy location of the copolymer at the interface under the processing conditions.
Shearing the polyblend is another variable making predictions still more hazardous, while the effect of block
copolymers on melt rheology of immiscible polymers is
still a pending question.
Not only molecular weight influences the interfacial activity of the PS-hPBD diblock copolymers but also the internal structure. Instead of a single covalent bond between
PS and hPBD, a more or less random copolymer sequence can
form the transition between the two blocks resulting in
the so-called tapered diblock copolymers. The effect of
the tapered structure (12,22) and its superiority over the
pure diblock (18) have been documented elsewhere and discussed in terms of a change in the interface itself and
the melt viscosity of the interfacial agent.
     On a cost-benefit basis, it is interesting to outline
that the best mechanical performances of the PE-PS polyblends can be reached using a minimum amount (ca. 2wt%) of
an appropriate diblock copolymer (19). Furthermore, not
only reproducible samples can be prepared under processing
conditions, but an apparent equilibrium of phase morphology and mechanical properties is obtained within half to a
few minutes depending on the melt viscosity of the blend
and especially the microstructure of the diblock copolymer.

Emulsification of polar commodity polymer blends

On the basis of the PE/PS polyblends discussed hereabove, it can be stated that the emulsification of immiscible polymer blends by a few percents of tailored diblock copolymers is a concept that should be extended to other systems, and especially to blends of more polar commodity polymers, like PVC, SAN and PMMA. This very attractive extension step is however limited by the capability of the polymer chemist to synthesize appropriate diblock copolymers. When it is difficult (or impossible) to synthesize in a controlled way copolymers containing blocks of the same structure as the blended homopolymers, one should consider blocks miscible with these homopolymers. It was proposed some time ago (33) that attention should be paid to pairs of polymers that have mutual interactions, such as H-bonding or dipolar interactions, strong enough to promote miscibility. In this regard, s.PMMA, as prepared by anionic polymerization in a polar solvent, is an interesting constitutive block due to its miscibility with a number of polymers, such as PVC, SAN and PVF_2 (polyvinylidene fluoride) (34). Furthermore, MMA can be block polymerized by sequential living anionic polymerization with styrene or 1,4-conjugated dienes, and the selective hydrogenation of polydienes can be performed in the presence of PMMA (35).

Among other targets, it is tempting to incorporate a preformed rubber into fragile thermoplastics (SAN,PVC) by melt-blending. If effective, this would be a direct way for the production of high impact materials using the available processing equipments. It is well-known that high impact performances strongly depend on the state of dispersion of the rubber particles in the thermoplastic matrix (36). Actually, they require a good interfacial adhesion, an optimum average diameter of the rubber particles (0,5µm) and the appropriate crosslinking of the rubber. Since it is difficult to control the crosslinking during the melt blending process, a thermoplastic elastomer, i.e. a physically crosslinked rubber has been purposely selected. Solprene 411 is a well-known thermoplastic elastomer containing polybutadiene as the major component and polystyrene as the minor one. Considering PS-s.PMMA and PBD-s.PMMA diblocks, the PS as well as the PBD blocks can interact with Solprene, while the s.PMMA block is selectively miscible with SAN. Actually the impact strength of SAN is dramatically improved when it is melt-blended with Solprene, in the presence of 9% of a PBD-s.PMMA diblock copolymer. The Charpy test indicates a tenfold increase in impact resistance (37). Quite interestingly, the impact resistance of that polyblend is substantially greater than that of a commercial ABS resin (Cycolac GSM Natural) - 35 against 19KJ/m^2 according to Charpy test, and 0,51 against 0.30 KJ/m according to Izod test - while all other mechanical features like modulus, are quite comparable (37). According to scanning electron

micrographs of fracture surfaces, the superiority
of the polyblend over the traditional ABS resin might be
attributed to the stronger interfacial adhesion promoted
by the diblock copolymer used in the polyblend in contrast to the graft copolymer generated during the production of the ABS resin (37). The rubber particles, stained
by OsO4 and observed by transmission electron microscopy,
are generally smaller than 1 μm - thus very close to the
required value - and the phase morphology has nothing to
do with the usual honey-combed structure of the traditional toughened plastics (36) supporting that the shape of
the rubbery phases has no critical effect on the impact
resistance.
        Finally, it has been stated that this new toughening
technique can be safely extended to other fragile plastics, such as PVC and PS (38). This observation supports
that a properly tailored diblock copolymer can help in
generating materials with highly improved properties.

Engineering polymer blends

Filling an expensive engineering polymer, like PVF2, with
a cheaper commodity polymer is certainly a valuable economic target. That approach has been successfully applied
melt-blending PVF2 with poly(α-methylstyrene) (PMS) and
polyolefins (39,40). The most convincing example of the
beneficial effect of a diblock copolymer on melt-blending
of PVF2 has to be found in the PE containing mixtures.
PVF2 and PE are indeed immiscible to such an extent that,
after milling and molding, their blends are quite heterogeneous and show a coarsely layered and non-cohesive
structure (39). Both optical and scanning electron microscopy (Figure 9) support that 10% (non optimized value)
of a poly(hydrogenated butadiene-b-MMA) copolymer prevent
efficiently any macrophase separation to occur and lead
to a valuable material. It means that the diblock copolymer acts as an effective processing aid and compatibilizing agent.
The same conclusion arises when PVF2-PMS blends are considered (40). For instance, optical microscopy shows that
upon the addition of a PMS-PMMA diblock copolymer, the
mean size of the phases decreases below one micron. Similarly, the two-phase morphology that is easily dislocated
by cryofracture is considerably improved by the diblock
copolymer, and accordingly the surface fracture of a
60/40 PVF2/PMS blend consists of small (200 nm), regular
in size beads firmly bonded to the matrix. Finally, the
emulsification of the blends leads to averaging values of
the Young modulus, while high ultimate tensile strengths
are observed when PMS is the minor component.
        Last but not least, PVF2 has also been melt-blended
successfully with another engineering polymer, like Noryl
(PPE + PS) using a PS-PMMA diblock copolymer (41), since
it is known that PPE and PS are miscible (6). Scanning

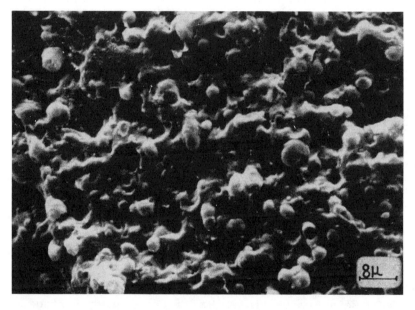

Figure 9. Scanning electron micrograph of a 60/40 blend of PVF$_2$ and LDPE added with 10% of a hPBD/PMMA diblock.

electron microscopy of fracture surfaces of a 80/20 PVF$_2$-
Noryl blend supports that 2% of the PS-PMMA emulsifier
promote an approximately 3 to 5 fold reduction in size of
the Noryl domains, while 12% (Figure 10) do not allow the
phases to be identified without resorting to some solvent
or etching techniques. The approximate additivity of the
ultimate tensile strength results from the addition of
only 2% of the diblock copolymer, while the same situation
is reported for the elongation at break provided the
interfacial adhesion is strong enough (12% diblock). That
improvement in ductility is dramatic when the extreme
brittleness of the unmodified polyblends is considered.
     The reported examples convincingly support that using
a purposely tailored diblock copolymer is a powerful tool
of generating new, or at least improved, multiphase polymeric materials. Quite interestingly, the interfacial activity of diblock copolymers is not exclusively confined
to immiscible polymer blends but it can be fruitfully
extended to dispersions of fine solid particles in either
a liquid phase or a polymeric matrix, as illustrated
hereafter.

Diblock copolymers as surface-active agents in the
dispersion of solid particles

Nowadays, there is a demand for dispersants of fine solid
particles in non-aqueous media, for the preparation of
inks, paints and liquid electrographic toners. Block copolymers are potential stabilizers provided their constitutive blocks are mutually repulsive and interact selectively with the particles and the dispersion medium,
respectively.
     Poly(styrene-b-stearyl methacrylate) copolymers have
been synthesized by the sequential living anionic polymerization of the comonomers and found to stabilize over
long periods of time dispersions of colloidal carbon black
in a hydrocarbon, like isododecane (42). The insoluble PS
block is selectively adsorbed onto the pigment surface,
while the soluble poly(stearyl methacrylate) block forms
an effective steric barrier around the particles. Depending on the total molecular weight and weight composition
of the diblock copolymer, particles of a diameter of 300
to 500 nm have been stabilized. Moreover, the experimental
adsorption of the copolymer on carbon black is consistent
with the Langmuir predictions. The area occupied by one
amphipatic macromolecule corresponds pretty well to the
diameter expected for the collapsed PS block.
The anionic block copolymerization of methyl methacrylate
and glycidyl methacrylate leads to very versatile surface-
active agents. The hydrophilic/hydrophobic balance of
these amphipatic macromolecules can be easily controlled
not only by the relative length of the constitutive blocks
but also by the chemical modification of the epoxy groups.
The poly(glycidyl methacrylate) block can be easily sulfo-

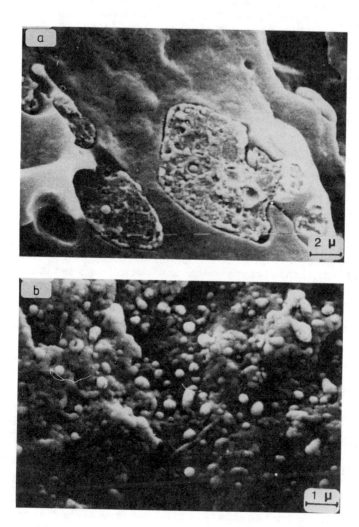

Figure 10. Scanning electron micrographs of a 80/20 blend of PVF$_2$ and Noryl, before and after addition of 12% of a PS/PMMA diblock.

nated by phase transfer catalysis (43) or reacted by pro-
tic organic functions (carboxylic acid, amine) with formation of more polar and/or adsorbing groups (44). In
aqueous media, the sulfonated diblock copolymer is a potential surfactant of emulsions provided the insoluble
PMMA block is irreversibly adsorbed onto the polymer particles, or coprecipitated within the dispersed polymer.
In non-aqueous pigment dispersions, proton transfer from
the acid groups of a polymeric dispersant to the surface
of basic particles is a process promoting effective
stabilization (45).
     Dispersion of solid particles in a polymeric matrix
is another target that can be reached successfully using a
purposely tailored diblock copolymer.
For instance, 25% of CaCO3 can be dispersed into low density PE, using a poly(hydrogenated butadiene-b-acrylic
acid) copolymer as an interfacial agent. The polyolefin
block is miscible to the LDPE matrix, while acid-base interactions are expected to occur between the polyacid
block and the CaCO3 particles. The upper scanning electron
micrograph of Figure 11 shows that the particles are not
expelled during the fracture of the filled polymer, and
the lower micrograph illustrates the adhesion between the
filler and the polymeric matrix. Ion pair bonding seems to
be an effective anchoring mechanism when an ion exchange
or a proton transfer process can occur between the anchoring groups of the diblock and groups available at the filler surface.

Mutually interacting groups in the control of the phase
morphology of immiscible polymers

The question might be addressed now to know whether phase
morphology and properties of immiscible polymer blends can
be modified by a way different from the previously described emulsification.
Examples are available that show that the degree of miscibility of two polymers depends on possible mutual interactions (46). Therefore, let us consider the proton transfer
from the acid-groups of polymer A to the aliphatic tertiary amines of polymer B (Equation 1). The ion pair formation is nothing but a block polymerization process, since
the extremities of the immiscible polyA and polyB are now
held together through an ionic bonding instead of a covalent bond as in the traditional block copolymers (47).

$$\sim polyA \sim\sim COOH + R_2N \sim\sim\sim polyB \sim\sim \longrightarrow$$
$$\{polyA \sim\sim COO^-R_2{}^+HN \sim\sim\sim polyB\} \sim \qquad (1)$$

Optical microscopy shows that the phase separation of immiscible polyisoprene and poly(α-methylstyrene) is controlled by the ammonium carboxylate pairs bridging the
chain extremities at a level that depends on the ion pair
content (i.e. the molecular weight of the blended teleche-

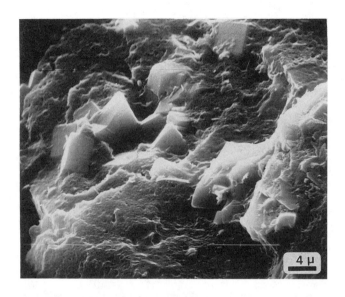

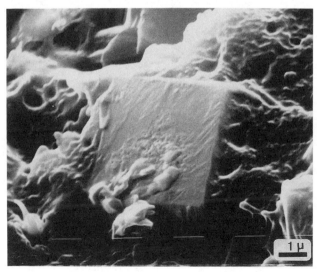

Figure 11. Scanning electron micrographs of LDPE filled with 25% CaCO$_3$, in the presence of a hPBD/poly
(acrylic acid) diblock.

lic polymers) (Figures 12b and c). The initial degree of
immiscibility of the mixed polymers has also an effect on
the mean size of the phases; according to Figure 12 (a
and b) polystyrene should be more miscible with polyisoprene than poly(α-methylstyrene) is. Another decisive parameter is the strength of the ion pairs: ammonium sulfonates are more effective in fighting the immiscibility of
the blended polymers than ammonium carboxylates are (48).
     Thermal dependence of the SAXS profile of a blend
comprising a dimethylamino telechelic polyisoprene (18K
molecular weight) and a sulfonic acid telechelic PMS (8K
molecular weight) gives an insight into the role played by
the ionic cross-interactions of the chain extremities
(Figure 13a). At room temperature, a strong reflection is
observed at 170Å which supports that a phase separation
has occurred with a characteristic periodicity. The phase
morphology improves on heating up to 180°C, and beyond
that temperature, the intensity of the scattering maximum
decreases in agreement with a gradual disordering due to
thermal fluctuations. At 260°C, no scattering is observed
anymore and a homogeneous system is formed as supported
by thermal analysis which shows only one T_g for the blend
annealed at 250°C and rapidly quenched. These observations
can be rationalized as follows. At 25°C, the phase morphology is controlled by two opposite trends: the phase separation of the immiscible PIP and PMS chains, and the electrostatic association of the ion pairs that link these
polymers to each other. As a result an organized phase
morphology occurs as tentatively schematized on Figure 14
(right handside, above). Above 180°C, the ion pair interactions are no longer strong enough to hold tightly the
mesophases and an order-disorder transition takes place
which is similar to that occurring in the covalently bonded block copolymers. The homogeneity of the blend observed
above that transition i.e. above ca.260°C (Figure 14;left
handside, above), can only be accounted for by the
"blocky" structure of the chains. It means that the ammonium sulfonate pairs are still stable at 250°C, otherwise
a macrophase separation should occur as observed for the
unfunctionalized PIP and PMS of the same molecular weight
(Figure 13,b).
When the molecular weight of PMS increases by a factor of
2, it has been shown that the blend (amine/acid molar
ratio = 1) is still microphase separated at room tempera-
ture(49). However, the extent of the phase separation upon
heating is so pronounced that a third order reflection
characteristic of a lamellar morphology emerges. The mesophases are observed to disappear near 210°C instead of
260°C as before. Thus, increasing the immiscibility of the
starting polymers reduces the range of stability of the
mesophase structures due to a decrease in the content of
the mutually interacting ion pairs. As a final difference,
a mixed state of PIP and PMS chains interconnected by
ammonium sulfonates goes unnoticed due to the early occur-

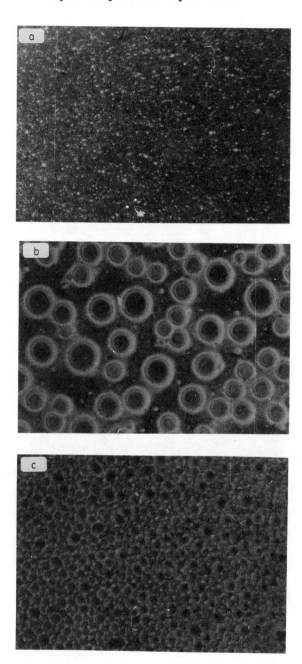

Figure 12. Optical micrographs of blends of amino terminated PIP (20K) with carboxylic acid terminated, PMS (10K): micrograph (a), PMS (20K): micrograph (b), and PS (30K): micrograph (c).

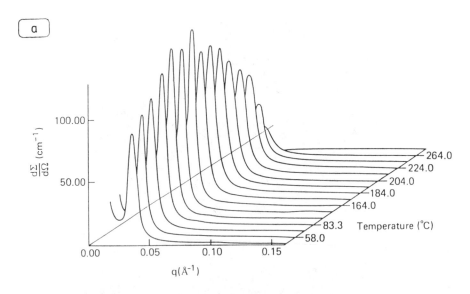

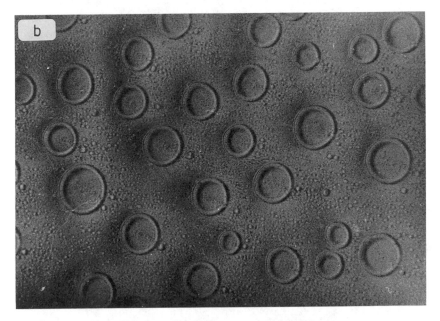

Figure 13. Thermal dependence of the SAXS profile of a solvent cast blend of PIP$(NR_2)_2$ (18K) and PMS $(SO_3H)_2$ (8K) (a); optical micrograph of the related unfunctionalized polymer blend (b). The scattering vector, $q = (4\pi/\lambda)\sin(\theta/2)$ where θ is the observation angle. $d\Sigma/d\Omega$ is the differential scattering cross-section per atom with respect to the solid angle, as normalized to a unit volume.

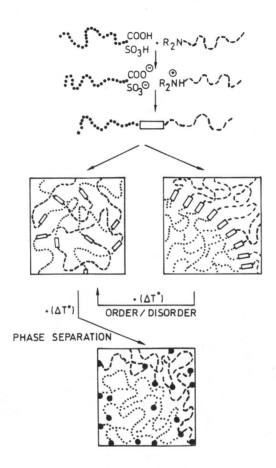

Figure 14. Schematic representation of the main phase situations observed when ion pairs associate the chain ends of immiscible polymers.

rence of the complete phase separation of PIP and PMS
(Figure 14; below). Only the thermal break up of the ammonium sulfonate ion pairs with release of sulfonic acid
and tertiary amine end-groups can account for that macrophase separation (Figure 14). In conclusion, increasing
the length of the blended telechelic polymers enhances the
quality of the mesophase structure while it has a deleterious effect on the thermal stability of the electrostatic
interactions of the ion pairs on one hand and the ion
pairs themselves on the other hand.

Figure 15 schematizes the expected phase diagram of
the $PIP(NR_2)_2/PMS$ $(SO_3H)_2$ polyblends $(NR_2/SO_3H = 1)$. By comparison with polybutadiene/polystyrene blends (50), it is
assumed that PIP/PMS mixtures exhibit a UCST curve. Since
the ammonium sulfonate pairs provide the blended macromolecules with a block copolymer structure, the UCST should
be shifted downwards in the temperature scale and desi-
gnated as the mesophase separation temperature (MST)
curve. T_g is the glass transition temperature of the hard
phases and T_i, the temperature at which the ion pairs are
disrupted, depends on the molecular weight of the blended
polymers. A defined composition $\emptyset_{PIP}$ and three distinct
values of T_i are now considered. In the first instance,
T_i is below T_g and the phase morphology is expectedly
stable up to T_i^1, above which (representative system A)
the block copolymer structure of the mixed telechelic
chains is definitely lost and the system phase separates
in agreement with the UCST curve. That behavior has been
actually observed when ammonium carboxylate ion pairs have
been substituted for sulfonate ones (49). Would T_i be above
T_g but below the MST curve, and the phase morphology can
improve itself above T_g on the impetus of the ion pairs
which tend to associate as extensively as possible.
However, above T_i^2, the ion pairs are broken up and the
phase separation occurs. Typically, this is the behavior
of the $(PIP(NR_2)_2-18K)/(PMS$ $(SO_3H)_2-14K)$ blend. Finally,
when T_i is above both T_g and the MST curve, an order-
disorder transition has to occur when the system goes
accross the curve leading to an uniform melt as long as
T_i^3 is not reached as shown by the $(PIP(NR_2)_2-18K)/(PMS$
$(SO_3H)_2-8K)$ blend.

SAXS results discussed hereabove show, for the first
time, that the blending of mutually interacting telechelic
polymers can promote a phase morphology very similar to
that seen in covalently bonded block copolymers. This is
a promising way to control the interfacial situation in
multiphase polymeric materials.

Conclusion

The whole set of experiments described in this paper provides evidence that the polymer chemist is able to improve
significantly any type of multiphase polymeric systems
that suffer from a poor interfacial adhesion. Today, two
strategies are available: either diblock copolymers of an

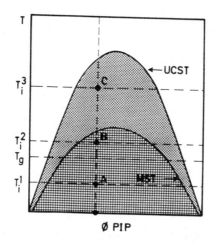

Figure 15. Proposed phase diagram of blends of amino terminated PIP and carboxylic acid or sulfonic acid terminated PMS .

appropriate interfacial activity are purposely tailored, or mutual interactions of the chain extremities of otherwise immiscible macromolecules are promoted.

Acknowledgments

The authors thank the "Services de la Programmation de la Politique Scientifique" as well as IRSIA (Brussels) for continuous financial support. They are indebted to all of their co-workers whose names appear in the literature citations, as well as to Mrs M. Palumbo for skilful technical assistance.

Literature cited

1. Mark, H. J. Chem. Educ. 1987, 64, 858.
2. Szwarc, M.; Levy, M.; Milkovich, R. J. Am. Chem. Soc. 1956, 78, 2656.
3. Holden, G.; Bischop, E.T.; Legge, N.R. J. Polym. Sci. Part C, 1969, 26, 37.
4. Utracki, L.A. Intern. Polymer Processing 1987, 2, 3.
5. Schreiber, P.; Latham, J. Polym. News 1987, 13, 19.
6. Olabisi, O.; Robeson, L.E.; Shaw, M.T. Polymer-Polymer Miscibility, Academic, New York, 1979, Chapter 7.
7. Attwood, D.; Florence, A.T. Surfactant Systems. Their Chemistry, Pharmacy and Biology; Chapman and Hall: New York, 1985; Chapter 8.
8. Paul,D.R.; Barlow, J.W.; Keskkula, H. In Encyclopedia of Polymer Science and Engineering; Mark, H.F.; Bikales, N.M.; Overberger, C.G.; Menges, G., Eds.; Wiley: New York; Second Edition, 1988; Vol. 12, p.399.
9. Locke, C.E.; Paul, D.R. J. Appl. Polym. Sci. 1973, 17, 2791.
10. Locke, C.E.; Paul, D.R. Polym. Eng. Sci. 1973, 13, 308.
11. Barentsen, W.; Heikens, D. Polymer 1973, 14, 579.
12. Heikens, D.; Hoen, N.; Barentsen, W.; Piet, P.; Ladan H. J. Polym. Sci. Symp. 1978, 62, 309.
13. Heikens, D.; Barentsen, W. Polymer 1977, 18, 69.
14. Barentsen, W.; Heikens, D.; Piet, P. Polymer 1974, 15 119.
15. Paul, D.R.; Newman, S. Polymer Blends; Academic, New York, 1978; Vol. 2, p. 40.
16. Fayt, R.; Jérôme, R.; Teyssié, Ph. J. Polym. Sci. Polym. Lett. Ed. 1981, 19, 79.
17. Fayt, R.; Jérôme, R.; Teyssié, Ph. J. Polym. Sci. Polym. Phys. Ed. 1981, 19, 1269.
18. Fayt, R., Jérôme, R.; Teyssié, Ph. J. Polym. Sci. Polym. Phys. Ed. 1982, 20, 2209.
19. Fayt, R.; Jérôme, R.; Teyssié, Ph. Makromol. Chem. 1986, 187, 837.
20. Sjoerdsma, S.D.; Dalmelen, J.; Bluyenberg, A.C.; Heikens, D. Polymer 1980, 21, 1469.
21. Fayt, R.; Jérôme, R.; Teyssié, Ph. J. Polym. Sci. Polym. Lett. Ed. 1986, 24, 25.

22. Hoen, N. Ph.D. Thesis, Eindhoven University of Technology, 1977.
23. Sjoerdsma, S.D.; Heikens, D. J. Mater. Sci. 1982, 17 2350 and 2609.
24. Lindsey, C.R.; Paul, D.R.; Barlow, J.W. J. Appl. Polym. Sci. 1981, 26, 1.
25. Paul, D.R. In Thermoplastic Elastomers; Legge, N.R; Holden, G.; Schroeder, H.E., Eds.; Hanser Publ., Munich, 1987; Chapter 12, Section 6.
26. Fayt, R.; Hadjiandreou, P.; Teyssié, Ph. J. Polym. Sci. Polym. Chem. Ed. 1985, 23, 337.
27. Gallot, B. Adv. Polym. Sci. 1978, 29, 85.
28. Green, P.F.; Russell, T.P.; Jérôme, R.; Granville, M. Macromolecules, to be published.
29. Roe, R.J. Polym. Eng. Sci. 1985, 25, 1103.
30. Zin, W.C.; Roe, R.J. Macromolecules 1984, 17, 183.
31. Roe, R.J.; Zin, W.C. Macromolecules 1984, 17, 189.
32. Rigby, D.; Roe, R.J. Macromolecules 1986, 19, 721.
33. Teyssié, Ph. In Polymer Science Overview: A tribute to Herman F. Mark: Stahl, G.A., Ed.; ACS Symposium Series N. 175; American Chemical Society: Washington, D.C., 1981; p. 311.
34. Olabisi, O.; Robeson, L.E.; Shaw, M.T. Polymer-Polymer Miscibility, Academic, New York, 1979, Chapter 5.
35. Fayt, R.; Jérôme, R.; Teyssié, Ph. J. Appl. Polym. Sci. 1986, 32, 5647.
36. Bucknall, B.C. In Polymer Blends; Paul, R.D.; Newman, S., Ed.; Academic, New York, 1978; Vol. 2, Chapter 14.
37. Fayt, R.; Teyssié, Ph. Macromolecules 1986, 19, 2077.
38. Teyssié, Ph.; Fayt, R.; Jérôme, R. Makromol. Chem., Macromol. Symp. 1988, 16, 41.
39. Ouhadi, T.; Fayt, R.; Jérôme, R.; Teyssié, Ph. J. Appl. Polym. Sci. 1986, 32, 5647.
40. Ouhadi, T.; Fayt, R.; Jérôme, R.; Teyssié, Ph. Polymer Commun. 1986, 27, 212.
41. Ouhadi, T.; Fayt, R.; Jérôme, R.; Teyssié, Ph. J. Polym. Sci. Polym. Phys. Ed. 1986, 24, 973.
42. Leemans, L.; Fayt, R.; Teyssié, Ph.; Uytterhoeven, H. Macromolecules, to be published.
43. Leemans, L.; Fayt, R.; Teyssié, Ph. J. Polym. Sci. Polym. Chem. Ed. to be published.
44. Webster, O.W.; Sogah, D.Y. In Recent Advances in Mechanistic and Synthetic Aspects of Polymerization; Fontanille, M.; Guyot, A. Ed.; Nato ASI Series C, Vol. 215; Reidel : Dordrecht, 1987; p. 3.
45. Jakubauskas, H.L. J. Coat. Technol. 1986, 58, 71.
46. Smith, P.; Hara, M.; Eisenberg, A. In Current Topics in Polymer Science; Ottenbrite, R.M.; Utracki, L.A., Ed.; Hanser Publishers: New York, 1987; Vol. 2, p. 255.
47. Horrion, J. ; Jérôme, R.; Teyssié, Ph. J. Polym. Sci. Polym. Letters 1986, 24, 69.

48. Horrion, J.; Jérôme, R.; Teyssié, Ph. J. Polym. Sci. Polym. Chem. Ed., to be published.
49. Russell, T.P.; Jérôme, R.; Charlier P.; Foucart, M. Macromolecules 1988, 21, 1709.
50. Russell, T.P.; Hadziioannou, G.; Warburton, W.K. Macromolecules 1985, 18, 78.

RECEIVED October 21, 1988

Chapter 3

Coreactive Polymer Alloys

M. Lambla, R. X. Yu[1], and S. Lorek

Ecole d'Application des Hauts Polymères, Institut Charles Sadron (CRM–EAHP), F–67000 Strasbourg, France

The sales of plastics continue to increase in a large part due to technical and economic advancements of polymer blends. Reactive blending is a useful technique for elastomers but, it appears that chemistry could also play an important role in the correct microstructure adjustment of thermoplastic alloys. Interfacial reactivity should be the focal point in maintaining the expected structure during subsequent stages of manufacture. Besides industrial examples, various kinds of polymeric co-reacting systems are also presented in order to emphasise the key factors of reactive blending.

Historically, reactive blending has been associated with the manufacture of elastomeric-based products. In the future, although elastomers or thermosets will continue to play an important role, reactive blending will expand to the production of thermoplastic alloys. The reactive processing may provide viable mechanisms for the creation and preservation of the desired blends with controlled structure and morphology, and assures that the microstructure is maintained throughout the subsequent stages of manufacturing. This paper will deal with a general overview of significant industrial developments in this field, whereas the key factors in reactive blending will be illustrated mainly with results from laboratory studies.

GENERAL CHARACTERISTICS OF BLENDS

Most polymer mixtures phase separate, as the entropy of mixing of the polymeric species is very low and the enthalpy of mixing is positive. Blends with a multiphase quenched structure are obtained by suitable melt-blending at high temperature and subsequent cooling below glass or crystallisation temperature. These blends may display a wide range of improved properties in comparison to pure homo or copolymers. The most important factors governing the mechanical properties of incompatible polymer blends are: the interfacial adhesion between separated

[1]Current address: University of Nanjing, China

phases and the morphology, i.e. the size and shape of the dispersed phase. The greatest toughness is achieved when the interparticle distance is smaller then a critical value (1), depending on the way the fracture energy is dissipated.

Morphology of the dispersed phase depends on complex shear and elongational deformations to which the materials are subjected during melt-blending. However, the size and shape of the dispersed phase are also controlled by interfacial tension (2) and viscoelastic properties of the components. Hence, by adding an appropriate emulsifying agent, the interfacial tension and particle size may be varied over a wide range. The use of block or graft copolymers is quite a common and versatile method for reducing particle size and enhancing interfacial adhesion in incompatible blends. A schematic representation of this feature is given in Figure 1 ; the emulsifying effects of block and graft copolymers have been previously demonstrated in several systems (3). A theoretical approach of this behaviour has been discussed recently by LEIBLER (4).

The block or grafted copolymers are generally prepared separately and introduced in low concentrations (1 - 3 wt %) in the melt-mixing system, which is based on two incompatible resins. It is also possible to create these amphiphilic agents in situ by chemical reactivity based on end-capping or grafting reactions. EISENBERG (5) has also shown that ionic interactions can be used to control the degree of miscibility of otherwise immiscible polymer systems. In some of these, the interactions are based on proton transfer from a donor site on one polymer to an acceptor site on another polymer, leading to ion-ion interactions. In others, the transfer of a metal cation from an ionomer to a polar polymer is the result of an ion-dipole interaction. More recently, JEROME and TEYSSIE (6) have confirmed the advantage of ionic-cross interactions between sulphon (or carboxyl)) and tertiary amine chain end-groups. The phase separation is controlled by the content and strength of the ion-pairs, as shown by SAXS characterization.

Thus, it appears that chemical reactivity or ionic-cross interactions could lead to in situ compatibilising or miscibility enhancement during melt-mixing. However, several questions remain. How does the reactivity modify the thermodynamic balance, the reciprocal miscibility or the rheological behaviour of the melt? Or, how the covalent or ionic bonding influence the interfacial adhesion processability and final mechanical properties of the immiscible blends ?

EXAMPLES FROM INDUSTRY

Among the numerous examples from industry, one of the first successful ones was based on reactive blending of polyamide and ethylene-propylene modified rubbers, containing grafted carboxylated or anhydride groups. In this case, approximately 1 wt% of carboxyl or anhydride groups was sufficient to create a graft product by condensation reaction with the amino-terminated polyamide, as described by CIMMINO and coworkers (7) and others (8-10). The important parameters are rubber concentration and particle size. An increasing rubber content and a decreasing particle diameter shift the brittle-tough transition to lower temperature, while the impact strength at very low temperature increases, as shown by BORGGREVE (11).

A similar idea of functional grafting was also applied by DOW Chemicals in a system based on styrenic copolymers containing oxazoline groups (RPS). These resins were mixed with carboxylated polyolefins. The reaction scheme is indicated in Figure 2, and Table 1 summarises the mechanical properties of pure homopolymers as compared to those obtained from a mixture of 50% by weight of the two reactive copolymers (12). It is interesting to note that most final properties of the reactive blends show the expected intermediate values. These kinds of reactive copolymers produced by classical synthesis or by grafting could incite future diversification in reactive mixing, even with olefinic products, as mentioned by GOTOH et al., (13).

Due to the excellent price/performance ratio, polypropylene should not only be a good base for unreactive blends, widely used in automotive industry, but also an attractive product for reactive blending. However, it is necessary to introduce polar reactive groups in the main chain which is difficult

Table 1. Reactive Polystyrene Blends

	RPS	LDPE	50/50 BLEND
TENSILE YIELD (MPa)	36.5	6.9	20
TENSILE MODULUS (MPa)	2890	124	690
ELONGATION (%)	2	600	75
NOTCHED IZOD IMPACT (FT-LB/ (INCH))	0.2	NB	13
VICAT (°C)	42	28	31
MELT FLOW (g/10 mn)	7	5	4
SPECIFIC GRAVITY	1.04	0.94	0.98

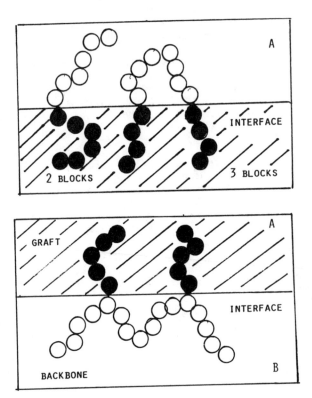

Figure 1: Models of polymeric compatibilizer at interface.

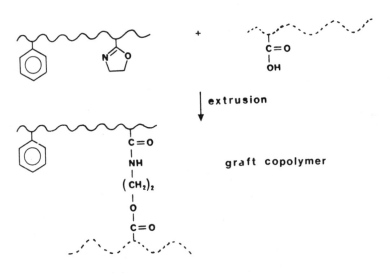

Figure 2: Reaction scheme of reactive polystyrene blends.

or impossible in synthesis by ZIEGLER NATTA catalysis or by
grafting. Monomer grafting onto propylene by radical reactions
is generally accompanied by chain degradation and related
decrease in mechanical strength but, as it will be shown,
considerable progress has been made recently in this field.

Polybutylene therephthalate (PBT) has been used as a blend
component to provide chemical resistance in various systems,
but the most interesting one results from a combination with
polycarbonate and, eventually, an impact modifier of the core-
shell type. Polyester blends containing polycarbonate exhibit
ester interchange chemical reactions, which add to the complexity
of property control of these materials. DEVAUX and co-workers
(14) have examined the transesterification reaction catalysed
by residual catalysts in PBT which can lead to the formation
of block and random copolymers. They have shown that allyl
or aryl phosphites inactivate the residual titanium catalyst
and minimise the transesterification reaction. HOBBS et al. (15)
reported a way of controlling miscibility behaviour, morphology
and deformation mechanisms, in order to obtain blends compatibilisation and excellent mechanical properties.

There are two further possibilities for preventing segregation
in complex immiscible mixtures. The first one implies crosslinking reaction of the dispersed phase, associated with interfacial bonding and leads to the well-known dynamic vulcanisation
(17). The second one results from mechanical interlocking
of the phases which are created in interpenetrated polymer
networks (IPN).

It is worthwhile to note that chemistry plays a major role
in the morphology and control of mechanical properties in
complex systems like PPE blends with crystalline polymers,
such as polyolefins, polyamides (PA) and polyesters (18).
The amount of copolymer formed during the reactive extrusion
between functionalised PPE and PA has a significant effect
on the impact-strength of blends. The latter levels off only
above 10% of copolymer.

Clearly, chemistry has been most helpful in the development
of industrial blends. Nevertheless, in many cases, the basic
reaction conditions are not completely understood. The following
is a description of laboratory studies carried out to examine
the interfacial reactivity and two kinds of condensation reactions
in immiscible polymer blends.

LABORATORY EXPERIMENTS AND FURTHER STUDIES

As mentioned above, direct compatibilisation by melt blending
could be obtained by functional end-capping or grafting reactions,
or again by interpolymeric bonding through a reagent which
is able to react with both polymeric species. This last system
will be examined first of all, in the case of a mixture based
on a polystyrene matrix containing dispersed polyolefinic

nodules, where the reactive product is bis-maleimide (BM). The first functional grafting studies were performed in a system containing a copolymer of methylmethacrylate and maleic anhydride as well as styrenic hydroxylated oligomers, both amorphous products. The second system was based on two semi-crystalline polymers : polypropylene and polyamide. To attain potential reactivity, polypropylene was first grafted with maleic anhydride and the results of this radicalar melt-grafting are presented. The final blend was obtained in one-step extruding of the two homopolymers and the reactive polypropylene.

INTERFACIAL REACTIVE BLENDING

Polystyrene (PS) and low density polyethylene (LDPE) are two incompatible polymers and their blends obtained by melt-mixing show very poor mechanical properties in comparison to the high impact polystyrenes obtained by direct synthesis in the presence of polybutadiene. In order to obtain a better quality of dispersion, i.e. regular and smaller particle size distribution, it is possible to add hydrogenated block copolymers of styrene and butadiene (SBS). Some of these products are supplied commercially by SHELL, under the tradename : KRATON G. While investigating the same model system, TEYSSIE et al., (19) have synthesised a large range of pure or "tapered" poly(styrene-b-hydrogenated polybutadiene)diblocks. Their main results can be summarised as follows :
- the dispersed phase mean size decreases to the sub-micron level, depending on concentration and structural characteristics (20) ;
- the bulk physico-mechanical properties of these blends are strikingly improved by the presence of the copolymer (21).

However, the polyolefinic dispersed phase cannot contribute to a large increase in impact performance, due to its crystalline structure, as shown previously by HEIKENS et al. (22). In order to verify this point, a low concentration of atactic polypropylene was introduced in certain blends in the polyolefin phase.

Studies of the interfacial and/or bulk reactivity were performed by adding BM, which is a difunctional highly reactive molecule. The two double-bonds are sensitive to radical attack and quite efficient in addition reactions on tertiary hydrocarbon sites. In a previous study we established that classical peroxydes could not play the major role in interfacial bonding. Bulk reactions, especially in polystyrene matrix, lead to chain degradation and related decrease of mechanical strength. Therefore, BM was not combined with peroxydes and it was reduced as much as possible (0.1. to 0.3%).

Final properties are summarised in Tables 2 and 3. Table 2 presents a comparison between two series of blends with different compositions. The system containing 25% polyolefin shows very poor mechanical properties, especially tensile strength at break

(σ_B), which is largely increased by the introduction of block copolymer (KRATON G 1650). The tensile strength is also increased in the presence of lower amounts of polyolefin by adding reactive BM to the mixture which does not contain the amphiphilic copolymer (KG). Inversely, the effect due to the presence of KG only leads to a higher performance in tensile strength and impact behaviour. Further addition of BM has the effect of reducing these characteristics somewhat.

Table 2 :   MECHANICAL PROPERTIES OF PS/PE BLENDS

Composition (%)				σ_B	ε_B	NOTCHED	ELASTIC
PS	PE	KG	BM	10.MPa	(%)	IS_2 (kJ/M^2)	MODULUS 10.MPa
75	25	–	–	94	75	1.8	2900
75	17,5	7,5	–	350	15	2.0	2950
85	15	–	–	330	6	2.1	3250
85	15	–	0.5	430	7	1.8	3400
85	15	2	–	470	8	2.7	3400
85	13	2	0.5	450	9	2.4	3400

Table 3 :   MECHANICAL PROPERTIES OF PS/PE - APP BLENDS

Composition (%)					σ_B	ε_B	IS[*]	ELASTIC
PS	PE	APP	KG	BM	10.MPa	(%)	(UNNOT-CHED) (kJ/m^2)	MODULUS 10.MPa
85	12	3	–	–	357	12.0	18	3500
85	12	3		0.2	386	14.4	20	3500
85	10	3	2	–	430	14.4	30	4400
85	10	3	2	0.2	430	14.2	42	4700

The results obtained with the last series of blends containing 3% atatic polypropylene are summarised in Table 3. It is interesting to note a limited, but general, increase in elongation at break, ε_B. Furthermore, the influence of both amphiphilic copolymer and reactive BM lead to satisfactory overall mechanical properties, especially modulus and impact strength. Even if it seems difficult to establish the real contribution of the reactive species, it is clear that interpolymeric bonding between PS and LDPE due to the presence of the added BM is not sufficient to create interfacial

efficient adhesion. However, the final compromise (last line in
Table 3) shows a positive influence of BM.

INTERPOLYMERIC CONDENSATION REACTIONS

a. PMMA-AM/PS BLENDS

Condensation reactions in the melt between the anhydride groups of
a copolymer containing maleic anhydride links and hydroxylated oligo-
styrenes were carried out in discontinuous or continuous mixers.
As shown previously (23), the progress of condensation reactions
are easily followed by the rising consistency in the melt,
if dihydroxylated oligomers are used (Figure 3). The speed
constants of these reactions were determined by rheological
studies (24).

The influence of reciprocal miscibility between the methacrylic
copolymer containing cyclic anhydride groups and the styrenic
oligomers was investigated in the case of monohydroxylated
oligomers, which were also prepared by anionic polymerisation
of styrene and further deactivation with ethyleneoxide.

The diffusion of these reactive smaller species is directly
influenced by their molecular weight ($\bar{M}_w$). Table 4 shows the
results obtained with three different oligomers of increasing $\bar{M}_w$ and
a copolymer synthesised on a laboratory scale, containing 10 wt%
cyclic anhydride.

Table 4 : INTERPOLYMERIC CONDENSATION IN THE MELT
PMMA/AM* - PS (OH)

PS (OH)		GRAFTING YIELD (%)
$\bar{M}n$ (kg/mol)	wt %	
2.45	46	55
	25	100
	12.5	100
6.85	75.5	46
	40	49
	24.5	56
9.0	73.5	0

$\bar{M}w$ = 72 kg/mol : MA ; 10 wt % B.W.

Table 5 : INTERPOLYMERIC CONDENSATION IN THE MELT
PMMA/AM* - PS (OH)

PS (OH)		GRAFTING YIELD (%)	
$\overline{Mn}$ (kg/mol)	wt %	SOLVENT EXTRACTION	GPC
	50	34	30
	37.5	52	39
3.8	25	67	47
	12.5	74	49

* HW 55 (ROHM) $\overline{Mw}$ = 130 ; MA : 20 wt %

The grafting yield, first evaluated by a solvent extraction technic, decreases as expected with increasing molecular weight, and it is also influenced by mixture composition. It appears that, with this first copolymer backbone of a rather low molecular weight ($\overline{Mw}$ = 72 kg/mole , there is no reactivity left with products where $\overline{Mw}$ equals 9kg/mol . Table 5 confirms these results with an industrial copolymer (HW 55, supplied by ROHM), of a higher molecular weight ($\overline{Mw}$ = 130). In this case, the limiting molecular weight of the poorly compatible oligomer is approximately 4 kg/mol. However, it was confirmed by GPC analysis that it is possible to attain high levels of grafting by mixing during approximately 10 minutes in the RHEOCORD discontinuous device. The grafted chains were stable at high temperatures and the reversible reaction mentioned by several authors (25-29) was not observed.

This point was confirmed recently by studying the reverse reaction with hemiester copolymers of various alcohol chain lengths. The reverse reaction is directly governed by the steric hindrance introduced by alcohol chain length and/or nature (30).

This grafting reaction was extrapolated to continuous-extruding systems but it was necessary to increase residence-time and mixing efficiency in order to obtain the optimal level of grafting (24). This study confirmed that interpolymeric condensation reactions are effective in the melt, but final reactivity is a function of composition and reciprocal miscibility of the two polymeric species. Further studies on crosslinking reactions in compatible reactive systems (31) have confirmed that diffusion plays an important role but surprisingly, the introduction of a tertiary amine, acting as catalyst, produces a multiplication of the rate constant by a factor of 100.

b. PP/PA 6 BLENDS: Radical grafting of maleic anhydride on PP

As indicated above, it is difficult to attach polar reactive
groups onto a polypropylene backbone. The grafting was first
carried out in solution and extrapolated to melt mixing (32,
33). It is necessary to define the compromise between grafting
yield and chain degradation in radical grafting of maleic anhydride.

We carried out experiments to define the optimal conditions
in a reactive extrusion process, and to obtain maleic anhydride
grafted polymers of polypropylene copolymers (3050 MN 4 supplied
by ATOCHEM). The previous studies were carried out in a batch
mixer (HAAKE RHEOCORD), at a temperature of 220° C and a mixing
speed of 64 rpm, during 20 minutes. Three different systems
were tested : i. with pure maleic anhydride ; ii. in the presence
of a solvent (toluene or chorobenzene) ; iii. by introduction
of a stoechiometric amount of maleic anhydride and styrene.
Dicumyl peroxyde was used as initiator. It is important to
note that all the ingredients were introduced simultaneously
in the mixing cell.

The principal results are summarised in Figure 4. The three
curves represent the variation of grafted versus added maleic
anhydride for three different systems, containing the same
amount of dicumyl peroxyde (0.48 wt %). Curve a, representing
pure maleic anhydride grafting, confirms the limited value
of radical grafting : 0.3%. In the presence of stoechiometric
amounts of butylacrylate or styrene, a linear variation of
grafted versus added cyclic anhydride is observed. It is important
to note that the best grafting yields are obtained with styrene,
leading to modified PP containing a few per cent of grafted
links. The melt-index characterisation of these products shows
a decrease of MFI versus grafted cyclic anhydride (Fig. 5).

In the presence of styrene as a comonomer, the grafting reaction
seems to be predominant and therefore, the chain degradation
by β -scission is reduced. This was confirmed by GPC analysis.
Cross-propagation of styrene and maleic anhydride, both in
solvent or in bulk, is well known. This behaviour is attributed
to a charge-transfer complex (CTC) between maleic anhydride
and styrene and can lead, in solution, to spontaneous copolymerisation (34,35).

Even if CTC has been confirmed by spectroscopy, it is characterised
in a few cases only, one of them being based on a stoechiometric
mixture of N-vinylpyrrolidone and maleic anhydride (36). However,
as mentioned by SEYMOUR et al. (37, 38), the stability of a
CTC decreases as polymerisation temperatures increase, furthermore,
for the styrene-MA pair the charge-transfer complex should
be non-existent above 130° C. However, alternating styrene-co-MA
and other MA copolymers have been grafted on a variety of other
polymeric materials (39,40,41). Our own results further confirm
the contribution of CTC to general activation of the grafting
reaction. It is important to note that the grafting efficiency

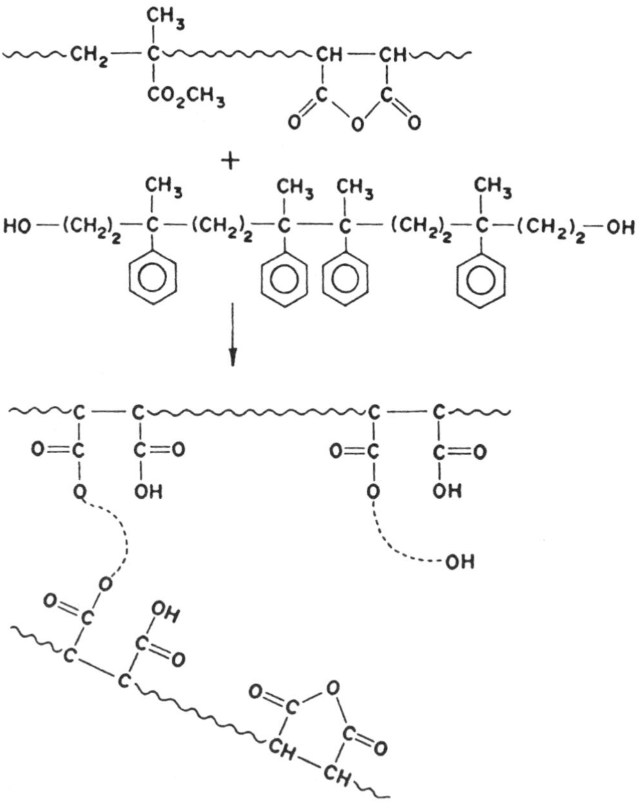

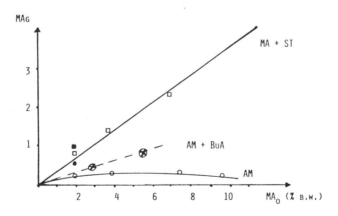

Figure 3: Interpolymeric condensation between dihydroxylated PS and copolymer PMMA/AM.

Figure 4: Grafting of maleic anhydride on PP: grafting yield versus monomer concentration.

observed when the reaction is carried out in a discontinuous mixer is also verified in a continuous process, using a twin-screw from WERNER PLEIDERER (ZSK 30). However, in this case it is essential to adjust feeding conditions of the various reagents, as shown in Figure 4 by the two points (black circle and black square), where grafting yield varies from 0.65% to 0.95% for two different ways of feeding. Further characterisation of the grafted PP before and after solvent purification have confirmed the alternating copolymerisation between styrene and MA.

c. PP/PA BLENDS: Interpolymeric reactivity

Polyamide-6 (PA-6) and polypropylene (PP) are both semi-crystalline polymers and the combination of an engineering plastic (PA) and the best commodity product (PP) could lead to new blends with interesting intermediate properties. We tested systems containing 50 wt% of each product and the ones obtained by addition of 3% of the reactive PP-g-AM resulting from previous continuous grafting in the extruder. The blends were prepared by simple mixing in the ZSK 30 twin-screw extruder and the samples for mechanical testing were molded by injection in a BILLION equipment.

Table 6 summarises the most important data resulting from stress-strain analysis. These four blends were obtained successively by simple melt-mixing (YMO), blends containing maleic anhydride grafted products in presence of a solvent: chlorobenzene (YM 1) and two products obtained by alternating grafting with styrene (YSM 2 - YSM 3). Due to the lower molecular weight of the grafted polypropylene, it seems easier to obtain a rapid and efficient mixing with YM 1 which results in a higher value of elastic modulus. The two blends containing PP-g AM/ST exhibit higher values both in elongation and yield-stress, which could be connected with better interfacial adhesion.

Table 6 : MECHANICAL PROPERTIES OF BLENDS

REFERENCE	E (MPa)	ε (%)	σy (MPa)
YSM 0	1900	6	35
YSM 1	2900	25	40
YSM 2	2180	29	43.8
YSM 3	2180	29	42

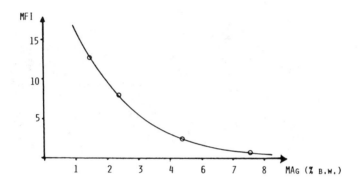

Figure 5: Melt index versus grafted MA.

The two Figures 6 and 7, scanning electron micrographs of fracture surfaces, show a great difference in particle size and interfacial adhesion between the 50/50 blend and the last one, YSM 3, containing 3% of the reactive PP-g AM/ST. The unreactive blend shows a large particle size (Figure 6), irregular shapes and poor interfacial adhesion. The reactive one (Figure 7) is characteristic of a dispersion of small spherical particles in a continuous matrix and apparently, satisfactory interfacial adhesion.

However, these results do not represent the optimised system. It appears that direct reactive blending between two homopolymers and the corresponding reactive product could lead to fair mechanical properties. The amphiphilic grafted copolymer, resulting from condensation reaction between cyclic anhydrides grafted onto PP and primary amino chain ends of PA (Figure 8) is produced during the melt mixing, leading to the expected decrease of interfacial tension and to higher cohesion of the blend.

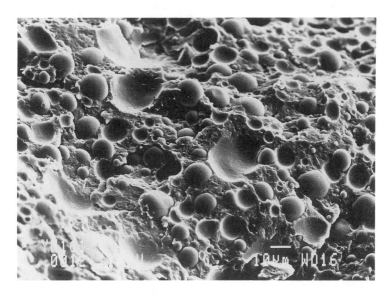

Figure 6: Scanning electron micrographs of fracture surface:
PP/PA without compatibilizer.

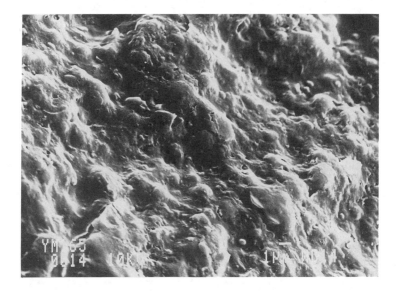

Figure 7: Scanning electron micrographs of fracture surface:
PP/PA with 3 wt % PP-g AM .

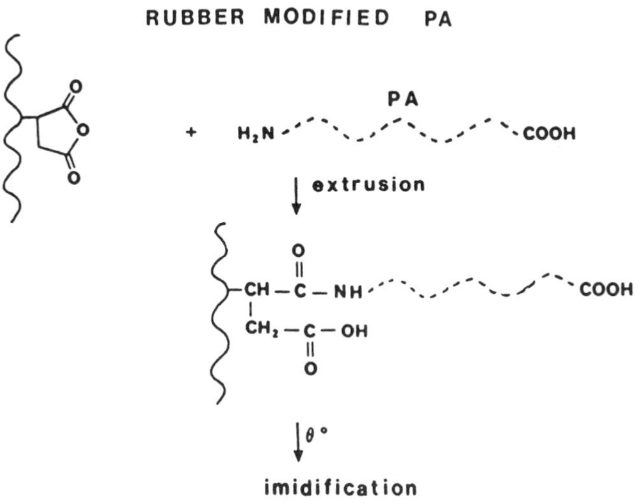

RUBBER MODIFIED PA

imidification

Figure 8: Reaction scheme of interpolymeric condensation between MA 6 and PP-g AM.

CONCLUSIONS

Industrial examples and further laboratory studies confirm that the performance of blended immiscible polymers not only depend on their mechanical nature and blend composition, but may be further improved by chemical reactions conducted during melt-mixing. The compatibiliser needed for particle size reduction and further interfacial adhesion may be formed by interpolymeric grafting through co-reactive micromolecular species or grafting and end-capping condensation reactions. Studies based on reactive systems containing copolymers of methyl methacrylate and maleic anhydride confirmed the efficiency of melt-mixing condensation reactions, taking into account the reciprocal partial miscibility of the polymeric species. The engineering blend obtained by mixing PP and PA-6, in the presence of a small amount of a reactive grafted polypropylene, seems to be of the greatest interest for intermediate materials with a satisfactory price-performance ratio.

LIST OF SYMBOLS

PMMA : Polymethylmethacrylate
PS : Polystyrene
PS (OH) : Hydroxylated polystyrene
PE : Polyethylene
LDPE : Low density PE
PP : Polypropylene
KG : Kraton G (block copolymer)
PA 6 : Polyamide 6
PP- g AM : Polypropylene grafted with maleic anhydride
PP-g AM/ST : PP grafted with maleic anhydride and styrene

M_n : Molecular weight (number average)
$\overline{M}_n$: Molecular weight (weight average)
wt % : Percent by weight
IS : Impact strength
σ_B : Tensile strength at break
ε_B : Elongation at break
MA^B: Maleic anhydride
ST : Styrene
BuA : Butylacrylate
MA_g : Grafted maleic anhydride

REFERENCES

1. Wu, S. Polymer 1985, 26, 1855; J. Appl. Polym. Sci. 1988, 35, 549.
2. Wu, S. Polym. Eng. Sci. 1987, 27, 335.
3. Teyssie, P.; Fayt, R.; Jerome, R. Makromol. Chem., Macromol. Symp. 1988, 16, 57.
4. Leibler, L. Makromol. Chem., Macromol. Symp. 1988, 16, 1.
5. Natansohn, A.; Murali, R.; Eisenberg, A. Makromol. Chem., Macromol. Symp. 1988, 16, 175.
6. Teyssie, P.; Fayt, J.; Jerome, R. Proc. ACS Meeting, Toronto 1988, 58, 622.
7. Cimmino, S.; D'Orazio, L.; Greco, R.; Maglio, G.; Malinconico, M.; Mancarella, C.; Martuscelli, E.; Palumbo, R.; Ragosta, G. Polym. Eng. Sci. 1984, 24, 48.
8. Flexman, E.A. Polym. Eng. Sci. 1979, 19, 564.
9. Wu, S. J. Polym. Sci., Polym. Phys. 1983, Ed. 21, 699.
10. Chen, D.; Kennedy, J.P. Polym. Bull. 1987, 17, 71.
11. Borggreve, R.J.M.; Gaymans, R.J.; Luttmer, A.R. Makromol. Chem. Macromol. Symp. 1988, 16, 195.
12. Sneller, J.A. Mod. Plast. Int. 1985, 42.
13. Gotoh, S.; Fujii, M.; Kitagawa, S. SPE ANTEC 1986, Techn. Pap., 32, 54.
14. Devaux, J.; Goddard, P.; Mercier, J.P. Polym. Eng. Sci. 1982, 22, 229.
15. Hobbs, S.Y.; Dekkers, M.E.J.; Watkins, V.H. Pol. Bull. 1987, 17, 341.
16. Hobbs, S.Y.; Dekkers, M.E.J.; Watkins, V.H. "Polyblends-88" NRC/IMRI Symp., Boucherville Que-Canada, April 5-6, 1988.
17. Coran, A.Y. Polym. Proc. Eng. 1988, 5, 317.
18. Juliano, P.C. US-France Workshop 1988, Williamsburgh.
19. Teyssie, P.; Fayt, R.; Jerome, R. Makromol. Chem., Macromol. Symp. 1988, 16, 41.
20. Fayt, R.; Jerome, R.; Teyssie, P. J. Polym. Sci., Polym. Phys. 1981, Ed. 19, 1269.
21. Fayt, R.; Jerome, R.; Teyssie, P. J. Polym. Sci. in press.
22. Heikens, D.; Hoen, N.; Barentsen, W.; Piet, B.. Ladan, H. J. Polym. Sci., Polym. Symp. 1978, 62, 309.
23. Lambla, M.; Killis, A.; Magnin, H. Europ. Polym. J. 1979, 15, 489.
24. Lambla, M.; Druz, J. Preprints IUPAC, Amherst 1982, 2, 487.
25. Muskat, I.E. U.S. Patent 3 085 986.

26. Whithworth, C.J.; Zutty, N.L. U.S. Patent 3 299 184.
27. Zimmerman, R.L. U.S. Patent 3 678 016.
28. Ardashnikov, A.Y.; Kardash, I.Y.; Praveonikov, A.N. Polym. Sci. USSR 1971, 13, 2092.
29. Barozier, G.; Niclo, A. French Patent 2 130 561.
30. Lambla, M.; Mazeres, F.; Druz, J. Preprints AICHE Meeting, New York 1987, 40B.
31. Lambla, M.; Satyarayana, N.; Druz, J. Makromol. Chem. 1988, 189, 0000.
32. De Vito, G. J. Polym. Sci. 1984, 22, 1335.
33. Ide, F. Kobunshi Kagaku 1968, 25, Nr 274.
34. Tsuchida, E.; Tomono, T. Makromol. Chem. 1972, 151, 245.
35. Vukovic, R.; Kurusevic, V.; Fles, D. J. Polym. Sci. 1977, 15, 2981.
36. Nikolaev, A.F.; Bondarenko, V.M.; Shakalova, N.K. Syn. High Polym. 1974, 1974, 81.
37. Seymour, R.B.; Garner, D.P.; Stohl, G.A.; Sanders, L.J. Am. Chem. Soc., Div. Polym. Chem. Prepr. 1976, 17, (2) 660.
38. Seymour, R.B.; Garner, D.P. J. Coat. Technol. 1976, 48 (612), 41.
39. Gaylord, N.G. Adv. Chem. Ser. 1973, 129, 209.
40. Gangarz, I.; Laskavski, W. J. Polym. Sci. Polym. Chem. 1979, Ed. 17, 3, 683.
41. Gangarz, I.; Laskavski, W. J. Polym. Sci. Polym. Chem. 1979, Ed. 17, 5, 1523.

RECEIVED April 27, 1989

Chapter 4

Blends of Polystyrene with Phenylene Oxide Copolymers

P. Padunchewit, D. R. Paul, and J. W. Barlow

Department of Chemical Engineering and Center for Polymer Research, The University of Texas at Austin, Austin, TX 78712

Blends of both atactic polystyrene, PS, and 90 mole% isotactic polystyrene, i-PS, with 2,6-dimethyl-1,4-phenylene oxide-co-2,3,6-trimethyl-1,4-phenylene oxide copolymers, PEC, containing up to 20 mole% trimethyl comonomer, were prepared by solution casting methods and found to be miscible by the presence of a single glass transition temperature, T_g. The equilibrium melting behavior of i-PS crystals in the blends with PEC was carefully determined by the Hoffman-Weeks method. Analysis of the observed depression in equilibrium melting temperatures suggests that the interaction between i-PS and PEC is exothermic and becomes increasingly exothermic with increasing PEC trimethyl comonomer content. These results are discussed in the context of recently developed theories for copolymer blends with homopolymers.

Owing to its considerable commercial significance, the miscible blend of poly(2,6-dimethyl-1,4-phenylene oxide), PPO, and atactic polystyrene, PS, has been extensively studied, as summarized in recent articles and reviews (1-9). An exothermic heat of mixing, ΔH_{mix}, between component repeat units is generally held to be necessary for formation of a miscible polymer blend (3), and this is thought to arise in PPO/PS blends from phenyl group coupling (8,9) between the aromatic ring of PPO and that of PS. Blends based on the miscible mixture of PS with poly(2,6-dimethyl-1,4-phenylene oxide-co-2,3,6-trimethyl-1,4-phenylene oxide), PEC, also have considerable commercial significance, although very little has been published concerning their thermodynamic behavior (10). One of the goals of this work is to characterize the behavior of PEC blends and to compare, where possible, the thermodynamic properties of these materials with those of PPO containing blends.

Since PEC is a copolymer, description of its interaction with PS is more complex than if it were simply a homopolymer. Binary interaction models have been presented which suggest that copolymer miscibility with a homopolymer can be enhanced by endothermic interactions between the unlike repeat units of the copolymer (11-13). In its simplest form (11), the binary interaction model for the heat of mixing, ΔH_{mix}, of a copolymer, A, containing repeat units 1 and 2, with a homopolymer, B, containing repeat units 3, is given by

$$\Delta H_{mix}/V = B \; \phi_A \; \phi_B \tag{1}$$

where ϕ_A is the volume fraction of copolymer, ϕ_B that of the homopolymer, V is the volume of the system, and B the interaction parameter for mixing A with B, given by,

$$B = B_{13}\Phi_1^A + B_{23}\Phi_2^A - B_{12}\Phi_1^A\Phi_2^A \tag{2}$$

where Φ_i^A is the volume fraction of repeat unit i in copolymer A. Equation 2 suggests that a variety of parameter combinations, leading to negative B and a miscible blend, are possible, depending on the natures of the polymers. Some of these are illustrated in Figure 1. Of most interest is the suggestion that exothermic or negative heats of mixing are possible, indicating miscibility of the copolymer with the homopolymer, even though all B_{ij} parameters are positive. Provided $(B_{12})^{1/2} > (B_{13})^{1/2} + (B_{23})^{1/2}$, there will be a region in copolymer composition, Φ_i^A, where B and ΔH_{mix} are negative when the copolymer and homopolymer are mixed, signifying the formation of a miscible blend. This simple result would appear to explain the "miscibility window" seen with variation of comonomer content in many different systems (14-17) where the homopolymer is not miscible with the homopolymers made from either comonomer, yet is miscible with the copolymer.

PEC, PPO, and PS are amorphous materials as normally melt processed, a fact which prevents the use of melting point depression analyses for experimentally determining ΔH_{mix} or B for the blends with PS. Isotactic polystyrene, i-PS, is able to crystallize, is miscible with PPO, and has been successfully used to determine the parameters related to ΔH_{mix} (7,12,18). This study shows that i-PS is also miscible with PEC copolymers which contain up to 20 mole% trimethyl comonomer. This fact permits the use of i-PS melting point depression analysis to determine the effect of comonomer content on ΔH_{mix} with i-PS.

Experimental Procedures

All of the materials used in this study have weight
average molecular weights between 28000 and 35000. The
PEC copolymers and the PPO were provided and characterized by Dr. Bill Pavelich, Borg-Warner Chemicals,
Parkersburg, West Virginia. The properties of these
materials are summarized in Table I. The i-PS, which
contained 90 mole% isotactic units, was purchased from
Scientific Polymer Products, Inc.

Table I. Physical Properties

Polymer	Abbreviation	M_w	T_g ($^\circ$C)	T_m ($^\circ$C)
Atactic Polystyrene	PS	28,000	96.3	----
Isotactic Polystyrene (90 mole% isotactic units)	i-PS	35,000	92.0	232*
Poly(2,6-dimethyl phenylene oxide)	PPO	30,000	212	256**
Poly(2,6-dimethyl phenylene-co-2,3,6-trimethyl-1,4-phenylene oxide), 1 mole% trimethyl comonomer	1%PEC	30,000	214	256
Phenylene copolymer, as above, 9 mole% trimethyl comonomer	9%PEC	30,000	220	264
Phenylene copolymer, as above, 15 mole% trimethyl comonomer	15%PEC	30,000	221	264
Phenylene copolymer, as above, 20 mole% trimethyl comonomer	20%PEC	30,000	222	266

*   After Hoffman-Weeks procedure.
**   Crystallized from solution, no annealing, for all
       phenylene oxide materials.

Blends were prepared by dissolving the components in
tetrahydrofuran, THF, at 75 $^\circ$C to form a solution containing about 10 wt.% solids. Films were prepared by
evaporating the THF from solution at room temperature for
1-2 days, followed by vacuum drying at 90 $^\circ$C for 24 hours
to remove final traces of the solvent. Films, prepared
in this way, were opaque due to crystallization of the

PEC, in agreement with previous observations (19,20), and
to the presence of crystalline i-PS. Amorphous, optically transparent, films could be prepared by heating the
solution cast films for a few minutes at 270 °C, followed
by rapid cooling to temperatures below the blend glass
transition temperature, T_g, to prevent crystallization of
the i-PS. Once the phenylene ether crystals are melted
out of the blend, crystallization of this material does
not recur. The i-PS material will crystallize readily
from the melt when the blend temperature is above T_g and
below about 200 °C for several minutes.

Blend T_g's were measured with a Perkin-Elmer DSC-7
differential scanning calorimeter by first heating 10 mg
samples of the blends at 20 °C/min to 270 °C, annealing
for 10 min to melt any crystallinity, then rapidly cooling to about 60 °C. The blend T_g was then obtained by
reheating at 20 °C/min.

Melting temperatures, T_m, of i-PS in the pure state
and in blends were measured after crystallizing the
blends at several crystallization temperatures, T_c, to
permit estimation of the equilibrium melting temperatures, via the Hoffman-Weeks method (21). Typically, a
10 mg sample was heated in the DSC-7 at 20 °C/min to
270 °C, held at that temperature for 10 min, then rapidly
cooled to T_c where it was held for at least 100 min. The
melting temperature, T_m, was then determined on reheating
the sample in the DSC-7 at 10 °C/min. The 100 min annealing time at T_c was found to be the minimum time required to achieve the maximum degree of crystallinity for
i-PS in either the pure state or in blends with phenylene
ethers. At annealing temperatures below 210 °C, the i-PS
degree of crystallinity was found to be the same after 1
hr as after 15 hrs, a result which is similar to that
determined by Overbergh, et al. (22).

Results and Discussion

The melting behavior of the i-PS material seems to indicate the presence of two crystalline forms, with two
melting temperatures marked H and L in Figure 2. The behavior of the multiple melting peaks in i-PS is actually
quite complex, and as the crystallization temperature,
T_c, is increased the temperature of peak L is found to
increase in direct proportion while that of peak H increases only slightly. As T_c is further increased, the L
endotherm increases in size while the H endotherm
diminishes and eventually disappears. These trends
generally confirm those reported previously (2), and suggest that the multiple peaks are most likely the result
of lamellar thickening processes, perhaps requiring melting and recrystallization (23).

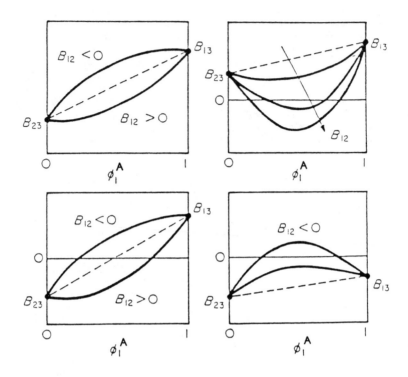

Figure 1. Illustration of ways the interaction parameter may
vary with copolymer composition for mixtures of a
copolymer with a homopolymer. (Reproduced with permission from ref. 11. Copyright 1984 Butterworth.)

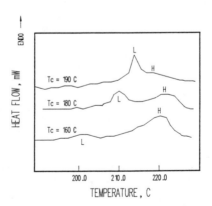

Figure 2. Variation in i-PS melting behavior with crystallization temperature.

A good straight line results when T_m of the L peak is plotted against T_c for i-PS materials crystallized between 150 - 180 °C, see Figure 3. Extrapolation of this line to the T_m = T_c line, suggested by Hoffman-Weeks (21), yields an equilibrium melting temperature, T_m^o, for i-PS equal to 232 °C. This value is approximately 8 °C lower than values reported by other investigators (4,23,24). This discrepancy is probably a consequence of our use of 90 mole% isotactic material instead of 100 mole% i-PS for this study.

Blend T_g Behavior. Blends of both atactic polystyrene with PEC and i-PS with PEC show the presence of a single T_g which varies monotonically with blend composition and with the percent trimethyl comonomer in the PEC. Typical results are shown in Figure 4 for the i-PS/PEC blends. The PS/PEC blends are not shown. The presence of a single T_g for each blend composition is indicative of the presence of a single amorphous phase (2,3), and on this basis we conclude that i-PS/PEC and PS/PEC blends are miscible for trimethyl compositions in the copolymer from 0 to 20 mole%. The T_g of pure PEC is raised by only 12 °C with respect to PPO by the addition of 20 mole% comonomer, and this does little to raise the blend T_g at intermediate PEC concentrations.

The lines in Fig. 4 are calculated by the Fox equation (25),

$$1/T_{g,b} = \Sigma_i (\omega_i/T_{g,i}) \qquad (3)$$

where ω_i is the mass fraction of species i, $T_{g,i}$ its glass transition temperature, °K, and $T_{g,b}$ is the glass transition temperature of the blend. While derived to explain the variation of T_g with copolymer composition, the Fox equation has successfully predicted the composition dependence of T_g in miscible blend systems (2,26). As found previously for the PPO/PS system (2), the agreement between this prediction and experiment is fairly good when the trimethyl comonomer level in the copolymer is low. As the trimethyl comonomer level increases, however, the observed T_g variation with PEC in the blend falls increasingly below the line predicted by the Fox equation. The reason for this behavior is unclear. The concave upward T_g behavior shown for these blends is relatively normal for miscible systems, and there is no evidence of strong interactions, such as hydrogen bonding, between the components which can lead to convex upward T_g vs. composition curves (26).

Blend Melting Behavior. The melting point depression shown by a miscible, high molecular weight, crystallizable blend component in a high molecular weight diluent can be analyzed by the Flory-Huggins equation (27),

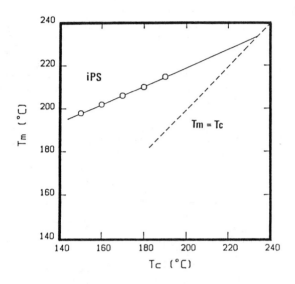

Figure 3. Hoffman-Weeks construction for i-PS crystallized between 150 and 180 °C.

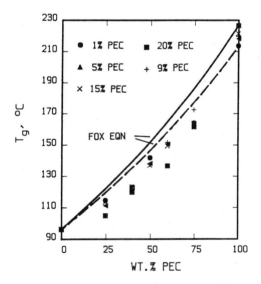

Figure 4. Glass transition behavior of i-PS/PEC blends.

$$\Delta T_m = T_m - T_m{}^o = - \{V_{2u} \cdot B \cdot T_m{}^o / \Delta H_{2u}\} \cdot \Phi_1{}^2 \qquad (4)$$

to obtain the interaction density, B, characteristic of
the polymer-polymer interaction. Accurate determination
of B requires accurate values for the volume fraction PEC
amorphous diluent in the blend, Φ_1, for the melting temperatures in both the blend, T_m, and in the pure state,
$T_m{}^o$, and for the heat of fusion per unit volume of i-PS
crystal, $\Delta H_{2u}/V_{2u}$. This latter value is taken to be
20.56 cal/cc (4). As discussed above, the melting temperature for i-PS is a sensitive function of T_c, and the
Hoffman-Weeks construction is necessary to accurately establish T_m and $T_m{}^o$ for this material, both in the pure
state and in blends with PEC.

The morphology of crystalline isotactic polystyrene,
i-PS, has been investigated by others, and they have concluded that i-PS normally crystallizes as stacks of
folded chain lamellae which are arranged in volume filling spherulites (2). The melting point of lamellar
polymer crystals depends on the lamella thickness, L, as
follows (28)

$$T_m = T_m{}^o \{ 1 - 2\sigma_e / (\Delta H_f \cdot L) \} \qquad (5)$$

where σ_e is the specific surface free energy of the end
face of the crystal and ΔH_f is the heat of fusion per
unit volume. Hoffman and Weeks (21) developed a theory
relating the observed melting point to the crystallization temperature,

$$T_m = T_m{}^o \{ 1 - 1/\beta \} + T_c /\beta \qquad (6)$$

Provided β is independent of temperature, a plot of
T_m vs. T_c is linear and should intersect the line $T_m = T_c$
at $T_m{}^o$, the equilibrium melting temperature of the in-
finitely large crystal. Figures 5 and 6 show Hoffman-
Weeks constructions for i-PS in PEC copolymers containing
1 mole% and 20 mole% trimethyl comonomer, respectively,
the limits of PEC compositions investigated. Similar
plots, not shown, were also obtained using the protocol
outlined above, for i-PS blends with the remaining PEC
materials, described in Table I. Deviations from
straight line T_m vs. T_c behavior are evident in all PEC
blend systems when the annealing temperature is raised
above 180 °C, indicating that the deviations are related
more to the i-PS and less to the PEC comonomer content.

Deviations of the sort seen in Figures 5 and 6 are
not rare and can result for a variety of reasons (28).
For example, crystallization-induced chemical reorganization has been reported (29-34) to cause such behavior for
a variety of materials, including polyesters and various
random and block copolymers. It would not be un-

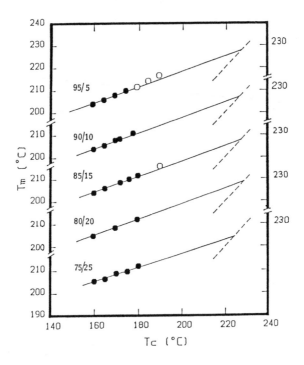

Figure 5. Hoffman-Weeks construction for i-PS crystallized from blends with 1%PEC.

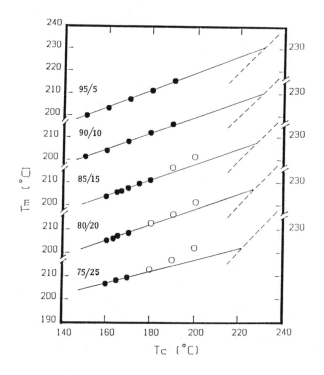

Figure 6. Hoffman-Weeks constructions for i-PS crystallized from blends with 20%PEC.

reasonable to speculate that crystallization-induced
chemical reorganization of the i-PS is responsible for
the increase in melting temperatures above the Hoffman-
Weeks line. The physical picture could simply be that
the atactic PS units in the initially random copolymer
are eliminated, through chemical rearrangements at high
crystallization temperatures, to produce a block
copolymer of isotactic strings which can crystallize at
higher temperature due to their improved perfection.
Such a mechanism, known to be operative in other systems,
would imply that annealing at high temperatures produces
an i-PS material that is physically different from that
annealed at lower temperatures where chemical rearrangement is less possible.

What is important to the present study is to examine the effect of PEC comonomer, not i-PS copolymer
structure, on the thermodynamics of interaction with
i-PS. Consequently, we limited crystallization temperatures to values less than 180 °C to inhibit the T_m
enhancement process. As can be seen from Figures 5 and
6, good straight lines, which appear to follow Equation
6, can be drawn through the solid data points corresponding to these lower crystallization temperatures, and
reasonable equilibrium melting temperatures result from
the intersection of these lines with the $T_m = T_c$ lines
for each blend fraction of PEC.

The ΔT_m values, computed for each PEC blend system
from the equilibrium melting temperatures of i-PS in the
blends and in the pure state are shown in Figure 7. As
suggested by Equation 4, good straight lines result for
each PEC blend system when ΔT_m is plotted against the
square of the volume fraction of PEC. Straight lines
suggest that the the interaction density, B, is independent of PEC concentration for blends containing less than
25% PEC, a result which contrasts somewhat with the extreme concentration dependence in B reported by Plans, et
al (4). These straight lines also show increasingly
positive slopes as the trimethyl comonomer content in the
PEC is increased, indicating that B, the interaction density, becomes more exothermic with increasing comonomer
content in the PEC. These results are summarized in
Figure 8. Error bars are included for each B value to
indicate that each has less than 10% relative uncertainty, based on the melting point depression data.

As shown in Figure 8, we estimate the interaction
energy density for i-PS/PPO blends to be B = -2.9 cal/cc
at 232 °C. This value is comparable to the value, B =
-1.4 cal/cc, calculated from χ = -0.17, reported by
Plans, et al (4), via

$$B = \chi RT/V_1 \qquad\qquad (7)$$

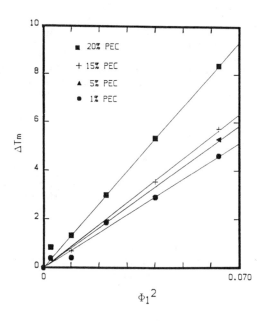

Figure 7. Melting point depressions for i-PS in various PEC copolymers.

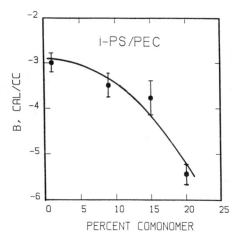

Figure 8. Interaction parameter vs. copolymer composition for i-PS in PEC copolymers.

where χ is the Flory-Huggins interaction parameter, V_1 the segmental molar volume of the PPO, taken in our calculation to be 121 cc/gmole. We believe that the difference between the Plans B value and ours is largely the result of a concentration dependency in their melting point depression data which were obtained at PPO concentrations up to 35%. If one restricts their data to PPO concentrations less than 25%, thereby eliminating one of the three data points used in their construction, their Fig. 6, one can easily bring their B parameter into agreement with ours for mixtures with the 1% comonomer containing PEC.

The difficulty of directly measuring the heats of mixing for two polymers is well known (35), but indirect routes which measure the heat of dilution (36,37) suggest that B for the PPO/PS system is -4 to -5 cal/cc at 35 °C. Gas sorption experiments (38) are also consistent with B values in this range.

Our measured B value for i-PS/1%PEC is generally more exothermic than the estimate obtained by Maconnachie, et al, (39) from neutron scattering in diluted atactic polystyrene/deuterated PPO systems. They obtain an equation,

$$\chi = 0.121 - 77.9/T \qquad (8)$$

which can be used with Equation 7 to obtain B = -0.3 cal/cc at 505 °K. The χ in Equation 8 contains an entropic contribution which can be evaluated from the temperature dependency (27). Doing so, allows one to evaluate the purely enthalpic portion and to obtain B = -1.5 cal/cc.

Given the various experimental and theoretical difficulties associated with measuring B or χ for this system, our value seems reasonable; however, it is important to note that we have measured B for blends containing i-PS not PS. Investigators of other blend systems (40-42) have consistently shown that isotactic and syndiotactic blend components interact less exothermically than their atactic counterparts, a situation which can lead to the immicibility of polymers with regular tacticity. On this basis, we could expect our B values for i-PS/PPO blends to be somewhat less exothermic than values obtained for PS/PPO blends.

Since only four copolymer compositions were studied and particularly since the maximum trimethyl comonomer content available was only 20%, it is difficult to precisely determine the three binary interaction parameters, suggested by the "copolymer model", Equations 1 and 2, from the observed variation of B with trimethyl comonomer content in the PEC, Figure 8. It is none the less interesting, however, to qualitatively assess the magnitudes and signs of the interaction parameters. Because B for i-PS/PEC is negative and becomes increasingly negative as the trimethyl comonomer content in the copolymer increases from zero, we can conclude that B_{13} for the interaction between the trimethyl comonomer and the i-PS repeat unit is more exothermic than that between the dimethyl comonomer and i-PS, B_{23}. While somewhat less certain, the downward concavity of the curve in Figure 8 also suggests that the self-interaction, B_{12}, between the trimethyl and the dimethyl comonomers in PEC, is also negative or exothermic.

The observation that B_{13} is more exothermic than B_{23} is not surprising when one considers the fact that blends of tetramethyl bisphenol-A polycarbonate with polystyrene are miscible while blends of bisphenol-A polycarbonate with polystyrene are immiscible (43). Apparently, substitution of methyl groups on the aromatic rings in the backbones of these materials enhances miscibility with the styrene repeat unit. The observation that B_{12} is also exothermic is quite surprising, considering the similarity of the trimethyl and dimethyl ether units. Using regular solution approximations, one could reasonably expect these similar materials to mix endothermically, but the negative B_{12} suggests they do not.

Studies of other blend systems have observed that cloud temperatures, associated with the onset of immiscibility on heating, generally increase as the interaction between blend components becomes increasingly exothermic (44-46). The results of Figure 8 might logically lead one to expect i-PS/PEC blends to show higher cloud temperatures than i-PS/PPO blends. This possibility, which was a partial reason for the research reported here, was examined by heating blends at approximately 10 °C/min. in a simple cloud point apparatus, described previously (47). No cloud points in any of the blends were observed at temperatures up to 380 °C, at which point decomposition became severe. These results contrast with reports from the field which suggested that PEC containing blends could be molded, without delamination (possibly associated with multiphase formation), at higher temperatures than PPO containing blends (48). One possible reason for the discrepancy between laboratory and field tests for cloud points is the rather low molecular weights of the components in the laboratory blends. By analogy with other systems (49), cloud temperatures generally rise as the molecular weights of the components are reduced.

Acknowledgments

The authors gratefully acknowledge an unrestricted grant by Borg-Warner Chemicals, Inc. for partial support of this research.

Legend of Symbols

Symbol	Typical Units	Description
B	J/m^3 , cal/cc	Binary interaction energy density between polymer repeat units
B_{ij}	J/m^3 , cal/cc	Interaction energy density between units i and j
ΔH_{mix}	J, cal	Enthalpy of mixing
ΔH_{2u}	J/gmole, cal/gmole	Enthalpy of fusion per mole pure crystal
ΔH_f	J/m^3 , cal/cm^3	Enthalpy of fusion per unit volume of crystal
L	m, cm	Lamella thickness
R	$J/gmole^\circ K$, $cal/gmole^\circ K$	Gas Constant
T	$^\circ K$	Absolute temperature
$T_m{}^\circ$	$^\circ K$	melting temperature of perfect crystal
T_c	$^\circ K$	Crystallization temperature
T_m	$^\circ K$	Observed melting temperature
T_g	$^\circ K$	Glass transition temperature
V	m^3 , cc	System volume
V_{2u}	m^3/gmole, cc/gmole	Molar volume pure crystal 2
V_1	m^3/gmole, cc/gmole	Molar volume of component 1, either solvent molecule or polymer repeat unit
β	___	Hoffman-Weeks parameter
σ_e	J/m^2 , cal/cm^2	Surface free energy of crystal end face
ϕ_i	___	Volume fraction i in blend
$\phi_i{}^A$	___	Volume fraction i in polymer A
ω_i	___	Mass fraction i in blend

Literature Cited

1. Hay, A. S.; Polym. Eng. Sci., 1976, 16(1), 1.
2. MacKnight, W. J.; Karasz, F. E.; Fried, J. R. In "Polymer Blends, Vol. I," Paul, D. R.; Newman, S., Eds.; Academic Press, New York, 1978, Chapter 5.
3. Paul, D. R.; Barlow, J. W.; J. Macromol. Sci. Rev. Macromol. Chem., 1980, C18, 109.
4. Plans, J.; MacKnight, W. J.; Karasz, F. E.; Macromolecules, 1984, 7(4), 810.
5. Prest, W. M.; Porter, R. S.; J. Polym. Sci., 1972, A2(10), 1639.
6. Fried, J. R.;Karasz, F. E.; MacKnight, W. J.; Polym. Eng. Sci., 1979, 19, 519.
7. Runt, J. P.; Macromolecules, 1981, 14(2), 420.

8.   A. F. Yee, A. F.; Engineering Science, 1977, 17,
     213.
9.   Baer, E.; Wellinghoff, S.;Polymer Preparation, 1977,
     18, 836.
10. Warren, R. I.; Polym. Eng. Sci., 1985, 25, 477.
11. Paul, D. R.; Barlow, J. W.; Polymer, 1984, 25, 487.
12. Kambour, R. P.; Bendler, J. T.; Bopp, R. C.; Macro-
     molecules, 1983, 16, 753.
13. Ten Brinke, G.; Karasz, F. E.; MacKnight, W. J.;
     Macromolecules, 1983, 16, 1827.
14. Barnum, R. S.; Goh, S. H.; Paul, D. R.; Barlow, J.
     W.; J. Appl. Polym. Sci., 1981, 26, 3917.
15. Keitz, J. D.; Barlow, J. W.; Paul, D. R.; J. Appl.
     Polym. Sci., 1984, 29, 3131.
16. Alexandrovich, P.; Karasz, F. E.; MacKnight, W. J.;
     Polymer, 1977, 18, 1022.
17. Pfennig, J. L. G.; Keskkula, H.; Barlow, J. W.;
     Paul, D. R.; Macromolecules, 1985, 18, 1937.
18. Plans, J.; MacKnight, W. J.; Karasz, F. E.; Macro-
     molecules, 1984, 17(4), 810.
19. Horikiri, S.; J. Polym. Sci., 1972, A2(10), 1167.
20. Butte, W. A.; Price, C. C.; Hughes, R. E.; J.
     Polym. Sci., 1962, 61, 528.
21. Hoffman, J. D.; Weeks, J. D.; J. Res. Nat. Bur.
     Stds., 1962, 66A(1), 13.
22. Overbergh, N.; Berghmans, H.; Smets, G.; J. Polym.
     Sci., 1972, C(38), 237.
23. Overbergh, N.; Berghmans, H.; Reynaers, H.; J.
     Polym. Sci.:Polym. Phys. Ed., 1976, 14, 1177.
24. Lemstra, P. J.; Kooistra, T.; Challa, G.; J. Polym.
     Sci., 1972, A2(10), 823.
25. Fox, T. G.; Bull. Am. Phys. Soc., 1956, 1, 123.
26. Olabisi, O.; Robeson, L. M.; Shaw, M. T.; "Polymer-
     Polymer Miscibility", Academic Press, New York,
     N.Y., 1979.
27. Flory, P. J.; "Principles of Polymer Chemistry",
     Cornell University Press, Ithaca, New York, 1953,
     Chapt. 8.
28. Wunderlich B., "Macromolecular Physics, Vol. 3 Crys-
     tal Melting", Academic, New York, 1980.
29. Barnum, R. S.; Barlow, J. W.; Paul, D. R.;J. Appl.
     Polym. Sci., 1982, 27, 4065.
30. Schulken, R. M.; Boy, R. E.; Cox, R. H.; J. Polym.
     Sci., Pt. C, 1964, 6, 17.
31. Lenz, R. W.; Martin, E.; Schuler, A. N.; J. Polym.
     Sci.: Polym. Chem. Ed., 1973, 11, 2265.
32. Lenz, R. W.; Schuler, A. N.; J. Polym. Sci.: Polym.
     Symp., 1978, 63, 343.
33. Lenz, R. W.; Go, S.; J. Polym. Sci.: Polym. Chem.
     Ed., 1973, 11, 2927; 1974, 12, 1.
34. Thiele, M. H.; Mandelkern, L.; J. Polym. Sci., 1970,
     A2(8), 957.
35. Robeson, L. M. In "Polymer Compatibility and Incom-
     patibility: Principles and Practice, Vol. 2"; Solc,

K., Ed.; M.M.I. Press Symposium Series; Harwood
Academic Press: New York, 1982; p.177.

36. Karasz, F. E.; MacKnight, W. J.; Pure App. Chem.,
1980, 52, 409.

37. Weeks, N. E.; Karasz, F. E.; MacKnight, W. J.; J.
Appl. Phys., 1977, 48(10), 4068.

38. Maeda, Y; Paul, D. R.; Polymer, 1985, 26, 2055.

39. Maconnachie, A.; Kambour, R. P.; White, D. M.; Rostami, S.; Walsh, D. J.; Macromolecules, 1984, 17,
2645.

40. Goh, S. H.; Paul, D. R.; Barlow, J. W.; Polym. Eng.
Sci., 1982, 22, 34.

41. Roerdink, E.; Challa, G; Polymer, 1980, 21, 1161.

42. Schurer, J. W.; de Boer, A.; Challa, G.; Polymer,
1975, 16, 201.

43. Fernandes, A. C.; Barlow, J. W.; Paul, D. R.;
Polymer, 1986, 27, 1788.

44. Bernstein, R. E.; D. C. Wahrmund, D. C.; Barlow, J.
W.; Paul, D. R.; Polym. Eng. Sci., 1978, 18(16),
1225.

45. Barnum, R. S.; Goh, S. H.; Barlow, J. W.; Paul, D.
R.; J. Polym. Sci. Polym. Lett. Ed., 1985, 23, 395.

46. Pearce, E. M.; Kwei, T. K.; Min, B. Y.; J. Macromol.
Sci. Chem., 1984, 21, 1181.

47. Bernstein, R. E.; Cruz, C. A.; Paul, D. R.; Barlow,
J. W.; Macromolecules, 1977, 10, 681.

48. Pavelich, W. A.; private communication.

49. Kwei, T. K.; Wang, T. T.; Chapt. 4 in Ref. [2].

RECEIVED April 27, 1989

Chapter 5

Fractionated Crystallization in Incompatible Polymer Blends

H. Frensch, P. Harnischfeger, and B.-J. Jungnickel

Deutsches Kunststoff-Institut, D—6100 Darmstadt, Federal Republic of Germany

The crystallization of the minor component in incompatible polymer blends starts sometimes at distinctly larger undercoolings than in the pure polymer, and proceeds in several separated steps. After a short survey on the history of the effect in the available literature, the several types and the origin of this "fractionated crystallization" as observed in some selected systems are described. The information on the blend which can be deduced from the effect is discussed, and the consequences for the blend processing and properties are investigated.

The properties of incompatible polymer blends depend to a large extent on the mutual dispersion of the components, on the supermolecular structure within the phase of a single component, and on the structure of the interface. These structural parameters, in turn, depend on the processing or mixing conditions and on the strength of the thermodynamic incompatibility of the components as well. These boundary conditions together with the cooling rate control also the solidification process of a melt.

Crystallization is a special solidification process. It is well known that the crystallization of a blend component can differ remarkably from that of the corresponding pure material. As in polymers in general, the course of crystallization in blends is governed by equilibrium thermodynamics and by kinetic boundary conditions as well. These boundary conditions change remarkably with blending, thus causing a lot of technically important and scientifically interesting effects. Depending on the blend components under consideration, they can origin in a large number of physical and physico-chemical phenomena, among them also some having non-equilibrium thermodynamic basis. For melt-compatible polymer blends, among other effects, a variation of the crystallization process due to altered nucleation and growth conditions has been reported (1-11). The usual mixing induced melting point depression has been observed too (1-5,11,12).

0097—6156/89/0395—0101$07.25/0
© 1989 American Chemical Society

For polymer blends with extended interfacial regions, these phenomena were reported to apply also to the bulk of the phase borders (12-16). The investigation of incompatible polymer blends revealed, e.g., the induction of specific crystal modifications (16), the rejection, engulfing and deformation of the dispersed component by the growing spherulites of the matrix material (4,17-19), and nucleation at the interface (4,20,23).

Polymer blends containing one component as finely dispersed droplet suspension exhibit sometimes the phenomenon of "fractionated crystallization" which originates in primary nucleation of isolated melt particles by units of different nucleating species (24-33). This phenomenon resembles to some extent the classical droplet crystallization in which crystallization is inhibited until homogeneous nucleation occurs (34-39). Fractionated crystallization proceeds stepwise at greatly different undercoolings, these steps sometimes being separated by more than 60°C (25,33). It is the aim of the present paper to give a short survey of the history and the available literature on this particular blend crystallization phenomenon, and to describe and interprete all kinds of manifestations of it as observed so far, considering particularly the authors work. It will turn out that fractionated crystallization exhibits occasionally a number of interesting and surprising new effects, the origin of which will be discussed. Finally, the qualitative and quantitative conclusions that can be drawn from the effects on the system under consideration are summarized.

History

Crystallization is a phase transition that is controlled by nucleation and growth processes (38). The nuclei which are necessary for the onset of crystallization can consist, on the one hand, of small crystals of the crystallizing material itself which are statistically created through thermodynamical fluctuations in the melt (homogeneous nuclei). On the other hand, heterogeneities due to the presence of impurities or other materials can act as nuclei too (heterogeneous nuclei). Each nucleating species, in particular those of the different impurities, is characterized by a specific undercooling ΔT_h (at non-isothermal crystallization), or induction time Δt_h (at isothermal crystallization) at which it induces remarkable crystallization. By experience, the undercooling at which homogeneously nucleated crystallization sets in (ΔT_{ho}) is usually much greater than those of the many heterogeneous nucleating agents ($\Delta T_{he}^{(i)}$). Real crystallization processes, therefore, are generally heterogeneously induced if the substances under investigation are not specially purified. If several different heterogeneities are present, however, only that with the lowest attributed undercooling ΔT_{he} is nucleatingly active: when started by the described process ("primary nucleation"), crystallization instantly spreads over the whole available material via "secondary nucleation" before another type of heterogeneity can become efficient. The dynamics of the latter process depends for a given substance only on the temperature. These necessary preceeding remarks are valid not only for polymers but for all kinds of substances.

It was in 1880 when the Dutchman Van Riemsdyk reported that small gold melt droplets solidified at much larger undercoolings than the bulk material (40). Similar observations were made later for many other metals, indicating this to be a basic crystallization phenomenon (41-43). Turnbull and Cech (44) investigated it in detail. They remarked that the delayed solidification of the droplets (with sizes between 10μm and 100μm) after slow cooling of their melt covered a broad temperature interval with a lower limit T_{ho} that was typical for the material under investigation. The approximate relation

$$T_c^{(o)} - T_{ho} \approx T_c^{(o)}/5 \qquad (1)$$

($T_c^{(o)}$: melt/crystal equilibrium temperature) held for most metals. The observation could be explained quantitatively by assuming an exponential law for the crystal induction rate, and by consideration of the distribution of the nucleating heterogeneities (all of which were assumed to be equally efficient) over the large number of droplets of different size (41,44). The heterogeneous nucleation activity is terminated at that temperature where homogeneous nucleation sets in. The course of the crystallization before the onset of homogeneously nucleated crystallization, therefore, reflects the number density of the heterogeneities and the size distribution of the droplets. From the homogeneous nucleation, finally, conclusions can be drawn on the interface parameters of the crystals (35-39,45).

The creation of comparably stable suspensions of sufficiently small polymer droplets is not easy. It was therefore only in 1959 when such investigations and observations were reported also for polymers. Price (34) found that droplets of polyethylene oxide (PEO) with an average diameter of 5μm crystallized only at an undercooling of 65°C. This value is to compare with the maximally attainable undercooling of 20°C in bulk. Turnbull, Price et al. (36,37,46) prepared rather stable suspensions (particle diameter some μm) of n-alkanes, polyethylene (PE) and polypropylene (PP) in a thermodynamically inert liquid. They observed the delayed crystallization at a low cooling rate of the molten polymer droplets in a light microscope and stated their course to be continuous. Their explanation was the same as for the time and temperature dependent course of the crystallization in the latter. Those droplets which do not contain at least one heterogeneous impurifying particle undergo homogeneous nucleation. The finer the dispersion, the more this nucleation type dominates. Some years later, Koutsky et al. (35) reported on another basic experiment in this field. They sprayed solutions of several polymers on a silicon oil film which was spread over a glass plate. During spraying, the solvent evaporated, and polymer droplets of the size of (1...100)μm arose. The observation of the crystallization during slow cooling (cooling rate: 0.1°C/min) of their melt in a microscope revealed clearly for the first time that the crystallization did not occur continuously but in distinct steps which were initiated at distinctly different undercoolings. Most of the material, however, crystallized again at a maximally possible undercooling which was typical for the polymer. This undercooling varied between about 50°C (for PE) and 100°C (for PP) and was attributed again to homogeneous nucleation (PE, PP), or to the nucleating efficiency of the interface between the silicon oil and the polymer melt (for PEO, polyoxymethylene (POM), isotactic polystyrene (iPS)), respectively.

The explanation of the observed crystallization temperature distribution with peaks at distinct temperatures is obvious: the spectrum of undercoolings at which the several crystallization steps occur reflects immediately the efficiency spectrum of the several nucleating heterogeneous species. It should be recalled that only that heterogeneous species with the lowest specific undercooling ΔT_h is active even when several different kinds of them are present. If the dispersion of the polymer is so fine that not every droplet contains at least one particle of the usually active species (that is, if the degree of dispersion is of the order of magnitude of the number density of those heterogeneous particles), that heterogeneous species with the second lowest specific undercooling ΔT_{he} can become active and so on.

The polymer droplets under investigation so far were immersed in a liquid that did not influence crystallization. The described effects, however, should occur also if they are inserted in a surrounding of a solid or molten second polymer, that is, in polymer blends where the polymer under investigation is the minor component, dispersed as finely as possible in the matrix polymer. Such materials should be advantageous in that sense that they enable the investigation of the crystallization by usual thermoanalytical techniques, e.g. by DSC. This, in turn, would allow a more precise description of the effects. It is surprising that such investigations have not been performed for a long time and that corresponding observations were made rather by accident. Only in 1969, Lotz & Kovacs (26), and O'Malley et al. (27) reported independently that the PEO component in a PEO/PS block copolymer crystallized in two steps as revealed by DSC measurements, one of them at about the same temperature at which the PEO homopolymer crystallizes, and the other at -20°C. The latter step occurred only if the PEO was distinctly the minor component and finely dispersed in the PS matrix, and was attributed to homogeneously nucleated crystallization in the sense of the foregoing considerations. Romankevich et al. (24) made a similar observation with respect to the POM component in a POM/PE block copolymer. The additional peak was about 13°C below that temperature at which POM usually crystallizes. This experimental result was confirmed later by Jungnickel et al. (32,47,48) who observed up to four crystallization peaks for POM in a blend with PE and who introduced the term "fractionated crystallization" for the effect. The relative intensity of the several peaks changed with the degree of dispersion of the POM in the PE matrix. Tsebrenko (31) found also up to three crystallization peaks for the POM component in blends with an ethylene/vinyl acetate copolymer, and explained it by the delayed crystallization of the POM in the interphase between the phase separated components. Hay et al. (29), and Baitoul et al. (30) investigated blends of PE and PS, and they found the crystallization of the PE component to proceed in two steps, one at about 100°C and another at about 80°C (29), or 70°C (30), provided this component was the minor phase and formed insertions in the PS with a diameter of typically 10μm. The additional peak was attributed to homogeneous crystallization as it did Turnbull, Price et al. (36,37) in their droplet experiments. Yip et al. (25,33) investigated the crystallization of PP blended with a SBS block copolymer. They found two additional crystallization peaks at 74°C and at 44°C if the PP was the minor component. They remarked, moreover, that the usual peak at 107°C disappeared completely when the PP content dropped sufficiently, and/or became finely enough dispersed. The peak at $74^{\circ}C$ is located at the same temperature where the main crystallization occurred in the droplet experiments of Turnbull, Price et al. (36,37), and Koutsky et al. (35). The authors, therefore, attributed it to homogeneous crystallization. The peak at $44^{\circ}C$ was assumed to be due to the solidification of PP in a smectic chain alignment. This "crystal modification" usually arises if the melt is quenched (49). This interpretation, however, is questionable since the homogeneously nucleated crystallization should occur at the highest attainable undercooling if the droplets have a monomodal size distribution. The experimental facts of Yip et al., on the contrary, hint at that the main crystallization step in the referred droplet experiments (35-37) is not due to homogeneous nucleation as the authors claimed but rather by a currently unknown heterogeneously nucleating species or surface.

This survey on the literature indicates that only few data are available on the droplet crystallization phenomena in incompatible polymer blends. Moreover, these observations are partly not completely explained, and, where explained, these explanations are partly not satisfying or contradictory. In the next chapter, therefore, experimental results for some selected systems as investigated by the authors are presented with the aim to show all faces and properties of fractionated crystallization in detail, and to contribute to a better understanding of the origin of the effect.

Theoretical Considerations

The overall crystallization kinetics in a dispersed system is determined by several processes. First, there is the time and temperature dependent nucleating efficiency of a certain heterogeneous species which, in turn, causes a time and temperature dependent course of the crystallization that it induces. This process depends on the number density of the nucleatingly active heterogeneous species under consideration (which must not necessarily be constant but in turn can be time dependent) and the dispersion of the crystallizing polymer. Its graph is smooth without maxima or shoulders if the mentioned distribution functions are monomodal. It can be characterized by an undercooling ΔT_h, e.g., by that temperature at which a given percentage of the material has crystallized, or by the temperature of the highest crystallization rate. It is difficult to derive analytical expressions for this function which take into account all relevant structural parameters, in particular the size distribution of the droplets, the distribution of the nucleating heterogeneities over the several droplets, the cooling rate (with the limiting value zero, i.e. isothermal crystallization), the reaction rate (turning into efficiency of the nucleating sites), and the crystal growth rate itself. Some considerations and calculations in this respect can be found, e.g., in (36,37) and (45). An evaluation of this function, basicly, allows estimation of the interface energies between the polymer melt and the heterogeneity and the polymer crystal faces, respectively.

If, additionally, several different nucleating heterogeneities are present, the overall crystallization is a complicated superposition of the contributions of them all. Frequently, the crystallization that is caused by a certain heterogeneity is completed within time or - for nonisothermal conditions - temperature intervals which are small in comparison to the differences in the induction times or specific undercoolings, respectively, of the several subsequently active heterogeneities. The superposition of their contributions, then, leads to a stepwise overall crystallization. These repeatedly mentioned specific undercoolings can be estimated in the following way.

The free energy $\emptyset^*$ of formation of a rectangular nucleus of critical size at an undercooling ΔT is proportional to a "specific interfacial energy difference" Δy_{ps} (38).

$$\emptyset^* \sim \Delta y_{ps}/(\Delta T)^2. \tag{2}$$

Δy_{ps} is defined by (38,50):

$$\Delta y_{ps} = y_p(m,c) - y_{ps}(m) + y_{ps}(c). \tag{3}$$

Here, the convention $y_{12}(1,2) = y_{substance"1",substance"2"}$ (state of substance"1", state of substance"2") is introduced; m = melt, c = crystal, s = substrate, p = polymer). $y_{ps}(m)$ and $y_{ps}(c)$, followingly, are the interfacial energies between the nucleating substrate and the polymer melt and crystals, respectively. $y_p(m,c)$ is the surface free energy parallel to the molecular chain direction between the crystal and its own melt. The relation

$$\Delta y_{ps} = 2y_p(m,c) \tag{4}$$

holds for homogeneous nucleation.

If one assumes that for the onset of crystallization, $\emptyset^*/kT$ must be smaller than a certain critical value which is independent on the material under consideration, and if one neglects that the crystallization rate depends also on the temperature dependent mobility of the crystallizable segments, then from Equation 2 the following approximate expression follows for the relation between Δy and the undercooling at which the nucleation by two different species "A" and "B" gives rise to crystallization:

$$\Delta y_A/\Delta y_B \approx (T_A/T_B)(\Delta T_A/\Delta T_B)^2. \tag{5a}$$

If, in particular, "B" designates the homogeneous nucleation,

$$\Delta y_B = 2y_p(m,c),$$

(cf. Equation 4), and if Equation 1 is valid,

$$\Delta T_B = T_c^{(o)}/5, \text{ and } T_B = (4/5)T_c^{(o)},$$

then Δy_A is given by

$$\Delta y_A/y_p(m,c) \approx 62.5(T_A/T_c^{(o)})(\Delta T_A/T_c^{(o)})^2. \tag{5b}$$

From Equations 5, the relative Δy-values of the heterogeneities can be determined from the corresponding undercoolings. The inversion of this statement is true too. The temperature T_A at which a certain heterogeneity "A" induces crystallization depends on its Δy-value (Figure 1). Usually, only that heterogeneity with the smallest Δy-value is efficient; via secondary nucleation at the created crystals, the crystallization process spreads over the whole volume once it has started, and it is completed before the undercooling of the heterogeneity with the second smallest Δy-value is reached. But exactly this effect is inhibited if the material volume is divided into many separated droplets.

Among a large number of small polymer droplets, each of volume v_D, the fraction of droplets which contain exactly z heterogeneities of the kind "A" that usually induce crystallization follows a Poisson distribution function (45):

$$f_z^{(A)} = ((M^{(A)}v_D)^z/z!)\exp(-M^{(A)}v_D) \qquad (6)$$

where $M^{(A)}$ is the concentration of randomly suspended heterogeneities and $M^{(A)}v_D$ is their mean number per droplet. The fraction of droplets which contain at least one heterogeneity of the kind "A" is given by $f_{z>0}^{(A)} = 1 - f_0^{(A)}$ and amounts to:

$$f_{z>0}^{(A)} = 1 - \exp(-M^{(A)}v_D). \qquad (7)$$

The consideration of a droplet size distribution may somewhat modify this equation. $f_{z>0}^{(A)}$ describes that part of the droplets and, therefore, of the material that crystallizes induced by heterogeneity "A". The remainder crystallizes at a greater undercooling induced by heterogeneity "B" and so on. For these further crystallization steps, the same considerations hold. From the relative intensity of the different crystallization steps, therefore, conclusions can be drawn on the concentration of the respective heterogeneities if the mean size of the droplets is known. The larger a particle is, the greater is the probability of containing units of a certain heterogeneity and, followingly, the probability to crystallize at the usual temperature.

It is easy to realize that the crystallization at the usual temperature which is induced by heterogeneity "A" is completely suppressed if the relation

$$M^{(A)}v_D \ll 1 \qquad (8)$$

holds.

It should be pointed out that there is no direct physical relation between the phenomenon of fractionated crystallization and the number and the size of spherulites in the pure polymer. Whereas the occurrence of fractionated crystallization is related to the ratio between the number densities of dispersed polymer particles and primary nuclei, the size and the number of spherulites are additionally influenced by the cooling rate and the crystallization temperature. There is, therefore, also no relation between the fractionated crystallization and the type of the arising crystalline entities (complete spherulites, stacks of lamellae,...) both in the pure and in the blended material. There is, finally, no relation between the scale of dispersion which is necessary for the occurrence of fractionated crystallization and the spherulite size in the unblended polymer.

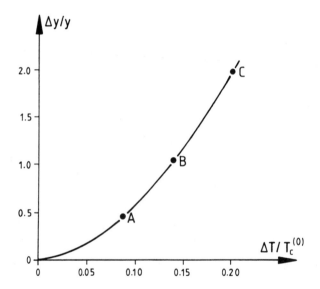

Figure 1. Plot (according to Equation 5b) of the specific interfacial energy difference Δy against the relative undercooling at
which a heterogeneity nucleates the polymer. A, B : Heterogeneous
nucleations; C : Homogeneous nucleation.

The dispersion level and the particle sizes are governed by the melt rheology of the blend components. Theoretical approaches as well as experiments give evidence that the minor component particle radius R during melt processing is related to the interfacial energy y between the components, to the matrix viscosity η and to the shear rate τ by (1,61-63)

$$R \sim y(\eta\,\tau\,)^{-1}. \tag{9}$$

Furtheron, the dispersed droplets are the smaller the closer to unity the viscosity ratio of the components is (62-64). Their sizes decrease also if the first normal stress difference of the dispersed phase becomes smaller than that of the matrix (61). The droplet size, moreover, is influenced by the tendency to further break down of elongated particles due to capillary instabilities (61) as well as by coalescence via an interfacial energy driven viscous flow mechanism. All these procedures and dependences affect the structure formation within their typical time scales (61,62).

Selected Systems

Sample Materials and Preparation. We want to present here the main results of our investigations of the blend systems PE/POM (32,47,48), poly (vinylidene fluoride)/ polyamide-6 (PVDF/PA-6), and PVDF/poly (butylene terephthalate) (PVDF/PBTP) (Frensch, H., Jungnickel, B.-J. Coll. Polym. Sci., in press). The PVDF, the PA-6, and the PE that we used were commercial materials, the latter a stabilizer free grade. The POM and the PBTP of our investigations were especially prepared and contained no additives apart from an unusually low amount of a stabilizer. Separate investigations revealed that all components degraded only negligibly during the several preparational steps and the measurements.

The dried components were melt mixed in a single-screw laboratory extruder. The extruded strands were regranulated and the extrusion cycle was repeated up to five times. After every cycle, samples were taken for the investigations. By the repeated extrusion, different degrees of dispersion were realized.

Investigation Techniques. DSC measurements were carried out under nitrogen atmosphere. In order to destroy the self-seeding nuclei in the components, the samples were preheated for 5 min at least 35°C above the melting points of the higher melting component. Then, several crystallization and reheating runs were performed at a standard rate of 10°C/min. In some cases, other rates were used too.

The dispersion structure of the blends both in the melt and in the solid state was imaged partly by light microscopy (LM), and partly by scanning (SEM) and transmission electron microscopy (TEM). Wide-angle X-ray scattering (WAXS), Infrared (IR) measurements, and torsional pendulum analysis at 1Hz were performed too.

Results/Morphology. LM micrographs of the melt structure of the PE-POM blends at 170°C are displayed in Figure 2. The components are phase separated. The diameter of the dispersed particles of the POM, that is the minor component in these figures, decreases from about 20μm after one extrusion run to about 5μm after five ones.

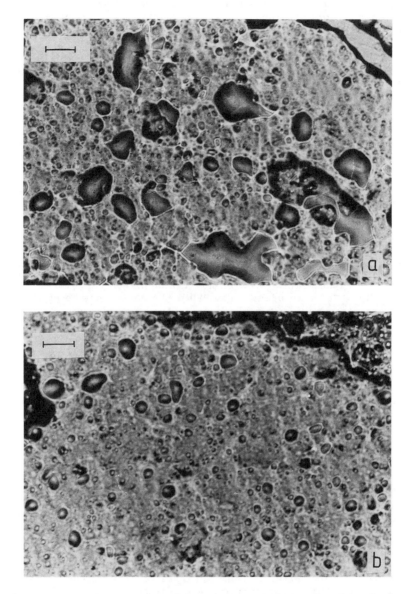

Figure 2. LM micrographs of a PE/POM = 85/15 vol.-% blend after one (a), and five (b) extrusion cycles. Scale bar corresponds to 30µm.

The TEM micrographs of the PVDF-PA-6 blends (Figure 3) reveal that these components are phase separated too. The particle sizes in the four times extruded samples vary between 0.1µm and 3µm for the PVDF component and 0.05µm and 2µm for the PA-6, depending on the composition. The greater particles have a composite structure since in their turn they contain particles of the matrix phase as a further level of dispersion. With increasing extrusion cycle number, the dispersion becomes finer, and the composite character, where initially present, is occasionally lost.

Torsional pendulum analysis exhibits discrete relaxations at the glass transition temperatures of PVDF and PA-6 at -45°C and 50°C, respectively. This also indicates incompatibility of the blend components in the amorphous phase after solidification.

SEM of fracture surfaces reveal phase separation in the PVDF-PBTP blends too (Figure 4). The particle sizes (PVDF: between 1µm and 20µm, PBTP: between 0.1µm and 2µm)) are considerably larger than those in the PVDF-PA-6 blends of comparable composition. The particle diameters again decrease with increasing extrusion cycle number.

Results/Crystallization Behaviour/PE-POM Blends. The PE component in this blend crystallizes in all samples at almost the same temperature and to the same extent. In contrast, the POM, if the minor component, crystallizes in up to three distinct steps ("fractionated", Figure 5) at 146°C, 138°C, and 132°C, the relative DSC peak area of which depends on the number of extrusion cycles, that is, on the degree of dispersion of that component (Figure 6). The relative intensity of the several peaks, moreover, changes with increasing DSC cycle number to that corresponding to a lower extrusion cycle number. This indicates a coarsening of the dispersion with increasing dwelling time in the melt. The overall degrees of crystallinity of both components, however, are almost independent of the extrusion and DSC cycle number, respectively. The POM, in particular, exhibits only one melting endotherm at an almost constant temperature for all samples.

Results/Crystallization Behaviour/PVDF-PA-6 Blends. The crystallization of several four times extruded blends (that is, for constant mixing conditions but variable composition), as studied by DSC, is displayed in Figure 7. The crystallization temperature T_c taken as the maximum of the crystallization curve at a cooling rate of 10°C/min is at 140°C for pure PVDF and at 178°C with a shoulder on the high temperature side for pure PA-6. The most striking result of these investigations has been found with the crystallization behaviour of the 85/15 blend. It is remarkable that it shows only one crystallization exotherm at 140°C, that is, at the T_c of PVDF whereas nothing happens at the usual PA-6 crystallization temperature. The WAXS analyses and the DSC melting runs, however, reveal that, nevertheless, the PA-6 crystallizes to the usual extent. In the 70/30 and 50/50 blends, the PA-6 crystallizes in two steps, partly at 184°C (that is, at a somewhat higher temperature than in the pure material) and partly at 140°C (that is, as in the 85/15 blend) as can be derived from comparison of the exotherm and the endotherm peak areas. In the 50/50 blend, also the crystallization of the PVDF component splits up into two steps at 140°C and 116°C. In the 15/85 blend, finally, the PVDF crystallizes merely at the low temperature step at about 112°C.

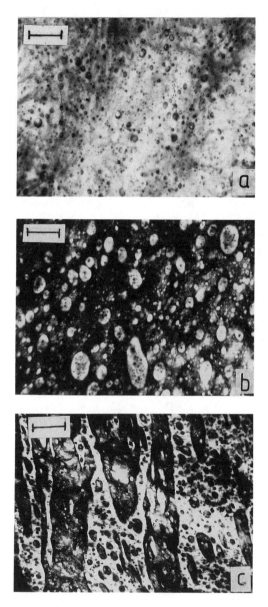

Figure 3. TEM micrographs of PVDF/PA-6 blends. PVDF is the dark
phase. Scale bar corresponds to 2µm (a,b), or 5µm (c).
a) PVDF/PA-6 = 15/85 vol.-%, four extrusion cycles;
b) PVDF/PA-6 = 75/25 vol.-%, four extrusion cycles;
c) PVDF/PA-6 = 75/25 vol.-%, one extrusion cycle.
(Reproduced with permission from ref. 68. Copyright 1988 Steinkopff-Verlag Darmstadt.)

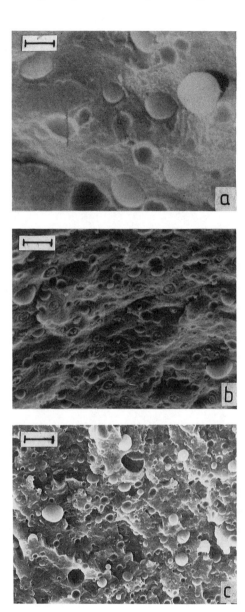

Figure 4. SEM micrographs of PVDF/PBTP blends. Scale bar corresponds to 2um (a,b), or 10µm (c).
a) PVDF/PBTP = 15/85 vol.-%, four extrusion cycles;
b) PVDF/PBTP = 85/15 vol.-%, four extrusion cycles;
c) PVDF/PBTP = 85/15 vol.-%, one extrusion cycle.
(Reproduced with permission from ref. 68. Copyright 1988 Steinkopff-Verlag Darmstadt.)

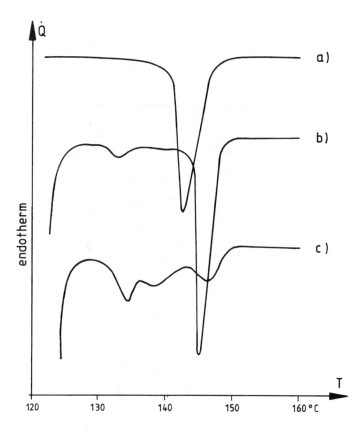

Figure 5. DSC cooling curves of a PE/POM = 85/15 vol.-% blend after one (b) and four (c) extrusion cycles. (a): pure POM.

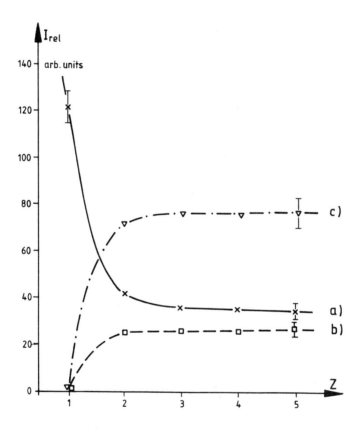

Figure 6. Relative area of the POM crystallization peaks at 146°C (a), 138°C (b), and 132°C (c) in dependence on the extrusion cycle number Z.

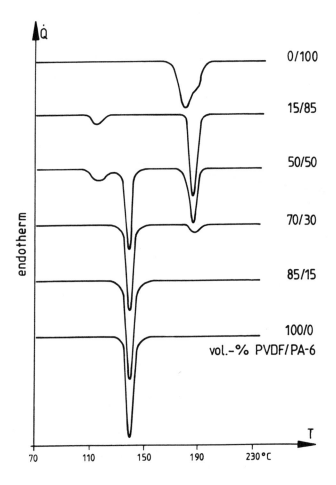

Figure 7. DSC crystallization curves of PVDF/PA-6 blends. Four extrusion cycles; parameter: blend composition.(Reproduced with permission from ref. 68. Copyright 1988 Steinkopff-Verlag Darmstadt.)

Variation of the cooling rate does not affect the number and the relative intensities of the crystallization peaks. In particular, the unusual, complete coincidence of the T_c in the 85/15 blend occurs at various cooling rates ranging between 0.5°C/min and 100°C/min.

The crystallization dependence on the extrusion cycle number, that is, its dependence on the mixing intensity, with a fixed composition is shown in Figure 8. The DSC cooling curve of the four times extruded 75/25 blend exhibits again the coincident crystallization of the PVDF matrix and the PA-6 droplets at 140°C as already the 85/15 blend did (cf. Figure 7). In the one time extruded blend, additionally, a part of the PA-6, possibly the larger domains, crystallizes at its usual T_c of 184°C and a part of the PVDF, probably the droplets dispersed inside the mentioned larger PA-6 domains, crystallizes at 113°C. The 85/15 blend exhibits the same behaviour. With increasing number of extrusion cycles, the PA-6 domains become smaller and loose their composite character. This causes the disappearance of the usual high temperature crystallization peak of the PA-6 as well as that of the low temperature crystallization peak of the PVDF. The latter, therefore, is assumed to represent the crystallization of those PVDF droplets which are inserted in PA-6 droplets. Obviously, the described effects, in particular the fractionation of the crystallization, depend to a large extent on the kind and on the degree of the dispersion of the minor component. With increasing dispersitivity of that component, the magnitudes of its additional crystallization peaks become stronger at the expense of the usual peak.

Despite the occasional fractionation of the crystallization or its suppression at the usual temperature, the DSC heating curves of all blends exhibit in all cases for both polymers a single melting endotherm (PA-6: one for both crystal modifications) at an almost constant temperature (variation range smaller than 6°C) of about 175°C for the PVDF and of about 217°C/223°C for the PA-6. The relative crystallinity of each component also does not change significantly with composition and extrusion cycle number. In fact, WAXS- and IR-analyses show that, in all samples, PA-6 crystallizes mainly in the γ -modification at an almost constant γ/α -ratio, and PVDF in the α -modification.

The basic character of all observations is independent on the number of the cooling/heating cycles.

Results/Crystallization Behaviour/PVDF-PBTP Blends. The pure PBTP (Figure 9) crystallizes at about 180°C. The crystallization temperature T_c rises remarkably to 189°C and 194°C for the one and for the four times extruded blends, respectively, after adding 15 vol.-% PVDF. In a similar way, the pure PVDF crystallizes at 140°C whereas the corresponding T_c in the blends is between 142°C and 148°C. The PVDF in the 15/85 blend, e.g., crystallizes at 147°C after one extrusion cycle whereas it crystallizes at 143°C after four ones. The PBTP crystallization in the four times extruded 85/15 blend is suppressed in a similar manner as already described for the PA-6. In place of that, the PBTP crystallizes at 147°C simultaneously with the PVDF as derived from comparison of the exotherm and the endotherm DSC peaks of the cooling and reheating runs.

Again, the melting curves show no significant differences to that of the pure materials, and all described effects occur also

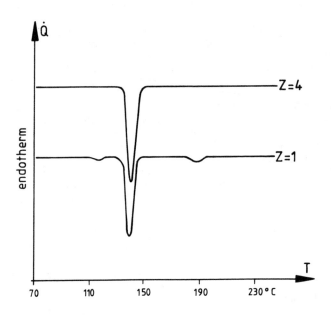

Figure 8. DSC crystallization curves of the PVDF/PA-6 = 75/25 vol.-% blend. Parameter: number of extrusion cycles Z. (Reproduced with permission from ref. 68. Copyright 1988 Steinkopff-Verlag Darmstadt.)

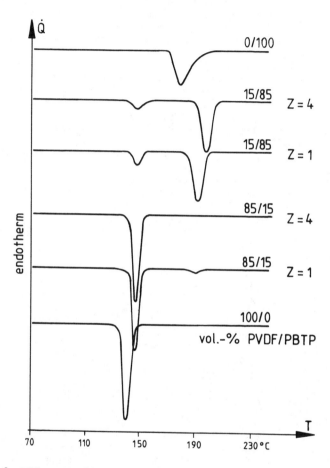

Figure 9. DSC crystallization curves of PVDF/PBTP blends. Parameters: blend composition and number of extrusion cycles Z.(Reproduced with permission from ref. 68. Copyright 1988 Steinkopff-Verlag Darmstadt.)

during further cooling/heating cycles. The PVDF, in particular, as proved by WAXS measurement, crystallizes in all samples in the α -modification.

Discussion. The crystallization of the investigated blends exhibits several differences to that of the pure components. An occasional occurrence of douple melting endotherms (that has not been been described hitherto) of the components during the reheating runs and the small melting point depressions of the minor component can be explained in the usual manner by imperfect crystal formation during the - possibly fractionated - crystallization, by molecular reorganization and lamellar thickening during heating, and by polymorphism (55-60). The considerable rise in crystallization temperature after adding a second component (PA-6 and PBTP in their PVDF blends) may be due to migration of nuclei across the interface, or an altered chain mobility in the interface (48). These issues will, therefore, not be considered in the following. Other effects, however, are related to the topic of this paper and have to be treated in detail:
-- split up of the crystallization of the dispersed component into several distinct steps (fractionated crystallization; POM blended with PE, both PVDF and PA-6 in their blends);
-- complete initial suppression of the crystallization of the dispersed component, and subsequent fully coincident crystallization with the matrix (both PA-6 and PBTP in their blends with PVDF).
These effects are clearly connected with the dispersion of the component under investigation into the other, and they are enhanced if this dispersion becomes finer.

The crystal growth rates of PVDF, PA-6, and POM amount to at least 10μm/min in the temperature range where their crystallization steps occur (6,52,67). A dispersed particle, therefore, once nucleated, crystallizes promptly and the primary rather than the secondary nucleation is the rate-controlling factor of the crystallization kinetics of the dispersed phase. Thus, the crystallization temperatures as observed in the DSC-cooling run agree roughly with the nucleation temperatures.

Discussion/Melt Rheology of PVDF-PA-6 Blends. Normally, the melt viscosity of a polymer decreases with increasing temperature T and with increasing shear rate τ. Since our samples were processed at T = 230°C with $\tau = (10^{-1}...10^{3})s^{-1}$, the viscosities amount to η (PVDF) = $(2x10^{3}...2x10^{2})$Pas (65) and η (PA-6) = $(7x10^{2}...2x10^{2})$Pas (66). A low interfacial energy between PVDF and PA-6 melts which may be of the order of 1mJ/m^{2}, and the close match of the blend component viscosities at higher shear rates may account for the rather small particle sizes of the dispersed component and for the occurrence of subdispersions inside a particle. Furtheron, the slightly greater viscosity of PVDF compared to that of PA-6 may, according to Equation 9, be the reason for that the PA-6 particles are somewhat smaller than those of the PVDF.

Discussion/Estimation of Nuclei Concentrations. The average volumes v_D of the dispersed PA-6 particles in the four times extruded PVDF/PA-6 85/15 and 75/25 blends as estimated from the electron micrographs amount to about $4x10^{-3}μm^{3}$ and $3x10^{-2}μm^{3}$, respectively.

Since the PA-6 crystallization at its usual temperature of about 180°C is suppressed, and since almost all PA-6 droplets crystallize in a low temperature step together with the PVDF, Equation 8 must be fullfilled with $M^{(A)}$ as the number density of nucleating impurities active in PA-6 above 140°C. Followingly, $M^{(A)}$ should be less than $10\mu m^{-3}$. This value agrees roughly with the number density $(0.2...2)\mu m^{-3}$ of spherulites and sheaves grown in pure PA-6 at 140°C. On the contrary, the PA-6 particles in the one time extruded 75/25 blend are much larger. Accordingly, the crystallization peak at the initial temperature of about 180°C is not completely suppressed; there is, however, a split up of the PA-6 crystallization.

The average volume of a PVDF particle in the PVDF/PA-6 = 15/85 blend after four extrusion cycles amounts to about $8 \times 10^{-3} \mu m^3$. Moreover, all of them crystallize at the low temperature step. Followingly, the number density $M^{(A)}$ of the nucleating impurities which are active in PVDF at 140°C is below about $50\mu m^{-3}$ since Equation 8 must be fullfilled also for this sample. In the 50/50 blend which exhibits two crystallization steps there are, on the contrary, also larger PVDF particles with a volume of about $4\mu m^3$, the crystallization of which most probably gives rise to the DSC peak at 140°C since the probability of containing a heterogeneity is greater for a large particle. Therefore, $M^{(A)}$ of the PVDF should amount at least to $0.2\mu m^{-3}$ in order to fullfill the condition $M^{(A)}v_D \approx 1$ (cf. again Equations 6-8).

The average volume of dispersed PBTP particles in the PVDF/PBTP = 85/15 blend after the first and the fourth extrusion cycles amounts to about $4\mu m^3$ and $5 \times 10^{-2}\mu m^3$, respectively. By the same arguments as above one has to conclude from the fractionation of the crystallization of the dispersed PBTP droplets even in the one time extruded blend that the number density of the heterogeneities which nucleate PBTP above 148°C is below about $0.2\mu m^{-3}$.

Discussion/Estimation of Specific Interfacial Energies for PVDF.

With $T_c^{(o)}(\alpha\text{-PVDF}) = 459K$ (53) and $T_c^{(o)}(\gamma\text{-PA-6}) = 500K$ (52), and using Equation 1, one gets $T_{ho}(PVDF) \approx 94°C$, and $T_{ho}(PA-6) \approx 127°C$. Both these values are well below the T_c observed for the components in the PVDF/PA-6 blends which, therefore, in every case crystallize nucleated heterogeneously. At a low cooling rate of 1°C/min, the temperatures of the fractionated crystallization of PVDF are 148°C and 119°C. With $y_{PVDF}(m,c) = 9.7mJ/m^2$ (6,54) we obtain with Equation 5b for the specific interfacial energy differences of the two nucleating heterogeneities $\Delta y_{PVDF,s}(148) = 3.8 mJ/m^2$ and $\Delta y_{PVDF,s}(119) = 11 mJ/m^2$.

Discussion/Coincidence of Crystallization Temperatures.
Let us first consider the PVDF/PA-6 blend. In view of the nonaltered T_c of PVDF, we suppose that the PVDF crystallization induces the PA-6 crystallization rather than vice versa. Hence, the just created crystals of the PVDF matrix act as nucleating heterogeneity for the PA-6. The Δ y-value between PVDF crystals and PA-6 melt, obviously, is smaller than that of all other heterogeneities which are present in PA-6 to a sufficient extent except, possibly, the species "A". Its associated specific undercooling, moreover, must be so small that the PVDF crystals can induce the crystallization of the PA-6 from the instant of their own creation.

The coincident crystallization of the PVDF matrix and dispersed PBTP particles in the 85/15 blend (Z = 4) takes place at (142...148)$^{\circ}$C, that is, above the T_c of pure PVDF. It is not clear whether the PVDF or the PBTP crystallizes first. In either case, the nucleation of the first crystallizing component may be induced either by a species of nucleating heterogeneities or by the molten second blend component. The newly created crystals of the one component, then, act immediately as nuclei for the crystallization of the other in the same manner as already described for the PVDF/PA-6 blends.

Summary, Conclusions and Outlook

It was the aim of the present paper to show that crystallization in incompatible polymer blends can exhibit a lot of peculiar effects beside the classical well known physical and physico-chemical phenomena. The effects considered here, in particular, are due to the dispersion structure of such blends, and to the changes in the crystallization nucleation conditions which are such caused. They are important from a physical, a material scientific, and a technological point of view as well.

The phenomena on which has been reported here are linked to a droplet-like dispersion of the component under investigation. Therefore, they are usually exhibited only by the minor component in incompatible polymer blends. Both components, however, can exhibit the effects simultaneously if the dispersion is composed such that, in turn, a part of the matrix material is included into the particles of the dispersed component. These mentioned effects are mainly
(1) split up of the crystallization into several distinct steps, the temperature of which can differ by several ten degrees;
(2) inhibition of the crystallization at the usual temperature;
(3) coincidence of the crystallization of both components at that temperature at which one of them usually crystallizes, or at another temperature;
(4) occasionally homogeneously nucleated crystallization.

Basicly, the effects are caused by the nucleating activity of different inhomogeneities. They can become more complicated if the second component, in particular their just created crystals, acts as a crystallization nucleating inhomogeneity. Such a - in some cases mutual - nucleating activity is hidden under usual conditions. Finally, some blends, e.g. that of PBTP and PVDF, exhibit a most complicated mutual nucleation behaviour: the molten first component acts as nucleating substrate for the second one and becomes then itself nucleated by the newly created crystals of that component.

The technical importance of the described effects is obvious. The lowering of the solidification temperature, or the broadening of the solidification temperature range by occasionally several ten degrees influences remarkably the rheological boundary conditions during processing. They deliver, further, an additional connection between the degree of dispersivity and the rheological material parameters. The changes in material properties which are such caused are an open question although the available investigations do not indicate a strong change in supermolecular structure by the delayed crystallization.

The links between the degree and the level of dispersivity, on the one hand, and the type and strength of the particular fractionated crystallization effect that it causes, on the other hand, allow an at least qualitative characterization of the first by the latter via suitable reference measurements. By the several effects, moreover, an estimation of specific properties of the components like the absolute or relative amounts of the different crystallization inducing heterogeneities, and their nucleating efficiency are possible. A determination of the interface energies of the faces of the crystals is also possible if only the undercooling can be reached at which, finally, homogeneously nucleated crystallization starts.

Acknowledgment

Financial support of the Arbeitsgemeinschaft Industrieller Forschungsvereinigungen, grant No. 6015 and 6697, is acknowledged.

Literature Cited

1. Paul, D.R.; Newman, S., Eds. Polymer Blends; Academic Press: New York, 1978.
2. Olabisi, O.; Robeson, L.M.; Shaw, M.T. Polymer-Polymer Miscibility; Academic Press: New York, 1979.
3. Nishi, T. CRC Crit. Rev. Solid State Mater. Sci. 1985, 12(4), 329.
4. Martuscelli, E. Polym. Eng. Sci. 1984, 24, 563.
5. Alfonso, G.C.; Russell, T.P. Macromolecules 1986, 19, 1143.
6. Wang, T.T.; Nishi, T. Macromolecules 1977, 10, 421.
7. Calahorra, E.; Cortazar, M.; Guzman, G.M. Polymer 1982, 23, 1322.
8. Martuscelli, E.; Sellitti, C.; Silvestre, C. Makromol. Chem. Rap. Comm. 1985, 6, 125.
9. Bartczak, Z.; Martuscelli, E. Makromol. Chem. 1987, 188, 445.
10. Marinow, S.; May, M.; Hoffmann, K. Plast. Kaut. 1983, 30, 620.
11. Paul, D.R.; Barlow, J.W.; Bernstein, R.E.; Wahrmund, D.C. Polym. Eng. Sci. 1978, 18, 1225.
12. Eder, M.; Wlochowicz, A. Acta Polym. 1984, 35, 548.
13. Morris, M.C. Rubber Chem. Techn. 1967, 40, 341.
14. Ghijsels, A. Rubber Chem. Techn. 1977, 50, 278.
15. Lipatov, Y.S.; Lebedev, E.V. Makrom. Chem. Suppl. 1979, 2, 51.
16. Kishore, K.; Vasanthakumari, R. Polymer 1986, 27, 337.
17. Keith, H.D.; Padden, F.J. Jr. J. Appl. Phys. 1963, 34, 2409.
18. Bartczak, Z.; Galeski, A.; Martuscelli, E. Polym. Eng. Sci. 1984, 24, 1155.
19. Tanaka, H.; Nishi, T. Phys. Rev. Lett. 1985, 55, 1102.
20. Hsu, C.C.; Geil, P.H. Polym. Eng. Sci. 1987, 27, 1542.
21. Chatterjee, A.M.; Price, F.P.; Newman, S. J. Polym. Sci., Polym. Phys. Ed. 1975, 13 2369, 2385, 2391.
22. Lotz, B.; Wittmann, J.C. J. Polym. Sci., Polym. Phys. Ed. 1986, 24, 1559.
23. Bartczak, Z.; Galeski, A.; Pracella, M. Polymer 1986, 27, 537.
24. Romankevich, O.V.; Grzimalovskaya, L.V.; Zabello, S.E. Sin. Fiz. Khim. Polim. 1976, 17, 22.
25. Ghijsels, A.; Groesbeek, N.; Yip, C.W. Polymer 1982, 23, 1913.
26. Lotz, B.; Kovacs, A.J. ACS Div. Polym. Chem., Polym. Prepr. 1969, 10(2), 820.

27. O'Malley, J.J.; Crystal, R.G.; Erhardt, P.F.
ACS Div. Polym. Chem. Polym. Prepr. 1969, 10(2), 796.

28. Robitaille, C.; Prud'homme, J. Macromolecules 1983, 16, 665.

29. Aref-Azar, A.; Hay, J.N.; Marsden, B.J.; Walker, N.
J. Polym. Sci., Polym. Phys. Ed. 1980, 18, 637.

30. Baitoul, M.; Saint-Guirons, H.; Xans, P.; Monge, P.
Eur. Polym. J. 1981, 17, 1281.

31. Tsebrenko, M.V. Int. J. Polym. Mater. 1983, 10, 83.

32. Klemmer, N.; Jungnickel, B.-J. Coll. Polym. Sci. 1984, 262, 381.

33. Yip, C.W. Plasticon '81: Symposium on Polymer Blends,
1981, Paper #30.

34. Price, F.P. IUPAC Symposium on Macromolecules, 1959,
Paper #1B2.

35. Koutsky, J.A.; Walton, A.G.; Baer, E. J. Appl. Phys. 1967, 38,
1832.

36. Cormia, R.L.; Price, F.P.; Turnbull, D. J. Chem. Phys. 1962, 37,
1333.

37. Burns, J.R.; Turnbull, D. J. Appl. Phys. 1966, 37, 4021.

38. Wunderlich, B. Macromolecular Physics; Academic Press: New
York/San Francisco/London, 1976; Vol. 2: Crystal Nucleation -
Growth - Annealing.

39. Barham, P.J.; Jarvis, D.A.; Keller, A.
J. Polym. Sci., Polym. Phys. Ed. 1982, 20, 1733.

40. Van Riemsdyk, A.D. Ann. Chim. Phys. 1880, 20, 66.

41. Jackson, K.A. Ind. Eng. Chem. 1965, 57(12), 29.

42. Perepezko, J.H.; Paik, J.S. Undercooling Behavior of
Liquid Metals; In Rapidly Solidified Amorphous and
Crystalline Alloys; Kear, B.H.; Giessen, B.C.; Cohen, M.; Eds.;
North Holland Publ. Comp.: New York, 1982.

43. Zettlemoyer, A.C. Nucleation; Marcel Dekker: New York, 1969.

44. Turnbull, D.; Cech, R.E. J. Appl. Phys. 1950, 21, 804.

45. Pound, G.M.; LaMer, V.K. J. Am. Chem. Soc. 1952, 74, 2323.

46. Turnbull, D.; Cormia, R.L. J. Chem. Phys. 1961, 34, 820.

47. Harnischfeger, P. Master Thesis, Technische Hochschule,
Darmstadt, 1985.

48. Rech, N. Master Thesis, Technische Hochschule, Darmstadt, 1988.

49. McAllister, P.B.; Carter, T.J.; Hinde, R.M.
J. Polym. Sci., Polym. Phys. Ed. 1978, 16, 49.

50. Price, F.P. In Nucleation; Zettlemoyer, A.C., Ed.; Marcel
Dekker: New York, 1969; Chapter 8.

51. Lovinger, A.J.; Davis, D.D.; Padden, F.J. Jr. Polymer 1985, 26,
1595.

52. Magill, J.H. Polymer 1962, 3, 655.

53. Chen, C.T.; Frank, C.W. Ferroelectrics 1984, 57, 51.

54. Mancarella, C.; Martuscelli, E. Polymer 1977, 18, 1240.

55. Lovinger, A. In Developments in Crystalline Polymers; Bassett,
D.C., Ed.; Appl. Sci. Pub.: London/New York, 1982; Vol. 2.

56. Osaki, S.; Ishida, Y. J. Polym. Sci., Polym. Phys. Ed. 1975, 13,
1071.

57. Yadav, Y.S.; Jain, P.C. J. Macromol. Sci. Phys. 1986, B25, 335.

58. Illers, K.H.; Haberkorn, H.; Simak, P. Makromol. Chem. 1972,
158, 285.

59. Weigel, P.; Hirte, R.; Ruscher, C. Faserf. Textilt. 1974, 25,
198.

60. Hirami, M. Macromol. Sci. Phys. 1984, B23, 397.

61. v.d. Vegt, A.K.; Elmendorp, J.J. Blending of Incompatible Polymers; In International Fundamentals of Polymer Science and Technology; Kleintjens, L.A.; Lemstra, R.J.; Eds.; Elsevier, New York/London, 1986

62. Tokita, N. Rubb. Chem. Techn. 1977, 50, 292

63. Wu, S. Polym. Eng. Sci. 1987, 27, 335

64. Utracki, L.A. Rheology and Processing of Multiphase Systems; In Current Topics in Polymer Science; Vol. II; Ottenbrite, R.M.; Utracki, L.A.; Inoue, S.; Eds.; Hanser Publ., München/Wien/New York, 1987

65. Solvay Corp.; Customers Data Sheet

66. Laun, H.M. Rheol. Acta 1979, 18, 478

67. Pelzbauer, Z.; Galeski, A. J. Polym. Sci., Polym. Symp. Series 1972, C38, 23

68. Frensch, J.; Jungnickel, B.-J. Coll. Polym. Sci., 1988.

RECEIVED November 11, 1988

RHEOLOGY

Chapter 6

Phase Transitions in Simple Flow Fields

J. Lyngaae-Jørgensen

IKI, The Technical University of Denmark, 2800 Lyngby, Denmark

A theoretical expression for the depression of the
melting point of polymers with low degree of crystallinity is derived for simple shear flow at constant
rate of deformation using a thermodynamically
approach. The last surviving crystallite aggregate
is treated as a body consisting of many single
polymer molecules held together by a crystalline
nucleus. The theory is tested with data for PVC,
copolymers of ethylene and polypropylenes. An ana-
logical approach has been applied to blockcopolymers
and polymer blends in order to predict transitions
from two-phase to one phase melt state during simple
flow.

Two-phase flow of polymer systems is of increasing importance
because of the increased application of blockcopolymers, blends and
alloys. Most polymers are immiscible with other polymers of different molecular structure (1) because the entropy of mixing
converges to zero with increasing molecular weight. The properties
of an immiscible blend depend on the two-phase structure or morphology of the blend, the interface structure etc. It is therefore
important to be able to control the two-phase structure of the
blend. This discipline: called structuring, which is governing of
the two-phase or multiphase structures by proper selection of processing prehistory, is of increasing industrial and academic
interest. One possibility for obtaining materials with special
morphologies after processing, often in a reproducible way, is to
provoke phase transitions in flowing systems.
   The purposes of this contribution are; 1. to briefly review
some of the papers dealing with interactions between, e.g. shearing
stresses in flow and phase behavior, and 2. to develop models for

0097–6156/89/0395–0128$07.25/0

phase transitions in simple flow fields. The focus here is on transitions to "homogeneous" melt states in simple flow fields.

Two cases for phase transitions in simple flow fields will be treated,

1. a transition involving melting of crystalline areas in polymers with low degree of crystallinity, and
2. a transition to a homogeneous state of an originally phase separated blend of two polymers. This case represents an attempt to extend the solution developed for case 1; consequently the ideas are presented with emphasis on melt theory in flow fields as well as an evaluation of the theoretical predictions. The published work on polymer blends (18) will be only briefly summarized in this paper.

<u>Earlier findings.</u> The subject is broad because interactions between stresses and, say, the melting temperature may be encountered in many different situations e.g. during necking of crystalline polymers (2). In simple flow fields as in capillary flow it is normally reported that crystallinity is provoked by flow (3). However, for polymers with low degree of crystallization it was assessed that the melting point was depressed at high shear stresses (4). Shear induced melting has been theoretically predicted from non-equilibrium molecular dynamic simulations (5, 6); at higher shear rates a transition to a new ordered state was predicted.

The question of interaction between shear flow and phase structure was studied by Silberberg and Kuhn (7) in 1952. They reported that a homogeneous one-phase solution results if a velocity gradient is maintained in a stable two-phase system of polystyrene, ethylcellulose and benzene.

Rangel-Nafaile, Metzner and Wissbrun (8) reviewed studies on solubility phenomena in deforming solutions and developed an expression for stress-induced phase separation in polymer solutions. Other recent studies in this field were published by Wolf et al. (9-13) by Mazich and Carr (14) and by Vrahopoulou-Gilbert and McHugh (15).

There is no general consensus as to the influence of stress during flow on phase equilibria at present, and in fact observations from different sources are conflicting. Most observations of dilute solutions with upper critical solution temperatures indicate that phase separation may be provoked by mechanical deformation (8) or, stated differently, that the critical solubility temperature is increased by flow. Wolf et al. (9-12), however, developed a theory which predicts that the solubility temperature may decrease by flow in accordance with Silberberg's (7) observations.

Wolf et al. calculated the equilibrium size of droplets formed in a phase-separated system. From a force balance, he derived an expression indicating that the equilibrium droplet size r is a decreasing function of shear rate and that when r approaches the radius of gyration of the polymer molecules, redissolution will have occurred. Recently Krämer and Wolf have generalized the approach and formulated simple criteria for solution, respectively demixing (16).

Only a few investigations on shear flow induced changes in solubility phenomena in polymer blends have been reported. However, Lyngaae-Jørgensen and Søndergaard (17, 18) did advance a hypothesis predicting that a homogeneous melt should be formed at sufficiently high shear stresses for nearly miscible blends. Experimentally, such a transition was observed for simple shear flow for blends of SAN and PMMA (19). Winter observed a transition to homogeneous state in uniaxial extensional flow for a blend of PS and PVME (20) at high extensional strains $\epsilon = \dot{\epsilon}t \simeq 44$ Hencky.

Melting of Crystallites in Polymers with Low Degree of Crystallinity in Simple Flow Fields

We considered a linear polymer with crystallizable sequences. The molecules of the polymer consist in principle of alternating blocks of crystallizable and non-crystallizable sequences. Examples of such chains would be nearly atactic homopolymers and random copolymers.

Poly(vinylchloride) produced at temperatures between $40^{o}C$ and $60^{o}C$ are only slightly crystalline and it may be assumed that only syndiotactic sequences crystallize. Polymers where only sequences with a stereospecific configuration can crystallize may, in the context of this paper, be considered as copolymers.

The melting point of a polymer is defined as the temperature where the last trace of crystallinity disappears. Only the most stable crystallites existing in the material exist in the melt at the melting point. It is assumed that in a shear flow field the last surviving "structures" showing crystallinity consist of a crystalline nucleus acting as a giant branching point with n branches (a crystallite aggregate). This assumption is based on the fact that the last surviving crystallite aggregate in dilute solution has this structure (21). A key concept used in this work is that of "entanglements" even though it is an intuitive and rather ill-defined one. Entanglements here are considered to represent regions where neighbouring molecules are looped together one after another and thereby offer high resistance to deformation for a time (22). A number of assumptions are made: random coiled linear polymer molecules, isothermal conditions, constant volume, molecular weights (M) much larger than the critical molecular weights (M_c) for entanglement formation, constant segment concentration, negligible inertial forces (etc.)

At isothermal steady state conditions, we assume that the melt behaves as a lightly crosslinked network with non-permanent network points. The effect of destroying crystalline aggregates is evaluated as the removal of "extra" entanglements in a continuous phase with non-permanent crosslinks (entanglements). That is, crystalline aggregates are considered and measured in entanglement units.

The free energy of a closed system without flow is a function of state and may be expressed as a function of intensive state parameters including temperature, T, pressure, P and composition.

by   The number of degrees of freedom, n, in the system is given

$$n = k + 3 - f \tag{1}$$

where f is the number of phases, k the number of components. Eq. 1 states that the melting temperature is therefore a function of e.g. shear rate (or shear stress) and pressure. If solvent is present we may express the melting temperature as a function of e.g. pressure, shear stress and solvent volume fraction.

A Theory for Transition to a Monomolecular Melt state at Constant Shear Rate (Simple Shear Flow).
A criterion for a transition from a two phase state to a monomolecular melt state may be formulated as follows. The total change in free energy (ΔG_T) by removing one crystallite aggregate from a melt at constant shear rate may be considered to consist of the contributions:

$$\Delta G_T = \Delta G_{melt} + \Delta G_{mix} + \Delta G_{el} \tag{2}$$

ΔG_{melt} corresponds to the change in the free energy observed when pure crystalline phases are melted. Since the systems considered are originally two-phase systems (for shear stress: $\tau = 0$), ΔG_{melt} is always positive. The action of a domain in a polymer melt is assumed to be equivalent to the action of a giant crosslink in a rubber. Removing one "crosslink" is accompanied by a negative free energy change (ΔG_{el}). If solvent is present in the mixture one need to include the term ΔG_{mix}, which corresponds to the change in the free energy observed when pure phases are mixed. At steady state a condition for equilibrium between a two-phase structure and a homogeneous melt structure is that the chemical potential of a repeat unit in the crystalline phase μ_A^{cr} is equal to the chemical potential of a repeat unit in the melt state μ_A:

$$\mu_A^{cr} = \mu_A \tag{3}$$

or

$$\mu_A^{cr} - \mu_A^{ref} = \mu_A - \mu_A^{ref} \tag{4}$$

where μ_A^{ref} is a suitably chosen reference state. For pure polymers μ_A^{ref} is taken as the chemical potential of pure amorphous polymer with the same molecular structure as the sample. If solvent is present the reference state may of convenience be chosen as the chemical potential of a homogeneous amorphous melt of the same composition in the mixture considered.

Expressions for the left and the right hand side of Eq. 4 are derived in appendix.

For pure polymer melts, insertion of Eqs. 5 and 12 from appendix into Eq. 4 gives:

$$\frac{1}{T_{dyn}} - \frac{1}{T_m} = \frac{\tau^2}{Q\,T_{dyn}^2} \tag{13}$$

where

$$Q = \frac{2\,c^4\,R\Delta H_A}{a^2\,\rho^2\,M_A\,H^2\,M_c}$$

or

$$T_{dyn} = \frac{T_m}{2}\left[1 + \sqrt{1 - \frac{4\tau^2}{T_m Q}}\right]$$

where T_m is the static melting temperature and T_{dyn} is the melting temperature at constant shear rate in simple shear flow, a is a constant, ρ is polymer density, M_A is the molecular weight of a repeat unit, $H \equiv \overline{M}_W/\overline{M}_n$ is the ratio between the average molecular weight by weight and by number, respectively, M_e is the average molecular weight between entanglements. In the calculations we use an estimate $M_c \simeq 2\,M_e$, M_c is the molecular weight at the intersection of two lines defined by the equation $\eta_o = K\,M_W^b$ where b changes from ~1 to ~3.5; c is polymer concentration, R is the gas constant, $\tau_{21} = \tau$ is the shear stress and ΔH_A is the enthalpy of melting per repeat unit. The static melting temperature is depressed by solvent. The change may be written (23):

$$\frac{1}{T_m} - \frac{1}{T_m^o} = \frac{RV_A}{\Delta H_A V_1}\,(v_1 - \chi\,v_1^2) \tag{14}$$

where T_m is the static melting temperature of polymer-solvent mixture, T_m^o is the melting temperature of the pure polymer, V_A and V_1 the molar volume of a repeat unit and solvent molecule, respectively, v_1 is the solvent volume fraction and χ the interaction parameter. The final expression for the dynamic melting temperature may again be cast in the form of eq. 13 if T_m is defined as the melting temperature of the polymer-solvent mixture.

The Gel Destruction Temperature T_d^{dyn} in a Shear Flow field below the Static Gel Destruction Temperature T_d. The gel destruction temperature may be defined as the temperature where enough crystalline areas (which act as crosslinks) are melted to prevent an effective infinite network of crystalline areas to exist in the melt (threshold percolation structure). In this respect, gel formation is analogous to the gel point during condensation polymerization. Below the gel destruction temperature T_d the material is in a rubberlike solid state.

If a rubberlike solid (like PVC particles below T_d) is sheared in a shear flow field it will deform until the adhesion to the walls of the equipment fails, the gel structure is broken or melted or a slippage mechanism involving e.g. a particle flow mechanism such as Mooney flow as proposed by Berens and Volt (25-27) is established.

The derivation of a relation between T_d and τ is equivalent to the derivation in the preceding section. Crystalline network crosslinks, will melt if one transfers differential amounts of repetition units from the crystalline to the amorphous state in such a way that the free energy change is always negative or zero, or

$$-(\mu_A^{cr}-\mu_A^o)) + (\mu_A-\mu_A^o)_{el} = 0 \qquad (15)$$

$$\mu_A^{cr}-\mu_A^o = -\Delta H_A \left(1-\frac{T}{T_d}\right) \qquad (16)$$

given by Flory.

The change of free energy by melting of a crystalline crosslink in a deformed rubbery state is:

$$(\mu_A-\mu_A^o)_{el} = -\frac{1}{\bar{X}_{cr}} \left[\frac{\partial \Delta G_{el}}{\partial N}\right]_{T,P,\gamma \text{ (or evt. } \tau)} \qquad (17)$$

where $\bar{X}_{cr}$ is the number of repetition units between crosslink at T_d and N the number of crystalline crosslink. If we consider a material being deformed at constant shear rate: $(\gamma = \dot{\gamma}t)$:

$$\Delta G_{el} = \frac{1}{2} NRT (\dot{\gamma}t)^2 \qquad (18)$$

$$\tau = NRT \dot{\gamma}t \qquad (19)$$

For a momentary transition

$$\overline{X}_{cr} = \frac{\overline{M}_{cr}}{M_o} = \overline{X}_w \; ; \quad N = \frac{c}{\overline{M}_w}$$

$$\left(\mu_A - \mu_A^o\right)_{el} = -\frac{1}{\overline{X}_w}\left[\frac{\partial \Delta G_{el}}{\partial N}\right]_{T,P,\gamma} = -\frac{RT}{2 \cdot \overline{X}_w}\gamma^2$$

$$\left[\left(\mu_A - \mu_A^o\right)_{el} = -\frac{RT}{2\overline{X}_w}\frac{\tau^2}{(NRT)^2} = -\frac{RT}{2\overline{X}_w}\frac{\tau^2}{\frac{c^2}{M_w^2}\cdot R^2T^2} = -\frac{M_o\,\overline{M}_w\,\tau^2}{2c^2\,RT}\right]$$

$$(20)$$

Insertion of Eq. 16 and Eq. 20 into Eq. 15:

$$T_d^{dyn} = \frac{T_d}{2}\left[1 + \sqrt{-\frac{4\tau^2}{T_d Q^1}}\right], \text{ where } Q^1 = \frac{2c^2\,R\Delta H_A}{M_A\,\overline{M}_w} \qquad (21)$$

Thus the gel destruction temperature may be evaluated as a function of shear stress by Eq. 21. According to Eq. 21, the stresses necessary to depress T_d are so large, (at least for rigid compounds) that melt fracture or particle slippage takes place before any change in T_d. Melt fracture in a capillary corresponds to constant shear stress at the wall. It can be seen by analogy with the derivation in the preceding section that at constant stress T_d is expected to increase. Thus Eq. 21 predicts that the original particle structure cannot be destroyed below the gel destruction temperature for rigid and concentrated compounds.

Experimental

Sample Materials. Vinnol H 60d, Vinnol Y 60, and Vinnol E 60g are commercial polyvinylchlorides from Wacker, produced by suspension, bulk, and emulsion polymerization techniques, respectively. All materials have nearly the same molecular weight distribution (MWD) as Solvic 226 which, (with our SEC calibration) gave: $\overline{M}_w = 74.000$, $\overline{M}_n = 35,000$. The samples used in this investigation in the Vinnol 60 series had molecular weights in the range $\overline{M}_w = 72,000 \pm 2000$ and $\overline{M}_n = 34,000 \pm 2000$. Vinnol H 70 d: $\overline{M}_w = 110,000$ and $\overline{M}_n = 54,000$. Vinnol H 80 F: $\overline{M}_w = 166,000$, $\overline{M}_n = 80,000$. Four parts of liquid tin

stabilizer, Okstan X-3 (from Otto Bärlocher GmbH, Munich), were
used in all compounds. The stabilizer was chosen because of its
high thermostabilizing effect and compatibility. 25, 50, 100 and
200 pph of plasticizer: DOP (Di-2-ethylhexyl phthalate) were used.
Alathon E/VA 3185 from Du Pont, an ethylene-vinyl acetate copolymer
with 33 wt % vinyl acetate $\overline{M}_w$ = 79800, $\overline{M}_n$ = 19800 as described in
Ref. 28.

Methods. The molecular weight distribution was determined by size
exclusion chromatography SEC as described elsewhere (4).

Rheometry. The following rheometers were used in this study: An
Instron capillary rheometer, a Rheometrics mechanical spectrometer,
used in both cone and plate mode as well as in the biconical mode,
and a Brabender Plastograph (4).

Results. It is seen from eq. 13 that the parameter Q can be calculated if corresponding values of T_{dyn} and τ are measured and T_m
is found either by independent measurements or from rheological
data. Q has been determined by 1. plotting $\log\tau$ against $1/T$ with $\dot{\gamma}$
as discrete variable and curve fitting. The viscosity of a homo-
geneous melt can be approximated by $\eta = A \exp(E_{\dot{\gamma}}^*/RT)(22)$, where $E_{\dot{\gamma}}^*$
is the activation energy at constant shear rate and A and R are
constants. A plot of $\log\tau$ against $1/T$ with $\dot{\gamma}$ as discrete variable
is thus a straight line for a homogeneous melt.
    A melt containing crystallites with a distribution of cry-
stallite sizes "melts" over a broad range of temperatures. Crystallite aggregates acts as very large branched structures.
Destruction of crystallite aggregates causes a decrease in the
viscosity. Thus below the melting temperature but above the gel
destruction temperature a temperature increase will involve two
contributions the "normal" descrease in viscosity given by Eq. 22
and a decrease caused by crystallite melting. So below the melting
temperature a curve through corresponding points in a delineation
of $\log\tau$ against $1/T$ reflects the more stable part of the crystallite size distribution (a melting curve). At the melting temperature the melting curve and the straight line representing homogeneous melt behaviour crosses. The intersection points represent
corresponding values of T_{dyn} and τ. T_m can be found as the limiting value of T_{dyn} for small τ values. Such plots are shown on Fig.
1 and 2 for PVC compounds, on Fig. 3 for EVA and on Fig. 4 for
ethylene propylene copolymers, respectively (see next section). On
these plots Eq. 13 is shown as the full drawn curve. This curve
represents the geometrical place for the points where the last
crystallites disappear; consequently the straight lines representing homogeneous melt behaviour should intersect the measured $\log\tau$ -
$\frac{1}{T}$ curves on the full drawn curve.
    2. alternatively Q can be estimated as shown in Fig. 5. If "recry-
stallization" is a slow process corresponding values of T^{dyn} and τ

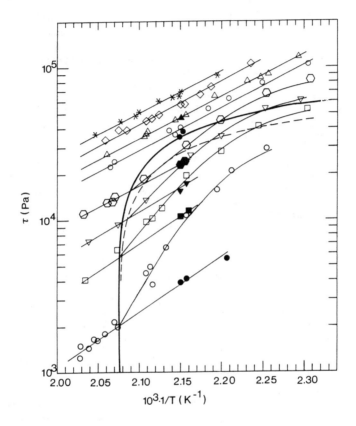

Fig. 1.   Delineation of logarithmic shear stress against reciprocal temperature with shear rate as discrete variable.   Shear rates: 1.8, 9, 18, 36, 72, 108, 180 and 270 sec^{-1}, respectively. Material: Vinnol H70/DOP/stab: 100/50/4.   Filled points correspond to data where samples were "premelted" at 210°C at $\dot{\gamma}$= 36 sec^{-1} to steady state.   Full heavy drawn curve: best estimate of Eq. 13 relation.   Dotted curve calculated from Eq. 13 with a = 7.7.

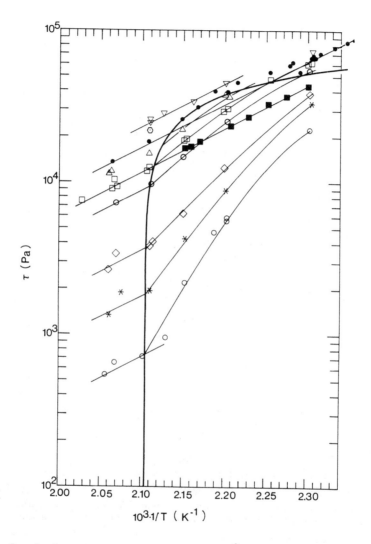

Fig. 2.   A plot of logτ against 1/T with $\dot{\gamma}$ as discrete variable for Vinnol H60/DOP/stab 100/50/4. $\dot{\gamma}$: 1.8, 9, 18, 36, 72, 108, 144, 180, 234 and 270 sec^{-1}.   Filled points: premelting at 205°C and 108 sec^{-1}.   Full drawn curve: best estimate.   Dotted curve: a = 7.7.

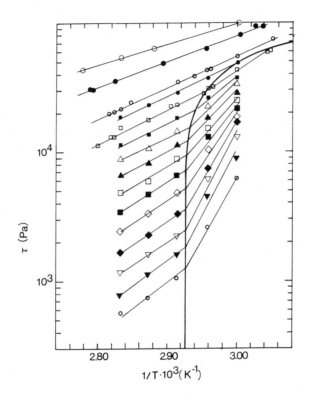

Fig. 3. Logτ against 1/T with shear rate as discrete variable for E/VA 3185 (33 (W%) vinylacetate). $\dot{\gamma}$:0.01, 0.016, 0.025, 0.04, 0.063, 0.1, 0.16, 0.25, 0.4, 0.63, 1.0, 1.6, 2, 5 and 10 sec^{-1}, respectively.

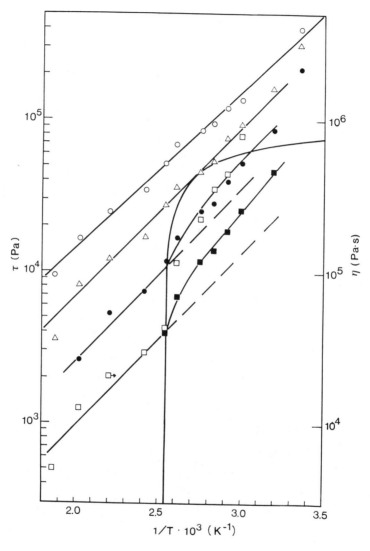

Fig. 4.   A plot of logτ against 1/T with $\dot{\gamma}$ as discrete variable
for EP copolymer with 66 (mol%) propylene.   Full drawn curve
calculated from Eq. 13 with a = 8.

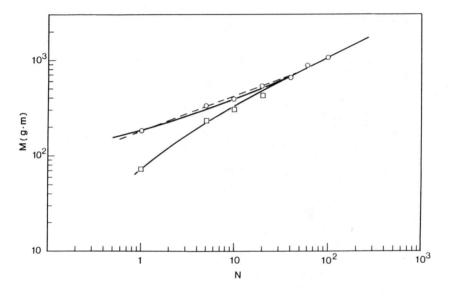

Fig. 5. Plot of shear stress against shear rate at 190°C for the sample used in Fig. 2. $\dot{\gamma} = 1.8 \cdot N$ (sec^{-1}) and $\tau = 60 \cdot M$ (Pa). The circles are measured starting with the lowest rates. The cubes are measured when the shear rates are decreased.

may be found from the intercept of the two curves found by first increasing and then decreasing the shear rate. These curves are measured again after annealing without flow.

Fig. 6 shows a double logarithmic plot of Q against concentration (c) for PVC samples. The slope is within experimental uncertainty equal to the theoretical value 4 given in Eq. 13. The average value of the constant a is found to be 7.7. $\cong$ 8.

Fig. 7 shows a delineation of Eq. 13 and Eq. 21 in a plot of logτ against reciprocal temperature for a rigid compound with

$\overline{M}_w$ = 72000. Values for stress giving melt fracture at the wall are shown too. For steady state measurements Fig. 7 shows that normal molecular melt flow, where the single polymer molecules constitute the flow units, is predicted in area A. In area B it is predicted that stable crystallite-aggregates exist in the melt. In area C the material is in a lightly crosslinked rubbery state.

Discussion (Comparison with Literature data). Fig. 1-4 show a rasonably accordance with eq. 13. From such plots T_m can be assessed as the limiting value of 1/T for small values of shear stress. From corresponding values of T_{dyn} and τ, Q may then be evaluated from Eq. 13.

The prediction that Q depends of the concentration to the fourth power is tested in Fig. 6. The uncertainty on a single estimate of Q is relatively large because Q reflects uncertainties on both T_m, T_{dyn} and τ (standard deviation on Q: ±15%). However, if we plot the average values for each concentration in DOP determined by either method 1 or 2, and a single point where tricresylphosfate (TTP) were used as plasticizer, we find a slope in a plot of logQ against logc equal to 3.97. This slope is definitely not significant different from the theoretical value 4. The concentrations were calculated at T_m. Some of the Q data have been published (4, 28 and 29). From the experimental Q values the constant a is found to be 7.7 $\simeq$ 8.

In Figs. 3 and 4, Eq. 13 is plotted as the heavy full drawn curve. The data used for the variables in Eq. 13 in Figs. 1-4 are shown in Table 1. The data for copolymers of ethylene-propylene copolymers (measured by capillary rheometry) are transplotted from Vinogradov (30).

Since the model describes different systems of polymers with low degree of crystallinity within experimental uncertainty, it is concluded that the model described in Eq. 13 cannot be rejected based on our test data. Dramatic changes in rheological response of PVC compounds taking place at characteristic temperatures (T_m) have been reported by many groups (4, 31-36).

The most elegant technique which permits an evaluation of Q by determination of corresponding values of T_{dyn} and τ is described by Villemaire and Agassant (37, 38). They constructed an ingenious piece of equipment which allows material to be processed in Couette flow under variable conditions and immediately thereafter to be extruded through a capillary rheometer. Their findings are indicated in Fig. 8 and 9 (by courtesy of Villemaire and Agassant).

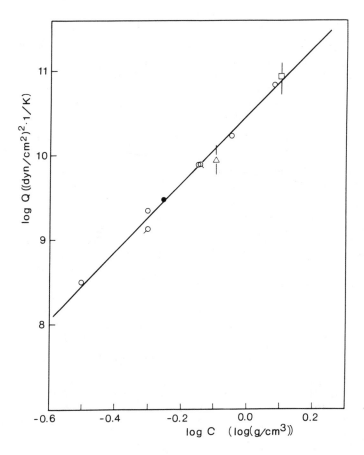

Fig. 6.   Log Q vs. log c for PVC compounds. O : Based on Vinnol
H60D/DOP/stab mixtures. Q : Vinnol H70d/DOP/stab: 100/50/4, ⚲ :
Vinnol H80F/DOP/stab: 100/100/4, ● : Vinnol H60d/Tricresylphos-
phate/stab: 100/100/4.

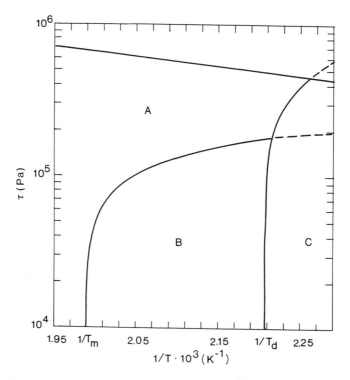

Fig. 7. Logτ against 1/T as predicted by Eq. 13 and Eq. 21 with melt fracture data plotted in the same delineation for a PVC sample with $\overline{M}_w$ = 72000 and PVC/stab: 100/4.

TABLE 1. Variables used in Eq. 14

Sample	Vinnol H600 PVC XOC / PPH 100 / 4	Vinnol H60d PVC DOP XOC / PPH 100 50 / 4	Vinnol H70d PVC DOP XOC / PPH 100 50 / 4	E/VA 3185 10-20% crystallinity	EPM Mol Fraction propylen (1) X_{pp}: 0,67
c (20°C) (kg/m³)	1325	792	792		
c (220°C) (kg/m³)	1208	708	708		
T_m °K	503.15	474.16	481.46	341	393
ΔH_A (kJ/mol)	3.285	3.285	3.285	7.533	9.207
ρ(20°C) (kg/m³)	1390	1390	1390	950	930
ρ(220°C) (kg/m³)	1280	1280	1280		
M_c (kg/mol)	11.5	11.5	11.5	5.7	6.4
M_c (g/mol)	11500	11500	11500	5700	6400 (2)
$V_A^{(4)}$ (m³/mol) · 10^{-4} (cm³/mol)	44.65	44.65	44.65	36.13	40.9
H = $\bar{M}_w/\bar{M}_n$	2.1	2.1	2.1	4.0	3.3 (3)
a	7.7	7.7	7.7	8	8
q calculated (eq. 14) (Pa²·K⁻¹·10⁻⁷)	54	4.9	4.9	60	7.8
q experimental ((dyn/cm²)² K⁻¹ 10⁻⁹) (Pa²·K⁻¹·10⁻⁷)	68	7.5	7.5	30	8

1. Estimated from the equation $\dfrac{1}{T_m} - \dfrac{1}{T_m^0} = \dfrac{R}{\Delta H_A} \ln X_{pp}$ with $\Delta H_A = 9.84 \dfrac{kJ}{mol}$, R = 8.317 $\dfrac{kJ}{mol \cdot K}$

2. From $M_c = 3.8 V_g^2$ where $V_g = M_g \cdot v_{sp}$ and M_g is the average molecular weight of a stretched chain which is 2.5 Å long. V_{sp} is the specific volume of polymer (44).

3. Estimated from viscosity - shear rate data (45).

   $\rho_{DOP}^{20°C} = 984 \dfrac{Kg}{m^3}$ $\rho_{DOP}^{220°C} = 856 \dfrac{Kg}{m^3}$ $\rho_{XOC}^{20°C} = 1070 \dfrac{Kg}{m^3}$

4. $V_A = \dfrac{M_A}{\rho}$

From plots like that in Fig. 8 and $T<T_m$ corresponding values of τ and T_m^{dyn} may be evaluated as the point of conversion of all curves irrespective of prehistory. Q can be calculated from Eq. 13. Villemaire and Agassant interpreted their results as caused by melting of crystallite aggregates. According to the gel destruction temperature hypothesis in Eq. 21 the prehistory effects can only be observed for $T_m>T>T_d$. This is documented by Villemaire and Agassant as shown in fig. 9; actually the figure allows estimation of both T_d and T_m. For a rigid sample with $\overline{M}_W \simeq 70000$, $T_d \simeq 180^oC$ and $T_m \simeq 230^oC$. The average Q values estimated from Villemaire and Agassant's data and an estimation of significance interval are shown on Fig. 6. The data are within experimental uncertainty in accordance with our data. Singleton et al. (39) have published data for PVC with 40 p.p.h DOP which may be similarily interpreted as those published by Villemaire and Agassant. The calculated Q data are shown in Fig. 6.

It is thus concluded that Eq. 13 is reasonable for samples where $T_{dyn} > T_d$. It can be shown (40) that the gel destruction temperature T_d is related to the melting temperature T_m by

$$T_d = \frac{(\xi_c - constant)}{\xi_c} T_m$$

where ξ_c is the sequence length of crystallites melting at T_d. Consequently, as the crystallizability increases, T_d and T_m converge. For polymers with regular stereospecific configuration or block copolymers with well-defined block lengths $T_d \cong T_m$ as may be shown for polypropylene model systems and Hytrel block copolymers. T_d predictions for PVC compounds are compared with experimental data in Ref. 40.

Phase Transitions in Polymer Blends and Block Copolymers

A derivation analogous to the derivation in this paper has been extended to diblock copolymer systems (41) and to immiscible (at $\tau = 0$) polymer blends (18). Theoretically, it is predicted that a homogeneous state may be formed during simple flow above a critical shear stress, which depends primarily on the miscibility of the two polymers. Light scattering measurements on a (nearly miscible) model system consisting of mixtures of SAN and PMMA were applied during flow in order to evaluate the prediction that a phase transition to a homogeneous state could be provoked by application of a critical shear stress. At conditions within the spinodal range of the phase diagram it was shown (19) that above the stress $\tau_c \cong 80$

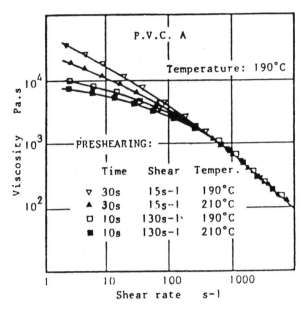

Fig. 8. Influence of mechanical and thermal history on the viscosity of PVC A at 190°C ($\overline{M}_w \simeq 70000$). (Reproduced with permission
from ref. 37. Copyright 1984 Elsevier.)

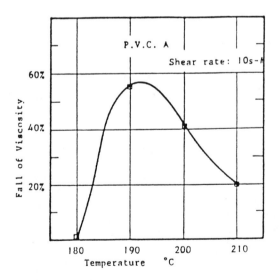

Fig. 9. Evolution of the fall of viscosity with measurement
temperature of PVC A for a shear rate of 10 s^{-1}. (Reproduced
with permission from ref. 37. Copyright 1984 Elsevier.)

kPa the total scattering is ~ zero and the anisotropic scattering pattern disappears. When the shearing is stopped the scattering pattern is independent of scattering angle and shows maxima in a plot of scattering intensity as a function of scattering angle(42). The structure formed is an interpenetrating cocontinuous structure. The melt is optically clear above τ_c and milky below.

Thus it has been documented that a phase transition to homogeneous state may be provoked by shearing. However, even though surprisingly good predictions of miscibility of polymer blends may be obtained based on e.g. group contribution models (43), such methods can at best give an order of magnitude estimate for a critical shear stress (τ_c). Consequently the predictive validity of expressions for τ_c in polymer blends is of limited value with the present accuracy of predictive schemes.

Acknowledgments

The author wish to express his gratitude to the Danish Council for Scientific and Industrial Research for financial support of the project.

Appendix

The expression:

$$\mu_A^{cr} - \mu_A^o = -\Delta H_A (1 - \frac{T}{T_m}) \qquad (5)$$

has been successfully used by Flory (23) and will be accepted here. ΔH_A is the enthalpy of melting per repeat unit and T_m is the static melting temperature (in contrast to the dynamic melting point introduced below).

A critical condition for a transition from a two-phase system to one single homogeneous phase is that we can transfer differential amounts of the components from the two-phase system to a homogeneous phase in such a way that the contribution to the fre energy is always negative or zero, or that:

$$(\mu_A - \mu_A^{ref})_{el} + (-(\mu_A^{cr} - \mu_A^{ref})) < o.$$

where $(\mu_A - \mu_A^{ref})_{el}$ is the chemical potential difference corresponding with ΔG_{el} (eq. 2).

The approach is based on the fact that (24):

a) the equations of state are valid in the steady state

b) the reversible processes are separable over definable steady state paths
c) the necessary thermodynamic functions are definable for the steady state.

In this evaluation, the final state is a state without crystallite aggregates and the starting state is the state with crystallite aggregates.
The change is defined as:

$$(\Delta\mu_A)_{el} \equiv (\mu_{A,final} - \mu_{A,start})_{el} \equiv -\frac{1}{A}\left[\frac{\partial\Delta G_{el}}{\partial N^*}\right]_{T,P,\alpha_x}$$

where A is the number of repeat units per chain between entanglements and N^* is the number of crystalline crosslinks measured in equivalent entanglement.
The elastic free energy function at steady state is assumed to be given by an equation of state of the form:

$$\Delta G_{el} = F_3(P,T,\alpha_x,N)$$

$$d\Delta G_{el} = \left[\frac{\partial\Delta G_{el}}{\partial P}\right]_{T,\alpha_x,N} dP + \left[\frac{\partial\Delta G_{el}}{\partial T}\right]_{P,\alpha_x,N} dT$$

$$+ \left[\frac{\partial\Delta G_{el}}{\partial\alpha_x}\right]_{P,T,N} d\alpha_x + \left[\frac{\partial\Delta G_{el}}{\partial N}\right]_{P,T,\alpha_x} dN$$

where P is the pressure, T is absolute temperature, α_x is a conformational tensor component α_{11}, and N is the number of entanglements per unit volume.

For ΔG_{el}, P and T constant:

$$\left[\frac{\partial\Delta G_{el}}{\partial N}\right]_{P,T,\alpha_x} = \left[\frac{\partial\Delta G_{el}}{\partial\alpha_x}\right]_{P,T,N} \left[\frac{\partial\alpha_x}{\partial N}\right]_{P,T,\Delta G_{el}} \tag{6}$$

In the following text, the case where $\alpha_x \gg 1$ is considered. α_x is the extension ratio of a chain between entanglements after a principal extension axis.

In first approximation:

$$\Delta G_{el} = NRT \; F_4(\underset{\sim}{\alpha}) \simeq 1/2 \; NRT\alpha_x^2 \qquad (7a)$$

and

$$\tau_{11} \cong NRT\alpha_x^2 \qquad (7b)$$

or

$$\Delta G_{el} \cong 1/2 \; \tau_{11} \qquad (8)$$

The tensile stress component τ_{11} is uniquely related to $\tau_{21} = \tau$.

Equation 6 can now be written:

$$\left[\frac{\partial \Delta G_{el}}{\partial N}\right]_{P,T,\alpha_x} = - \left[\frac{\partial \Delta G_{el}}{\partial \alpha_x}\right]_{P,T,N} \left[\frac{\partial \alpha_x}{\partial N}\right]_{P,T,\tau} \qquad (9)$$

where x, y and z are the principal axes of deformation of the chains between entanglements. x represents the largest deformation.

If crystalline entanglements and crosslinks are assumed equivalent, we have:

$$\left[\frac{\partial \alpha_x}{\partial N}\right]_{P,T,\tau} = \left[\frac{\partial \alpha_x}{\partial N^*}\right]_{P,T,\tau}$$

Near the transition to homogeneous phase, we assume:

$$(N^* + N)RT\alpha_x^2 \cong NRT \; \alpha_x^2$$

where N is the number of entanglements per unit volume, i.e. entanglement density. From the theory of rubber elasticity:

$$\Delta G_{el} = \frac{NRT}{2} \left[\alpha_x^2 + \alpha_y^2 + \frac{1}{\alpha_y^2 \alpha_x^2} - 3\right] \qquad (10)$$

For $\alpha_x \gg 1$ $\Delta G_{el} = \frac{NRT}{2} \alpha_x^2$ and:

$$\left[\frac{\partial \Delta G_{el}}{\partial N}\right]_{P,T,\alpha_x} = 1/2 \; RT\alpha_x^2 \qquad (11)$$

In order to introduce directly measurable quantities in the final expressions, the same derivation as in Ref. 18 is used to find a relation between α_x and τ_{21}.

The final expression (18):

$$(\mu_A - \mu_A^o)_{el} = - \frac{a^2 \, \rho^2 \, M_A \, H^2 \, M_c}{2 \, c^4 \, RT} \, \tau_{21}^2 \qquad (12)$$

a is a constant, ρ is polymer density, M_A is the molecular weight of a repeat unit, $H \equiv \overline{M}_w/\overline{M}_n$ is the ratio between the average molecular weight by weight and by number, respectively, M_e is the average molecular weight between entanglements. In the calculations we use an estimate $M_c \simeq 2 \, M_e$, M_c is the molecular weight at the intersection of two lines defined by the equation $\eta_o = K \, M_W^b$ where b changes from ~1 to ~3.5; c is polymer concentration, R is the gas constant, $\tau_{21} = \tau$ is the shear stress.

The derivation of Eq. (12) is based on a force balance on a single polymer molecule of the form

$$v^* \, F_1(\underset{\sim}{\alpha}) = v^* \, F_2(\underset{\sim}{\alpha}) \dot{\gamma}$$

where v^* is the number of entanglements per polymer molecule, $\underset{\sim}{\alpha}$ is a conformational tensor assumed to be expressible in α_x, F_1 and F_2 are functions and $\dot{\gamma}$ is the shear rate. It is furthermore assumed that shear stress is proportional to the force per polymer molecule times the number of polymer molecules per unit volume. Constant α_x is therefore equivalent with constant shear rate.

Literature Cited

1.   D.R. Paul and S. Newman, "Polymer Blends", I, Academic Press N.Y., 1978.
2.   N.J. Christensen, "Investigation of the Necking Phenomenon in Polymer Materials", Ph.D. Thesis, The Technical University of Denmark, 1985.
3.   A.K. Van der Vegt and P.P.A. Smit, S.C.I. Monograph No. 26, p. 314, (1967).
4.   J. Lyngaae-Jørgensen, Polym. Eng. & Sci., 14, 342, (1974).
5.   S. Hess, Intern. J. Thermophysics, 6 (6), 657, (1985).
6.   S. Hess, J. Physique, Col. C3 (S n°J), T 46, (3-191, (1985)).
7.   A. Silberberg and W. Kuhn, Nature, 170, 450, (1952).
8.   C. Rangel-Nafaile, A.B. Metzner and K. Wissbrun, Macromolecules, 17, 1187, (1984).
9.   B.A. Wolf, Macromolecules, 17, 615, (1984).
10. H. Kraemer, J.R. Schmidt, B.A. Wolf, Proc. 28th IUPAC, Macromol. Symp., p. 778, (1982).
11. B.A. Wolf, Makromol. Chem., Rapid Commun., 1, 321, (1980).

12. B.A. Wolf, J. Polym. Sci., Polym. Letters. Ed., 18, 789, (1980).

13. J.R. Schmidt and B.A. Wolf, Colloid & Polymer Sci., 257, 1188, (1979).

14. K.A. Mazich and S.H. Carr, J. Appl. Phys., 54, 5511, (1983).

15. E. Vrahopoulou-Gilbert and A.J. McHugh, Macromolecules, 17, 2657, (1984).

16. H. Krämer and B.A. Wolf, Makromol. Chem., Rapid Commun., 6, 21, (1985).

17. J. Lyngaae-Jørgensen, "Phase Transitions in Two Phase Polymer Blends During Simple Shear Flow", Proceedings 8th Scandinavian Rubber Conf., Copenhagen, June 10-12, 1985, p. 525-562.

18. J. Lyngaae-Jørgensen and K. Søndergaard, Polym. Eng. Sci., 27, 344, (1987).

19. Ibid, 27, 351, (1987).

20. J.D. Katsaros, M.F. Malone and H.H. Winter, Polymer Bulletin, 16, 83, (1986).

21. J. Lyngaae-Jørgensen, Macromol. Chem. 167, 311, (1973).

22. W.W. Graessley, The Entanglement Concept in Polymer Rheology, Adv. Polymer Sci., Vol. 16, (1974).

23. P.J. Flory, "Principles of Polymer Chemistry", Cornell University Press, Ithaca N.Y., 1953.

24. H.H. Hull, "An Approach to Rheology through Multi-variable Thermodynamics", Soc. Plast. Eng., 1981.

25. A.R. Berens and V.L. Folt, Trans. Soc. Rheol., 11, 95, (1967).

26. A.R. Berens and V.L. Folt, Polym. Eng. Sci., 8, 5, (1968).

27. Ibid, 9, 27, (1969).

28. J. Lyngaae-Jørgensen and A.L. Borring, Proceedings VII Int. IUPAC Congress on Rheology, p. 174, (1976).

29. J. Lyngaae-Jørgensen, J. Macromol. Sci.-Phys., B14(2), 213, (1977).

30. G.V. Vinogradov, Rheol. Acta., 6, 209, (1967).

31. E.A. Collins, Pure Appl. Chem., 49, 581, (1977).

32. L.A. Utracki, J. Poly. Sci. Phys., 12, 563 (1974).

33. L.A. Utracki, Polym. Eng. Sci., 14, 308 (1974).

34. L.A. Utracki, Z. Bakerdjian and M.R. Kamal, Trans. Soc.Rheol., 19 (2), 173 (1975).

35. A. Santamaria, M.E. Munoz and J.J. Pena, J. Vinyl Tech., 7, 22 (1985).

36. M.E. Munoz, J.J. Pena and A. Santamaria, J. Appl. Polym. Sci., 31 (3), 911 (1986).

37. J.P. Villemaire and J.F. Agassant, Mater. Sci. Monogr., 21, 271, (1984).

38. J.P. Villemaire and J.F. Agassant, Polymer Process Eng., 1, 223, (1983-84).

39. C. Singleton, J. Isner, D.M. Gezovich, P.K.C. Tson, P.H. Geil and E.A. Collins, Polym. Eng. Sci., 14, 371, (1974).

40. J. Lyngaae-Jørgensen, Makromol. Chem. Macromol. Symp., accepted for publication.

41.   J. Lyngaae-Jørgensen, Chapter 3, p. 75-123, in ed. M.J.
       Folkes, "Processing Structure and Properties of Block
       Copolymers", Elsevier Appl. Sci. Publ., London 1985.
42.   J. Lyngaae-Jørgensen, Proceedings ACS, Div. PMSE, 58, 702
       (1988).
43.   J. Holten Andersen, "Group Contribution Model for Phase
       Equilibria of Polymer Solutions", Ph.D. Thesis., Instituttet
       for Kemiindustri, Technical University of Denmark, 1985.
44.   M. Hoffman, Makromolecules Chem., 153, 99, (1972).
45.   J. Lyngaae-Jørgensen, "Advances in Rheology", 3, 503, 1984.

RECEIVED April 6, 1989

Chapter 7

Melt Flow of Polyethylene Blends

L. A. Utracki

Industrial Materials Research Institute, National Research Council of Canada, Boucherville, Québec J4B 6Y4, Canada

The rheology of polyethylene blends is discussed with an emphasis on those containing the linear low density polyethylenes, LLDPE. Flow of LLDPE with other types of LLDPE's, with low density polyethylene, LDPE, and with polypropylene, PP, was studied in steady state shear, dynamic shear and uniaxial extensional fields. Interrelations between diverse rheological functions are discussed in terms of the linear viscoelastic behavior and its modification by phase separation into complex morphology. One of the more important observations is the difference in elongational flow behavior of LLDPE/PP blends from that of the other blends; the strain hardening (important for e.g. film blowing and wire coating) occurs in the latter ones but not in the former.

Polyethylenes, PE, constitute an important part of the plastics market (see Table I). Since their discovery in 1933 they have seen continuous rise in consumption to the present level of 25M tons per annum, or 42% of all plastics (1). This extended period of growth originates in continuous development and modification of these resins, resulting from a widening range of polymerization techniques.
The history of PE can be divided into three periods: 1. the initial, characterized by predominence of the radical polymerization of ethylene, C_2, at high temperature and pressure, 2. development of coordination copolymerization of C_2 with other α-olefins, and 3. development of polymer blending technology. It is interesting that new methods have been developed without the older ones becoming obsolete. Thus it is difficult to put firm dates on transitions between these three periods. Development of Ziggler-Natta catalysts resulted in commercialization of high density polyethylene, HDPE, which had to be "toughened" by copolymerization with butene, C_4. Next was development of the linear low density polyethylenes, LLDPE, by DuPont Canada in the late 1950's. The polymer was prepared by coordination polymerization in solution of C_2 with 10 to 20 mol% of C_4, C_6 or C_8. In 1979 Union Carbide patented the gas phase fluodized

Table I. Global Consumption of Polyethylene (PE)

No.	Type	Code	Process	Density[a]	Consumption [b]	
					1986	1990
1.	High density PE	HDPE	low pressure; gas phase, solution or suspension	940-970	8.95	9.96
2.	Linear low density PE	LLDPE	low pressure; gas phase, solution or suspension	915-940	3.4	7.5
3.	Low density PE	LDPE	high pressure; pipe or autoclave reactor	915-930	12.3	10.0
4.	Very low density PE	VLDPE	low pressure; gas phase or solution process	900-915	0.03	0.2
5.	Ultra low density PE	ULDPE	low pressure; gas phase or solution process	~885	?	?

Note: (a) in kg/m^3; (b) in million tons per annum.

bed polymerization process.   It not only made LLDPE more popular around the world but also led to an ingress of blending methods as well as to development of new PE copolymers with very low density, butene based, VLDPE, and ultra low density, octene based, ULDPE. Both VLDPE and ULDPE are basically LLDPE copolymers with low crystallinity.

It is estimated that 60 to 70% LLDPE (including VLDPE and ULDPE) enters the market as blends.   One may distinguish several categories of PE-blends.

(i)   PE   blends with a small quantity of "external lubricant": fluoro-polymers, siloxanes, PE-waxes, etc.   These blends are primarily formulated for improvement of processability without affecting the PE performance (2).

(ii)   PE blends with high concentration of rigid polymer.   To this category belong blends of engineering resins with up to 10% of non-compatibilized   PE   acting   as   a   toughening   agent. Development of PE-ionomers and maleated PE allows for increase of PE content, generating a new class of materials.

(iii)   PE blends with up to 30 wt% of a rigid polymer.   The additional polymer plays the filler role increasing both the modulus and the heat deflection temperature.   At low concentration, say below 5% of rigid polymer, the compatibilization is seldom necessary, but it is a must for blends at higher loadings.

(iv)   LLDPE blends with other polyolefins or elastomers.   These blends are mainly designed for improved processability via increase of the melt strength in film blowing or wire coating applications.

(v)   PE blends with polypropylene, PP, and/or with one of their copolymers, EPR, EPDM, etc. constitute a large and important segment of the plastics market.

In this chapter only blends (iv) and (v) will be discussed.

PART I. LITERATURE SURVEY

The   methods   of   PE   polymerization   and   characterization   of   PE structures can be found in recent publications (3-6).   The melt flow of PE's and their blends has been reviewed by Plochocki (7-9) and Utracki (10, 11).   In the following text, recent information on PE/PE and PE/PP melt rheology will be outlined.

Polyethylene/Polyethylene Blends

Since the homologous polymer blends are known to be miscible it is not surprising that mixtures of **HDPE** with **HDPE** or **LDPE** with **LDPE** are miscible as well (12, 13).   However, due to the diversity of polymerization methods and the variety of resulting molecular characteristics **LLDPE/LLDPE** systems are not always miscible (10, 14-15).   In our laboratory three series of blends were prepared by identical procedure of mixing the same LLDPE with two other LLDPE resins and with LDPE.   The zero-shear viscosity vs. composition dependence, η_0 vs. w_2, of these systems is presented in Fig. 1.   Only the LLDPE's prepared with the same Ti-catalyst were found to be miscible (curve 2).   Neither blend of LLDPE with LDPE (curve 3) nor LLDPE prepared with a vanadium catalyst LLDPE (curve 1) were miscible.   There are indications in the literature (8) that

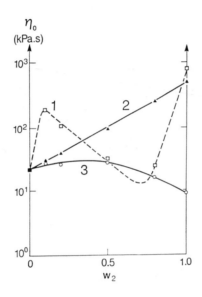

Figure 1. Compositional dependence of the zero-shear viscosity for blends of a linear low density polyethylene (LLDPE) with: (1) and (2) different LLDPE resins, and (3) with low density polyethylene, LDPE.

variations of η_0 with composition such as shown in Fig. 1 are a reflection of similar density changes caused by the thermodynamic interactions.

Dumoulin et al. (16) studied morphology and properties of HDPE blended with another HDPE resin of nearly identical density but ten times higher molecular weight. The melt-blending was carried out by three different methods. The polymers were found to be immiscible; the ultra high molecular weight component, UHMWPE, partially dissolved in the normal molecular weight (M_w = 280 kg/mole) polymer with the rest behaving as reinforcing filler particles. These observations were confirmed by Vadhar and Kyu (17) who concluded that only by dissolving UHMWPE and LLDPE in a common solvent, then casting, could the uniform, miscible blend be obtained.

Since dissolution of UHMWPE strongly affects the molecular weight distribution, MWD, of the matrix polymer, any rheological function sensitive to MWD can be used to follow the dissolution process. The dynamic viscoelastic data provided a simple and easy tool (10, 11). As will be discussed later, to accomplish this the dynamic data were examined using either: (i) Zeichner-Patel cross-point coordinates, (ii) Cole-Cole plot, or (iii) the relaxation spectrum. The stress growth or relaxation functions in shear or extension also depend on MWD. These as well were used in studies of blend miscibility by rheological means (11).

Use of rheological functions for detecting the state of PE-blend miscibility in the melt is particularly attractive. The only other available method is the direct and costly, Small Angle Neutron Scattering, SANS (18). The latter method requires deuteration of one component which may affect miscibility (19). All available information on miscibility of PE blends indicates that in the absence of strong thermodynamic interactions both the dissolution and the phase separation are diffusion controlled. Frequently, one can generate homogeneous melt via dissolution in a common solvent (20) while the melt blending will result in a phase-separated material. There is direct evidence of true miscibility in homologous polymer blends, while even a small change in polymerization method, polymer composition or structure may lead to apparent immiscibility (14, 15). Since for the high molecular weight, industrial PE the conformational entropy of mixing is negligibly small and there are no specific interactions between two polyethylenes of different structure, one can appreciate that in these systems the miscibility is controlled by small departures from zero of the free energy of mixing.

Blends of **HDPE with LDPE** were studied by Dobrescu (12) by means of a capillary viscometer. The constant stress viscosity, log $\eta(\sigma_{12}$= const) vs. w_2 plot indicated a strong positive deviation, PDB, from the log-additivity rule:

$$\log \eta = \sum w_i \log \eta_i \qquad (1)$$

where $\eta = \eta(\sigma_{12})$ is the constant stress viscosity of the blend and η_i, i = 1, 2 that of the components. The blends were reported immiscible. The higher the viscosity ratio η_i/η_j, the larger was the PDB deviation. Kammer and Socher (21) studied the two-phase HDPE/LDPE blends using the cone-and-plate, steady state rotational rheometer. Plots of the zero shear viscosity, η_0, the first normal stress difference coefficient ψ_{10} and the principal relaxation time, τ vs. w_2, all indicated PDB. In spite of immiscibility the shear viscosity data in a temperature range from 160 to 200°C could be superimposed on a master curve ([22, 23]):

$$\eta/\eta_0 = f(\dot{\gamma}\eta_0/T) \tag{2}$$

The compositional dependence of both $\eta = \eta(\sigma_{12})$ and the extrudate swell, $B = B(\sigma_{12})$ was PDB type. In conclusion, the current rheological information consistently reports that HDPE/LDPE systems are immiscible.

Santamaria and White ([24]) studied melt flow behavior and spinnability of several polyolefin systems including HDPE/LDPE blends. Defining the steady state compliance as:

$$J_e = N_1/2\sigma_{12}^2 \tag{3}$$

where N_1 is the first normal stress difference and σ_{12} the shear stress, the authors observed that J_e vs. w_2 shows a strong negative deviation from additivity. Furthermore, the extrudate shrinkage was nearly constant, independent of composition. Since the critical drawdown ratio was a linear function of the shrinkage, the net spinnability of HDPE and its blends with LDPE was low. The phase morphology was not studied.

The most important commercial blends of PE are those of **LLDPE with LDPE** ([25, 26]). The capillary flow data $\eta = \eta(\sigma_{12})$ and $B = B(\sigma_{12})$, indicated (similar to HDPE/LDPE) PDB-type behavior ([27-29]). The latter authors also reported a PDB relation between melt strength and composition. Recently ([14, 15]) these blends were studied under the steady state and dynamic shear flow as well as in uniaxial extension. A more detailed review of these results will be given in part 3 of this chapter. Like HDPE/LDPE blends, those of LLDPE/LDPE type are also consistently reported as immiscible.

Polypropylene/Polyethylene blends

PP/PE blends are of major commercial interest. Patents on PP with HDPE, LDPE and more recently with LLDPE constitute well over 40% of all patents on polyolefin blends. The main reason for blending is improvement of: processibility, low temperature toughness and impact strength (PP with HDPE or LLDPE), film clarity, drawability, orientability (PP with LDPE), etc. ([8]). The blends find application in automobiles, appliances, houseware, furniture, sporting goods, toys, packaging, chemical processing equipment and industrial components. The PE/PP systems are immiscible. Depending on the type and molecular parameters of the components the degree of immiscibility may be such that compatibilizing copolymers, viz. EPR or EPDM have to be added ([8, 30, 31]). The ultimate mechanical properties, i.e. impact strength, the stress and maximum strain at break, are the most sensitive to the extent of compatibility.

The capillary viscosities of **HDPE/PP** systems at 180 to 210°C were reported ([32, 33]) to superimpose on the Equation 2 master curve, with η_0 following the Arrhenius dependence:

$$\ln \eta_0 = A + E_\eta/RT \tag{4}$$

where A is a parameter, E_η the activation energy of flow and R the gas constant. The plot of shear viscosity at constant shear stress, $\eta = \eta(\sigma_{12})$, vs. volume fraction of the second component, ϕ_2, showed a negative deviation from the log additivity rule, Equation 2, NDB. Defining the viscosity ratio as:

$$\lambda = \eta_1/\eta_2 \tag{5}$$

where subscripts 1 and 2 represent the dispersed and the matrix phases respectively, the authors reported large differences in morphology for blends with $\lambda > 1$ and for those with $\lambda < 1$; in the first case the dispersed phase existed in a form of fibrillas, in the second as small droplets. The difference in structure originated in the kinetics of breakage of fibrillas created at the entrance to capillary. For systems whose $\lambda < 1$ the fluidity additivity equation (34) was found to be obeyed:

$$1/\eta = \beta_1 \sum w_i/\eta_i \tag{6}$$

with w_i replaced by ϕ_i and the slip coefficient $\beta_1 \approx 1$. For $\lambda > 1$ at concentrations $w_2 < 0.5$ Equation (1) yielded good approximation; at higher PP content, Equation 6 was obeyed.

The extrudate swell, $B = B(\sigma_{12})$, for HDPE and high PP-content blends, $w_2 \geqslant 0.75$, was independent of temperature. The plot for $w_2 = 0.25$ and 0.5 was temperature sensitive, indicating T-dependent morphology (32, 33). At high PP content and low strains the lower viscosity HDPE drops became deformed at the capillary entrance. Retraction of these fibers caused large extrudate swell. At high strains both phases were strained, yielding average, smaller B values. For high HDPE content B decreased with temperature, T, probably due to increased interlayer slip caused by lowering the viscosity.

Blends of three PP and two HDPE resins having different values of η_0 were studied in a capillary viscometer (35). Independently of the viscosity ratio, λ, at 200°C all blends showed small negative deviation from the log-additivity rule, NDB. The non-equilibrium extrudate swell at low stress (measured after quenching) showed small positive deviation from the additivity rule while at higher stresses the additivity. The NDB tendency for the viscosity combined with the constant critical shear stress for melt fracture, $\sigma_{MF} = 2$MPa, indicate a better extrudability of the PP/HDPE blends than that of the neat resins.

Blends of **PP with LDPE** have been of industrial interest for years (24, 35-38). Plots of $\eta = \eta(\sigma_{12})$ vs. w_2 invariably show NDB (35, 36, 38) whose magnitude varies with the method of blend preparation (37) and stress level (24). The negative deviations indicate that the mechanism responsible for NDB is interlayer slip. According to Lin (34) the interlayer slip factor in Equation 6:

$$\beta_1 = 1 + (\beta_{12}/\sigma_{12}) (w_1 w_2)^{1/2} \tag{7}$$

where β_{12} is the characteristic slip factor of the blend, i.e. the NDB should be, as observed, more pronounced at small σ_{12}. The extrudate shrinkage was also reported (24, 38). Blends with $\lambda \approx 1$ were extruded, then placed in an oil bath at 190°C for about one hour. The shrinkage was calculated as

$$\varepsilon_\infty = \ln (L_0/L_\infty) \qquad (8)$$

where L_0 and L_∞ indicate the initial and final extrudate lengths, respectively. The plot of ε_∞ vs. composition is presented in Fig. 2. It can be shown that while the absolute magnitude of ε_∞ depends on the initial aspect ratio of the extrudate, the form of ε_∞ vs. w_2 does not. It is apparent that for the system with $\lambda \approx 1$ the maximum shrinkage occurs at $w_2 \approx 0.5$ where the fibrillation affects both co-continuous phases.

The flow properties of **LLDPE/PP** blends were studied by Dumoulin et al. (39-41). This subject will be discussed in the last part of this chapter.

Concluding the literature review the following remarks should be made: (i) the NDB behavior of $\eta = \eta(\sigma_{12})$ vs. w_2, a large extrudate shrinkage near $w_2 = 0.5$ and a sensitivity of the rheological functions to methods of blend preparation all indicate immiscibility (10, 11, 42, 43), (ii) the single phase flow of PE blends was reported only for homologous PE blends when the components' molecular weight $M_w < 1000$ kg/mol; the homologous HDPE/UHMWPE blend was found to form two phases, probably caused by the slow rate of UHMWPE dissolution in the mechanically compounded mixtures, (iii) blends of HDPE with LDPE were found to be immiscible, although the immiscibility may be marginal, disappearing at high temperatures for low molecular weight components, and (iv) blends of PE with PP show strong, nearly antagonistic immiscibility.

PART II. LLDPE BLENDS WITH PE

A standard commercial film blowing LLDPE resin, LPX-30, was blended at different ratios with either other LLDPE's or a LDPE polymer. The characteristic properties of these materials are listed in Table II. The resins were generously donated to the project by Esso Chem., Canada. Prior to blending the polymers were thoroughly characterized by SEC, SEC/LALLS, solution viscosity, ^{13}CNMR, Atomic Absorbance, and their rheological behavior was characterized in steady state and dynamic shear flow as well as in the uniaxial extensional deformation (44-46).

Experimental
Blending was done using a Werner-Pfleiderer co-rotating twin screw extruder, model ZSK-30 at 140 rpm. The pelletized resins were pre-dry-blended and fed to the extruder using a volumetric metering feeder, Incrison model 105-C. The following temperature profile was used: (feeder) 158, 192, 214, 196 and (die) 196°C. The extrusion rate was Q = 6 to 8 kg/hr, indicating extensive backmixing. The extrudates were granulated, dried and formed into shapes suitable for rheological testing. The composition, code and molecular weight

Table II.   Characteristic parameters of neat PE resins [44 to 46]

No.	Parameter (units)	LPX-30	LLDPE-10	LDPE	LPX-24
1.	Density (kg/m^3)	918	955	922.5	951
2.	Melt index (190°C, 2.16 kg, g/10 min)	1.0	0.3	6.5	0.3
3.	Molecular weights (kg/mol) M_n	41 ± 2	17 ± 1	16 ± 1	28 ± 1
	M_z	272 ± 27	593 ± 11	225 ± 8	719 ± 30
4.	Polydispersity index (-) M_w/M_n	3.4 ± 0.2	11 ± 2	4.1 ± 0.1	7.6 ± 0.2
5.	Comonomer/concentration (wt%)	C_4/7.4	C_6/1.1	-	C_4/4.5
6.	Catalyst based on:	Ti	V	-	Ti
7.	Branches C_n 10 per 1000 C-atoms	0	0	1.2	0
8.	Zero-shear viscosity η_0 at 190°C (kPas)	34	645	9.3	8120

averages of the three types of blends are listed in Table III. Antioxydant, a mixture of Irganox 1010 and Irganox 1029, was used at a concentration 0.2 wt %.

For determination of the steady state shear viscosity the Instron capillary viscometer model 3211 was used at 190°C. Six capillaries were used, three each of diameter d = 747 and 1273 μm. The length to diameter ratio in each series varied from L/d = 0.6 to 60. The standard Bagley and Rabinowitsch corrections as well as that for the pressure effects (45) were applied. The extrudate swell was determined on air-cooled extrudates, 5 cm in length.

Rheometrics Mechanical Spectrometer, Model 605, RMS, was used in the dynamic mode at 150°C with the parallel plates geometry. To examine the thermal stability of the resins, first the two hours long time sweep at strain γ = 15% and frequency ω = 10 rads/s was carried out; the test samples were found to be stable for at least 60 min. The frequency sweep at γ = 15% was performed in two directions: for two samples from ω = 0.1 to 100 (rads/s) and for two others from 1 to 0.01 (rads/s). Two diameters of platens were used: 25 and 50 mm. The results were accepted if the maximum difference between the results of these four sweeps in the common range: ω = 0.1 to 1.0 did not differ by more than 5%. All tests were done under a blanket of dry nitrogen. The measurements were recorded using the RMS Columbia terminal, Model 964, then the data were transferred to a Hewlett-Packard, HP-85, mini-computer for analysis and mathematical manipulation, viz. calculation of the zero shear viscosities, η_0.

Rheometrics Extensional Rheometer, Model 605 (RER) was used at 150°C in a constant strain rate ($\dot{\varepsilon}$) mode. To facilitate storage and manipulation of data the RER was interfaced with a Hewlet-Packard model HP-85 micro-computer. For the test, the samples were transfer molded under vacuum, annealed, and then affixed with epoxy (24 hours curing) to the aluminum ties. Prepared in this manner, specimens did not show any shape change on immersion in hot (T ≃ 150°C) silicone oil. The specimens were mounted in the instrument and immersed in Dow 200 silicon oil at 150°C. The test started after a 7 to 10 minute temperature equilibration period. During the measurements the maximum temperature difference, as read by three thermocouples placed at different heights in the sample chamber, did not exceed 0.5°C. For each sample the uniformity of the cylindrical shape was verified at three different heights; in all cases the average diameter was found to be d = 5.55 ± 0.05 mm. The sample length used in this study was L_0 = 22 mm ± 0.1 mm.

It is convenient to discuss results of this work under separate sub-titles, starting with the capillary, then with dynamic and finally the extensional flow behavior.

Pressure correction

It was observed that Bagley plots of pressure drop P vs. L/d were not always linear (45, 47):

$$P = \sum_{i=0}^{2} C_i \ (L/d)^i \qquad (9)$$

Table III. Composition and Molecular Weights (in kg/mol) of the Three Series of LLDPE Blends

Code	Polymer Content (wt%)				M_n	M_w	M_z
	LPX-30	LLDPE-10	LDPE	LPX-24			
I. 0	100	0	0	0	41	133	272
I. 10	90	10	0	0	33	114	275
I. 20	80	20	0	0	36	144	357
I. 50	50	50	0	0	25	123	397
I. 80	20	80	0	0	25	151	525
I.100	0	100	0	0	17	152	593
II.0 = I.0	100	0	0	0	41	133	272
II.10	90	0	10	0	36	112	252
II.20	80	0	20	0	34	114	254
II.50	50	0	50	0	29	107	242
II.80	20	0	80	0	19	66	248
II.100	0	0	100	0	16	64	225
III.0 = I.0	100	0	0	0	41	133	272
III.10	90	0	0	10	39	141	340
III.20	80	0	0	20	38	150	401
III.50	50	0	0	50	34	174	548
III.80	20	0	0	80	30	199	659
III.100	0	0	0	100	28	216	719

where C_i's are parameters: $C_0 = P_e$ (the Bagley entrance and exit correction), $C_1 = 4\sigma_{12}$ and

$$C_2 \simeq C_1^2 \, a_1 b_1 / 2P^* \tag{10}$$

where: a_1 is the temperature and pressure sensitivity parameter defined by the dependence:

$$\ln \eta_0 = a_0 + a_1/(f + a_2) \tag{11}$$

with f being the free volume fraction computed from the Simha-Somcynsky equation of state (48); b_1 is the free volume pressure coefficient defined by the relation:

$$1/f = b_0 + b_1 P/P^*; \quad \text{at} \quad \tilde{T} \equiv T/T^* = \text{const.} \tag{12}$$

with P^* and T^* being the characteristic (for a given liquid) pressure and temperature reducing factors, respectively. Note that according to Equation 12 $b_1 = b_1(\tilde{T})$ is a universal constant for all liquids at the reduced temperature $\tilde{T}$, i.e. both b_1 and P^* are determined by the thermodynamic properties of a system with only a_1 remaining unknown. For HDPE or LLDPE at 190°C, Equation 10 can be rearranged to:

$$a_1 \simeq 154 \, \overline{(C_2/C_1^2)} \tag{13}$$

with the bar indicating the average value.

The values of a_1 were computed for a series of LLDPE and HDPE samples with a wide range of molecular weights. The data were found to follow the linear dependence:

$$a_1 = -0.1065 + 4031 \, M_w/M_n; \quad M_w/M_n \leqslant 35 \tag{14}$$

with the correlation coefficient squared, $r^2 = 0.9945$.

Computation of a_1 resulted in the values plotted vs. composition in Fig. 3. According to Equation 14, a_1 depends on polydispersity which for a blend can be calculated from molecular weight averages of the neat polymers assuming their miscibility:

$$M_n^{-1} = \sum w_i / M_{n_i} \tag{15}$$

$$M_w = \sum w_i M_{wi} \tag{16}$$

$$M_z = \sum w_i \, M_{zi} \, M_{wi} / \sum w_i \, M_{wi} \tag{17}$$

$$\therefore \, M_w/M_n = \sum (w_i/M_{ni}) \sum w_i \, M_{wi} \tag{18}$$

The broken line in Figure 3 was computed from Equations 14 and 18 for blends of Series-I. The agreement between experimental results and the theoretical predictions confirms assumed miscibility. On the other hand, a_1 vs. LDPE content for blends of Series II show

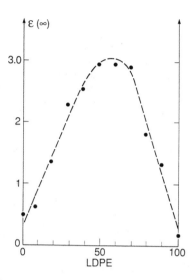

Figure 2. Extrudate shrinkage (in Hencky strain units) vs. composition of polypropylene/low density polyethylene blends. (Adapted from ref. 38.)

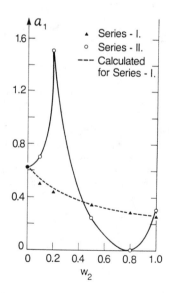

Figure 3. The temperature and pressure sensitivity coefficient for LLDPE 1 blended with either LLDPE 2 (triangles) or LDPE (circles). The broken line is theoretical, computed from variation of polydispersity in (assumed) miscible blends.

behavior incongruous with monotonic Equations 14 to 18 giving evidence of immiscibility. The extrema at II.20 and II.80 may indicate limits of miscibility. However, it is difficult to comprehend why crossing spinodals from the high LLDPE side results in a system with strong sensitivity of rheological behavior to T and P while crossing it from the high LDPE side generates blends with rheology insensitive to these independent variables.

Extrudate Swell

The extrudate swell, $B = D/d$, (where D and d are the extrudate and die diameter respectively) is plotted as B vs. w_2 in Fig. 4 for blends of Series I and II. In both cases B follows a simple parabolic equation:

$$B \equiv B_1 w_1 + B_2 w_2 + B_{12} w_1 w_2 \tag{19}$$

with $B_{12} = 0.32 \pm 0.06$ and 0.91 ± 0.05 for Series I and II respectively. The large B_{12} value for the latter blends suggests that here swelling is due to strain recovery of fibrillated drops rather than to extrudate swelling as in homologous polymer melts. Note that for system II the B vs. w_2 dependence resembles the extrudate shrinkage in Fig. 2.

Entrance flow

The entrance to capillary flow was studied by a flow visualization method (49). The flow pattern for Series II blends is presented in Fig. 5 (since the flow is axisymmetric only the left hand side part of the entrance region is shown). There is a large difference in flow pattern between that of LLDPE (0%) and of LDPE (100%). In the latter case the classical wine-glass shape with large vortices was observed. Upon a decrease of LDPE content both the glass-stem and vortices slowly decreased. It was noted that as little as 2% of LDPE was sufficient to introduce significant changes in LLDPE flow pattern. The mathematical modeling of the entrance flow allows correlation between the vortex size and the strain hardening in extensional flow.

The frequency dependence

The dynamic flow data can readily be evaluated in a three step process: (i) examination of the data for presence of an apparent yield stress, σ_y, and subsequently its subtraction, (ii) fitting the data to either the generalized Carreau model (42):

$$\eta' = G''(\omega)/\omega = \eta_0 \ [1+(\omega\tau_1)^{m_1}]^{-m_2} \tag{20}$$

or to the Havriliak-Nagami dependence (50):

$$\eta^* \equiv G^*/\omega = \eta_0 [1+(i\omega\tau_2)^{1-\alpha}]^{-\beta} \tag{21}$$

and (iii) generalization of the results by deriving other linear viscoelastic dependencies by means of the intermediation of the Gross' frequency relaxation spectrum:

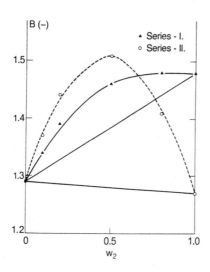

Figure 4. Extrudate swell at 190°C versus weight fraction of the second component in Series-I and II; curves - Equation 19.

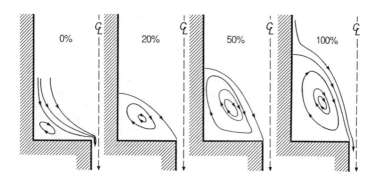

Figure 5. Flow pattern at the entrance to capillary for linear low density/low density polyethlene blend (LLDPE/LDPE) at 190°C and $\dot{\gamma}$ = 125 (s^{-1}). From the left hand side: 0, 20, 50 and 100% of LDPE. (Reproduced from ref. 49.)

$$H_G(\omega) = \pm (1/\omega\pi)\mathrm{Im}G^*(\omega e^{\pm i\pi}) = (2/\omega\pi) \mathrm{Re}G''(\omega e^{\pm i\pi/2}) \qquad (22)$$

In Equations 20 to 22: η', η^* are dynamic and complex viscosity; G', G'', G^* are the storage, loss and complex shear modulus; ω is the frequency; τ_i the primary relaxation time; m_1, m_2, α, β are parameters: $i = \sqrt{-1}$; and Im, Re indicate the imaginary and real parts of the complex function, respectively.

The convenient way of data examination for presence of the apparent yield stress is by means of the modified Casson plot (51):

$$F^{\frac{1}{2}} = F_y^{\frac{1}{2}} + \underline{a}\, F_m^{\frac{1}{2}} \qquad (23)$$

where F, F_m and F_y are rheological functions of the blend, the matrix phase and its yield value, respectively; $\underline{a}$ is a parameter representing the square root of the relative F-function. Note that Equation 23 looses its sense in systems where identification of the dispersed and matrix liquid is becoming ambiguous. For F any rheological functions, e.g. σ_{12}, G', G'', can be used. The yield phenomenon originates in a stable three-dimensional structure. When the stress exceeds its yield value, $\sigma_{12} > \sigma_y$, the structure is destroyed and the material behavior changes from solid-like to liquid-like. In this classical description the rate is not considered. In blends at high concentration of dispersed phase the droplets either form interactive clusters or they coalesce into a co-continuous network. The relaxation time of these interactive entities, τ_y, is rather long. When an experiment is carried out slowly enough, allowing the cluster to reform, $F_y \to 0$ can be found. On the other hand, when the experiment is conducted on a time scale comparable to τ_y, then the network response is analogous to that in the classical concept and $F_y \neq 0$ is observed. Formally one may express this idea in the form of the following relation:

$$F_y = F_y^\infty [1-\exp \{-\tau_y \omega\}]^u \qquad (24)$$

where τ_y is the relaxation time of interactive cluster and u = 0.2 to 1.0 is an exponent. At a given frequency ω and when $\tau_y \to 0$ and ∞ the apparent yield stress $F_y = 0$ and $F_y = F^\infty$, respectively. Similarly for τ_y = const. (specific polymer blend) F_y reaches these limits for $\omega \to 0$ and $\omega \to \infty$. Experimental verification of dependencies 23 and 24 was recently published (52, 53). After the apparent yield stress is calculated, the experimental values, F_a, should be corrected by subtracting F_y:

$$F(\omega) = F_a(\omega) - F_y(\omega) \qquad (25)$$

Once the values of $F(\omega)$ are known either Equation 20 or 21 can be used. The advantage of Equation 20 is that only $F_a = G_a''$ needs to be corrected for the apparent yield stress. If Equation 21 is to be used both G_a' and G_a'' must be corrected independently and then:

$$G^* = (G'^2 + G''^2)^{\frac{1}{2}} \tag{26}$$

calculated. (Note: since $G'_y \neq G''_y$ an attempt to use $F = G^*$ usually leads to nonsensical results).

Fig. 6 shows the curve-fit of η' vs. ω dependence for Series I and II blends by means of Equation 20. The fitting procedure generated the numerical values of the four parameters of the equation: η_0, τ, m_1 and m_2. It was found that the zero shear viscosity of homopolymers and blends followed the relation:

$$\eta_0 \propto M_\eta^a \tag{27}$$

where $a \simeq 3.5$. The molecular weight M_η, for polymers with log-normal distribution of molecular weights, can be expressed as (45):

$$M_\eta = M_z(M_w/M_n)^{0.2} \tag{28}$$

The dependence is shown in Fig. 7.

From Equation 20 at $\omega \gg 1$ the power law exponent $n = 1 - m_1 m_2$ i.e. $0 \leqslant m_1 m_2 \leqslant 1$ are predicted and experimentally verified.

The last parameter of Equation 19, the principal relaxation time τ_1, is a complex function dependent on sample polydispersity which can not be readily correlated with a single molecular or rheological function (45). However, the correlation can be obtained through H_G as intermediary.

Using principles of complex algebra H_G can be expressed in terms of either Equation 20 or Equation 21 parameters:

$$H_G(\omega) = (2\eta_0/\pi)r_1^{-m_2}\sin m_2\theta_1 = (\eta_0/\pi)r_2^\beta \sin \beta\theta_2 \tag{29}$$

where:

$$r_1 \equiv [1+2(\omega\tau_1)^{m_1} \cos (\pi m_1/2) + (\omega\tau_1)^{2m_1}]^{\frac{1}{2}}$$

$$\theta_1 \equiv \arcsin \{(\omega\tau)^{m_1} r_1^{-1} \sin (\pi m_1/2)\}$$

$$r_2 \equiv \{[1-(\omega\tau_2)^{1-\alpha} \sin (\pi\alpha/2)]^2 + (\omega\tau_2)^{2(1-\alpha)} \cos^2 (\pi\alpha/2)\}^{\frac{1}{2}}$$

$$\theta_2 \equiv \arctan \{[(\omega\tau^2)^{1-\alpha} \cos (\pi\alpha/2)] / [1-(\omega\tau_2)]^{1-\alpha} \sin (\pi\alpha/2)\}$$

Furthermore, from Equation 21 the real and imaginary parts of η^* can be expressed as:

$$\eta' = (\eta_0/r_2^\beta) \cos \beta\theta_2 \tag{30a}$$

$$\eta'' = (\eta_0/r_2^\beta) \sin \beta\theta_2 \tag{30b}$$

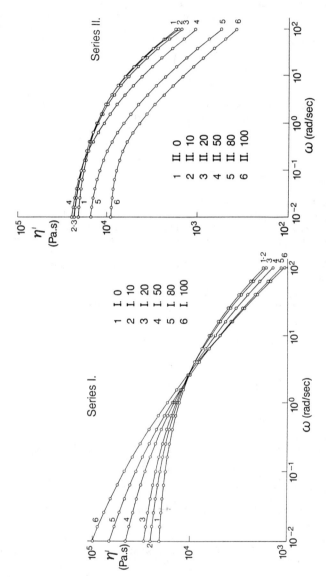

Figure 6. Frequency dependence of dynamic viscosity at 150°C for Series–I (left) and Series–II (right). Points – experimental, lines–computed from Equation 35b.

Comparing Equations 29 and 30 it is evident that for systems whose η^* vs. ω dependence can be described by Equation 21 there is a simple equivalence: $\eta''(\omega) = \pi H_G(\omega)$.

The molecular weight dependence enters Equation 29 through η_0. For this reason it is convenient to define the reduced frequency relaxation spectrum as:

$$\tilde{H}_G(\omega) \equiv H_G(\omega)/\eta_0 \tag{31}$$

An example of this function is presented in Fig. 8.   Since the integral:

$$\int_{-\infty}^{+\infty} \tilde{H}_G(s) \, d\ln s \equiv 1 \tag{32}$$

the coordinates of $\tilde{H}_G(\omega)$ maximum (ω_{max}, $\tilde{H}_{G,max}$) should correlate with the molecular parameters of the melt.   Indeed, it can be shown that in single phase systems ω_{max} depends on η_0 and $\tilde{H}_{G,max}$ on sample polydispersity (11, 14, 54).   These dependencies are presented in Fig. 9.   It is evident that for Series I blends there is a linear correlation between $\log \omega_{max}$ and $\log \eta_0$ as well as between $\log \tilde{H}_{G,max}$ and M_n/M_w but for LLDPE/LDPE, Series II, the linearity is not observed.

The linearity of dependencies in Fig. 9 can be predicted assuming miscibility in a blend composed of polymers with similar molecular weights.   For such systems the general, third order mixing rule for relaxation spectrum (55, 56):

$$H(\tau) = \sum_{ijk} P_{ijk} H_{ijk} (\tau/\Lambda_{ijk}) \tag{33}$$

can be reduced to simple additivity:

$$H(\tau) = \sum w_i H_i(\tau) \tag{34}$$

In Equation 33 P_{ijk} is the third order intensity function (replaced in Equation 33 by the weight fraction w_i) and Λ_{ijk} is the third order interactive relaxation time factor, which originates in the intermolecular interactions between dissimilar molecules.   In homologous polymer blends where the molecular weight of one polymer is lower than the entanglement molecular weight, M_e, and that of the second is higher than M_e, Equation 34 must be extended to accommodate a second order term (57).

The proposed method of data treatment has two advantages: (i) it allows assessment of the status of blend miscibility in the melt, and (ii) it permits computation of any linear viscoelastic function from a single frequency scan. Once the numerical values of Equation 20 or Equation 21 parameters are established the relaxation spectrum as well as all linear viscoelastic functions of the material are known. Since there is a direct relation between the relaxation and the retardation time spectra, one can compute from $H_G(\omega)$ the stress growth function, creep compliance, complex dynamic compliances, etc.

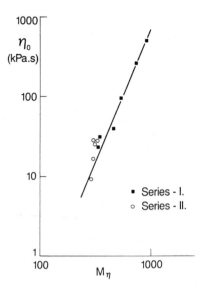

Figure 7. Molecular weight dependence of the zero-shear viscosity for Series-I and II. The line represents the dependence for neat LLDPE resins. (Reproduced with permission from ref. 45. Copyright 1987 SPE.)

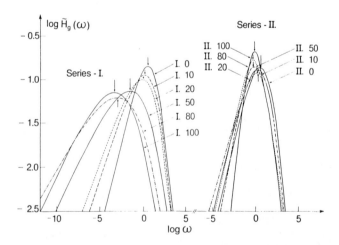

Figure 8. Reduced Gross frequency relaxation spectrum for Series-I and II. The arrows indicate computed coordinates of the maximum ($\tilde{H}_{G\ max}$, ω_{max}).

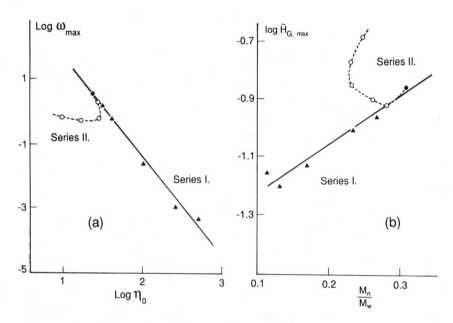

Figure 9. (a) Correlation between the abscissas of relaxation spectrum maximum and the zero-shear viscosity for Series-I and II; (b) Correlation between the ordinate of relaxation spectrum maximum and the polydispersity factor M_n/M_w.

(58). However, one must keep in mind that validity of the computed functions is only assured within the range of independent variables $(\omega, T, P, t, \ldots)$ used in determining H_G. For example:

$$G'(\omega) = \int_{-\infty}^{+\infty} \left\{ s \, H_G(s)/[1+(s/\omega)^2] \right\} \, d\ln s \qquad (35a)$$
calc

$$G''(\omega) = \int_{-\infty}^{+\infty} \left\{ \omega \, H_G(s)/[1+(\omega/s)^2] \right\} \, d\ln s \qquad (35b)$$
calc

$$\eta^+(t) = \int_{-\infty}^{+\infty} H_G(s) [1-\exp\{st\}] \, d\ln s \qquad (36)$$
calc

where η^+ is the stress growth shear viscosity. In Fig. 10 the experimental values of G' for blends Series I and II (points) are compared with G'_{calc} (lines) computed from H_G which in turn was determined using the η'-data. The agreement is quite satisfactory with residuals $|G' - G_{calc}| < 5\%$ for both series. The concluded immiscibility for LLDPE/LDPE blends of Series II make this observation particularly interesting.

Cole–Cole plot
The η'' vs. η' dependence is often used while discussing the dynamic data of polymer blends. Since for simple liquids the plot has the form of a semi-circular arc, deviation from a semi-circle is sometimes identified with immiscibility. However, as frequently demonstrated for homopolymers (16) and homologous polymer blends (59) the form of the η'' vs. η' plot is determined by the shape of the relaxation spectrum; the molecular polydispersity can modify the Cole–Cole plot as much as the presence of the interphase. As a rule, the presence of (unsubtracted) yield stress appears in the plot as a sudden departure from a semi-circular pattern at higher η' level.
From Equations 29 and 30 it is evident that both $H_G(\omega)$ and the Cole–Cole plot contain the same information; in principle these data treatments are equivalent. The advantage of H_G is the straight-forward interpretation and access to other linear viscoelastic functions. The advantage of the Cole–Cole plot is its simplicity – it can be constructed simply by plotting one experimental function versus another.
In accordance with Equation 30 η' and η'' can be normalized by dividing by η_0, i.e. η'/η_0 and η''/η_0. Such a normalized Cole–Cole plot has frequently been used for polymer blends (60, 61). In Fig. 11 the normalized Cole–Cole plot is shown for blends of Series I and II. The data used for constructing the Figure are the same as those used for computation of $\tilde{H}_G$ in Fig. 8.

Cross–point coordinates
The coordinates defined as $G_x \equiv G' = G''$ at $\omega = \omega_x$ were reported (62, 63) to be related to the molecular parameters of melts: $G_x \propto (M_w/M_w)^\alpha$ and $\omega_x \propto M_w^\beta$ with α and β negative. From Equations 34 and 35 it follows that $G' = G''$ at $\omega = s = 1/\tau$, i.e. when

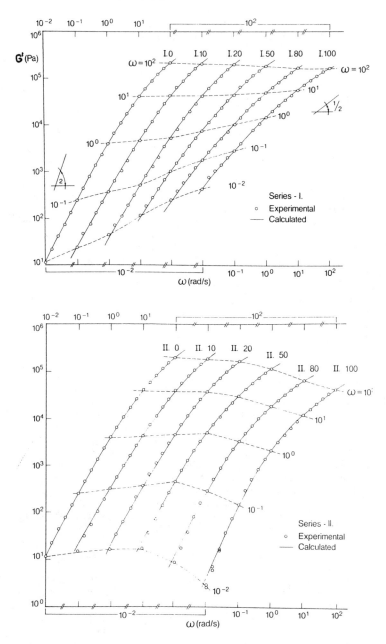

Figure 10.   Storage modulus vs. frequency for Series I LLDPE/LLDPE (top) and for Series II LLDPE/LDPE blends (bottom) at 190°C.   Points – experimental, solid lines computed from $H_G(\omega)$, broken lines connect the constant – ω data points.   Concentration of the second component is given in wt%.   The results for each blend is displaced horizontally by one decade.

the test frequency equals the inversed relaxation time of the system. Substituting this condition into the relation:

$$G'(\omega) = A\eta_0\omega^2\tau/[1+(\omega\tau)^2] \qquad (37)$$

derived for simple liquids with a single relaxation time by Maxwell ($A=1$) and by Rouse ($A = 6/\pi^2$), one gets:

$$\eta_{0M} = 2G_x/A\omega_x \qquad (38)$$

In Fig. 12 η_{0M}, calculated with $A = 1$, is plotted vs. η_0 for LDPE and LLDPE resins and for their blends. For $\eta_0 \leqslant 60$ kPas the totally unexpected agreement $\eta_{0M} = \eta_0$ was obtained. Above this limit the values diverged into $\eta_0 \simeq 7\eta_{0M}$ for $\eta_0 > 200$ kPas, where Equation 38 requires that $A \simeq 1/7$. From the general Rouse theory (64) A decreases with an increase of polydispersity of the relaxation times but the influence is not large enough to allow such a low A-value. As will be shown in part 4 the dependence presented in Fig. 12 was duplicated by the results of LLDPE/PP blends indicating a broad applicability of Equation 38 (41, 65).
    For LLDPE blends the relations reported by Zeichner et al. (62, 63) were not obeyed. For example the plot of G_x vs. w_2 showed an opposite trend from the predicted negative deviation from the additivity rule.
    For these systems the activation energy of flow $E_\sigma \equiv R(\partial\ln \eta/\partial 1/T)_\sigma = \text{const} = E_{\dot\gamma}/n = 30 \pm 2$ kJ/mol was found (45).

The uniaxial extensional flow
The behavior of LLDPE blends at constant rate of stretching, ε, was examined at 150°C. The results are shown in Fig. 13 for Series I and II as well as in Fig. 14 for Series III. The solid lines in Fig. 13 represent $3\eta^+_{calc}$ values computed from the frequency relaxation spectrum by means of Equation (36), while triangles indicate the measured in steady state $3\eta^+$ values at $\dot\gamma = 10^{-2}$ (s^{-1}), i.e. the solid lines and the points represent the predicted and measured linear viscoelastic behavior respectively. The agreement is satisfactory. The broken lines in Fig. 13 represent the experimental values of the stress growth function in uniaxial extension, η_E^+. The distance between the solid and broken lines is a measure of nonlinearity of the system caused by strain hardening, SH.

    The SH plays a positive role in film blowing (15, 44) or in wire coating (66). LPX-30 does not show this effect and its processability index is low. On the other hand, SH of LDPE is quite high and so is the processability index. The accepted mechanism for improvement of film bubble stability is self healing of a potentially weak spot by a local increase of extensional viscosity caused by SH. It is important to note that adding to LPX-30 either 10 wt% LDPE, 20% LLDPE-10 or 10% LPX-24 was sufficient for generation of SH. The subsequent trial on a production line confirmed these results; increased bubble stability allowed for higher production rates of the blends.

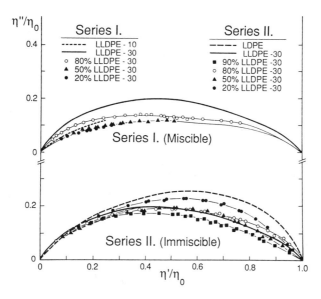

Figure 11. Cole-Cole plot for miscible (Series-I) and immiscible (Series-II) polyethylene blends.

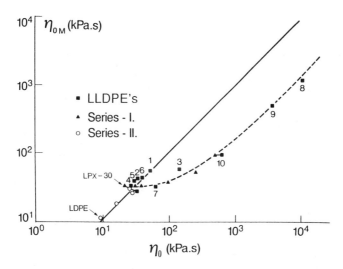

Figure 12. Maxwellian zero-shear viscosity computed from cross-point coordinates versus η_0 for Series-I and II as well as for neat LLDPE resins.

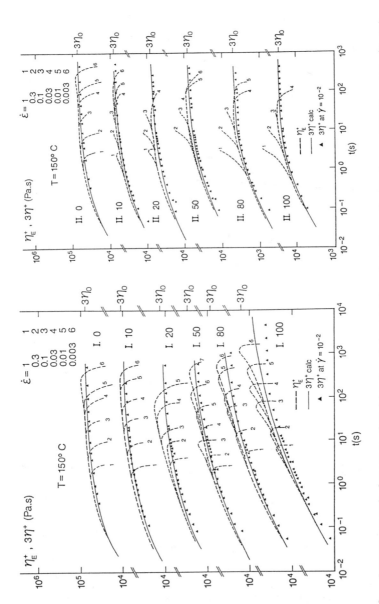

Figure 13. Stress growth function for blends of linear low density polyethylene, LLDPE, with (left) another type of LLDPE (miscible, Series-I) and (right) with low density polyethylene (immiscible, Series II). Broken lines: experimental data in elongation; triangles: experimental data in steady state shearing at $\dot\gamma = 0.01s^{-1}$. Solid lines were computed from the frequency relaxation spectrum. (Adapted from ref. 14.)

Figs. 13 and 14 also provide information on sample poly-dispersity. The initial slope of the stress growth function can be expressed as:

$$S = d\ln \eta_E^+/d\ln t\big|_{low\ t} \qquad (39)$$

For a series of LLDPE's the relation presented in Fig. 15 was found. An explanation for this empirical correlation can be found in Gleissle's mirror image principle (67):

$$\eta(\dot\gamma) = \eta^+(t); \qquad t \propto 1/\dot\gamma \qquad (40)$$

Accordingly, S is equivalent to:

$$S \simeq d\ln \eta/d\ln \dot\gamma\big|_{high\ \gamma} = n-1$$

where n is the power law index. Both (n-1) and S are related to polydispersity (68) Since for miscible blends the polydispersity can be calculated from Equations 15 to 18, the dependence in Fig. 15 provides a means for detecting miscibility in samples under uniaxial extensional flow. The data in Fig. 16 show the experimental S vs. w_2 dependence (points) as well as that predicted from variation of the polydispersity with composition (broken lines) for blends of Series I and II. The data in Fig. 16 confirm the previous results suggesting that while LLDPE/LLDPE Series I blends are miscible, those of LLDPE with LDPE, Series II blends, are not.

The S vs. w_2 dependence for Series III shown in Fig. 17 is particularly interesting. Apparently, upon addition of a small quantity of LPX-24, $w_2 \leqslant 20\%$, the polydispersity increases, as it should in a miscible system. However, for $w_2 > 20\%$ the poly-dispersity seems to decrease. Phase separation in the vicinity of 20 wt% of LPX-24 could provide a mechanism for such behavior.

Curve No. 1 in Fig. 1 shows the $\eta_E = \eta_E(\dot\varepsilon = 0.1\ s^{-1})$ dependence on w_2 for Series III blends. The points are experimental, the curve is a third order polynomial. Similarity of the shape of this dependence and that shown in Fig. 17 allows postulating that both are caused by the phase behavior. It is worth pointing out that the extrema on the η_E vs. w_2 plot represent compositions which may be of particular interest for processors: at the maximum the effect of LPX-24 on the stress hardening is most pronounced and the composition is most suitable for film blowing. At the minimum the η_E, and parallel with it the shear viscosity, is the lowest. The extrudability of LPX-24 with about 10 to 20 wt% LPX-30 is most efficient and stable. The usefulness of blends with compositions corresponding to extrema in the property-concentration dependence was recognized by Plochocki, who labeled these as the "rheologically particular compositions", or RPC (8). According to that author the shape of a rheological function vs. composition dependence is paralleled by a similar relation between the density and concentration. Judging by information in Fig. 17 it is the phase separation which provides the mechanism responsible for variation of density of Series III blends.

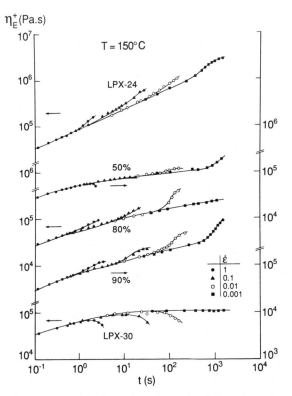

Figure 14. Stress growth function in uniaxial extension vs. time for LPX-30/LPX-24 blends (Series-III) at T=150°C and $\dot{\varepsilon}$ = 0.001, 0.01, 0.1 and 1 (s^{-1}).

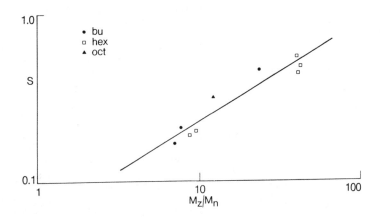

Figure 15. Initial slopes of the stress growth functions of LLDPE's (containing butene, hexene or octene comonomer) as function of resin polydispersity.

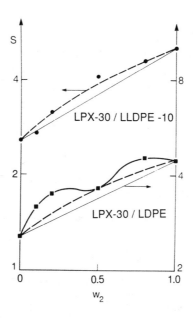

Figure 16. Compositional dependence of the initial slope of the elongational stress growth function for LPX-30/LLDPE 10 (top) and LPX-30/LDPE (bottom). The broken lines were computed assuming miscibility.

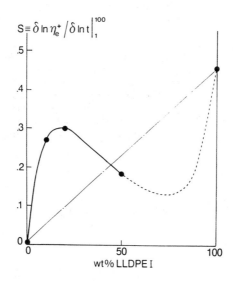

Figure 17. Initial slope of the stress growth function for LPX-30/LPX-24 blends vs. composition.

From the dependence: $\eta_E^+ = \eta_E^+ (\dot{\varepsilon},t)$ in Fig. 13 the values of the equilibrium extensional viscosity, η_E, were determined and plotted in Fig. 18. Operationally, η_E was taken as the maximum value of η_E at a given $\dot{\varepsilon}$. For specimens with maximum strain at break ε_b larger than the rheometer limit, $\varepsilon_L \simeq 3.2$, the arrow on a side of the data point was placed. The η_E vs. $\dot{\varepsilon}$ dependences for Series I are simple, resembling the classical shear flow curves, η vs. $\dot{\gamma}$. The few arrows indicate that for most of these blends $\varepsilon_b \leqslant 3.2$. By contrast, the plot for Series II is quite complex. Note that for several blends η_E as well as $\varepsilon_E > 3.2$ are larger than those for neat polymers. The complex shape of η_E vs. $\dot{\varepsilon}$ dependence for II.50 blend is particularly noticeable. Apparently the co-continuous blend morphology, expected at this concentration, resulted in self-reinforcing behavior, more sensitive to the rate of deformation than that observed for II.10 (LPX-30 with 10% LDPE) where the dispersed LDPE drops were transformed by uniaxial stress into reinforcing short fibers.

In Fig. 19 the steady state uniaxial elongational viscosity, $\eta_E/3$, is compared with the steady state shear, η, as well as dynamic, η', and complex, η^*, viscosities. It is evident that some strain hardening, evident in I.100 (LLDPE-10) is systematically diluted by the increasing amount of LPX-30. Thus, Series I behaves as a truly miscible system. By contrast, addition of LDPE (II.100) to LPX-30 (I.0) is generating more a complex variation of properties. The strain hardening, already visible at 10% of LDPE, reaches its maximum not at 100% LDPE but rather at 50:50 composition. Note that at $\dot{\varepsilon} \geqslant 0.1$ (s^{-1}) the maximum strain at break for II.50 is $\varepsilon_b > \varepsilon_L = 3.2$. The blends behave as immiscible.

The strain hardening, SH, causes the polymer to show a dual nature; at low strains it behaves as a linear viscoelastic liquid, whereas at high ones the non-linearity suggests a network-like behavior. For this reason the η_E vs. w_2 dependence may be misleading – it is not known where the linear behavior ends and SH begins. This is why in Fig. 20 only the linear viscoelastic part of η_E^+ is shown, the part computed from H_G by means of Equation 36. A significant difference in behavior between Series I and Series II is again visible at all deformation rates. For Series I the values of $3\eta_{calc}^+$ show simple additivity, whereas for Series II, there is a positive deviation from the log-additivity rule, PDB. While the PDB behavior per se is insufficient to call immiscibility, it is consistent with the expected behavior of emulsion-like immiscible polymer blends.

Conclusions

(i)   Blends of LLDPE with other LLDPE's or LDPE may show widely varying behavior, dependent on small changes in molecular structure engendered by e.g. different catalyst, polymerization method or composition. The LLDPE/LLDPE mixture may be miscible, as all rheological tests indicated for Series I, or partially miscible as for Series III blends. LLDPE with LDPE is immiscible (viz. Series II).

(ii) Blending two different types of LLDPE resulted in improved strain hardening at about 20 wt% of loading. However, it was more efficient to blend LLDPE with low molecular weight, MW,

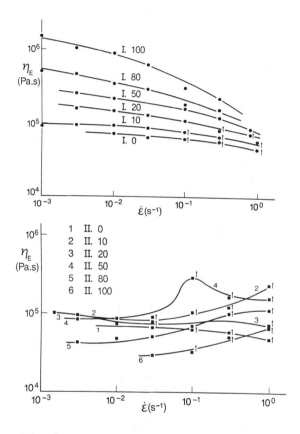

Figure 18.   The elongational viscosity in steady state flow at T = 150°C vs. strain rate for Series-I LPX-30/LLDPE-10 blends (top) and Series-II, LLDPE/LDPE blends (bottom).   The data points marked with arrow on a side indicate that equilibrium value has not been reached.

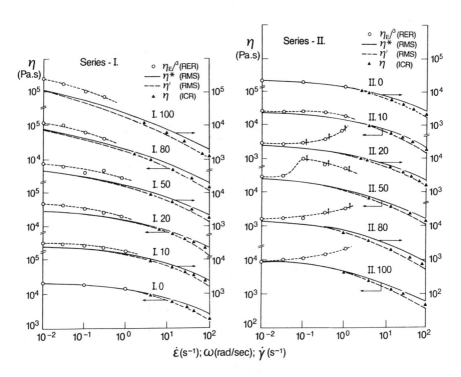

Figure 19. Comparison of rate of deformation dependence of the steady state (capillary) viscosity, η, (points) with the complex, η*, (solid line) dynamic, η', (broken line) and extensional viscosity, $\eta_E/3$ (open circles). The data are reduced to common temperature 150°C. Left, Series-I, LLDPE/LLDPE, right, Series-II LLDPE/LDPE blends. For clarity, the data of each blend is vertically displaced by one decade.

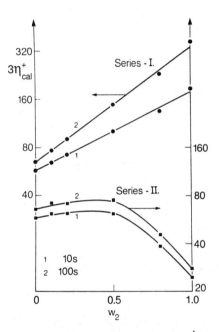

Figure 20.   Compositional dependence of $3\eta^+_{calc} \simeq \eta^+_{E,lin}$ at t = 10 and 100(s) for Series-I (top) and Series-II (bottom).

LDPE. High MW LLDPE and low MW LDPE have similar values of η_E which guarantees fine droplet dispersion. Addition of 10% LDPE was sufficient for engendering SH in LLDPE and increasing its maximum strain at break.

(iii) There is a direct correspondence between SH and processability on a film blowing line. The SH can be generated by long chain branching, blending, increase in MW or MW distribution.

(iv) The rheological studies of LLDPE and its blends led to development of several new rheological dependencies, viz. Equation 10 which allows calculation of the pressure correction in shear viscosity; Equation 24 expressing the rate dependence of the apparent yield stress; Equation 28 a new expression for melt-viscosity average molecular weight; Equation 29 allowing computation of the frequency relaxation spectrum from which all other linear viscoelastic functions can be derived; Equation 38 allowing estimation of η_0 from the cross-point coordinates; Equation 39 defining the initial slope of the stress growth function, which depends mainly on MW distribution.

PART III. LLDPE BLENDS WITH PP

Commercial PP and LLDPE resins were used. Their properties are listed in Table IV. For the first stage of the program PP/LLDPE = 50:50 blends were examined with or without 10 wt% of compatibilizing ethylene-propylene block copolymer EP-1 or EP-2 (39). Surprisingly, even at -40°C, the mechanical properties of the two-component blends were found to be either additive or synergistic. Addition of EP did not significantly affect them. For this reason, in the more extensive studies which followed only two component LLDPE/PP blends were investigated (40, 41, 69).

The mixtures were prepared in a Werner and Pfleiderer corotating twin-screw extruder, ZSK-30, at 350 RPM and T = 170°C in the feed zone increasing linearly to 240°C at the die. To prevent oxidation 0.1 wt% of Irganox 1010/1024 blend was added. The extruder screws were optimized for maximum mixing efficiency at low output rate, Q = 4 kg/hr. The granulated and dried extrudates were compression or transfer molded into specimens for rheological testing.

The rheological testing was carried out in RMS and ICR at 190°C and in RER at 150°C. The instruments and procedures were described in the preceeding part.

Preliminary tests

Both the theoretical and experimental studies indicated that the highest degree of dispersion can be obtained blending two liquids with similar viscosities at the same stress level as that expected during mixing:

$$\lambda_{opt} \equiv \eta_1/\eta_2 = 0.3 \text{ to } 1.0$$

where subscript 1 indicates the dispersed polymer and 2 indicates the matrix (11, 42, 43). For this reason PP-1 and LLDPE-A were chosen as the most suitable pair (39). The samples containing 0, 50 and 100

Table IV. Characteristic parameters of PP/PE neat resins

No.	Parameter	PP-1	PP-2	LLDPE-A	LLDPE-B	LLDPE-C	EP-1/EP-2[a]
1.	Manufacturer	Hercules	Hercules	Esso	Exxon	Dow Chem.	Hercules
2.	Designation	6704	6701	LLDPE-1	LPX-24	Dowlex 2517	--
3.	Density (kg/m^3)	904	900	918	951	917	901/897
4.	Melt index (g/10 min)	4.0	0.8	1.0	0.3	25	5/0.5
5.	Molecular weights: M_n	250	106	97.5	28.4	36	-
	(kg/mol) M_w	325	392	261	216	108	-
6.	Zero shear viscosity, η_0 190°C, (k Pas)	8.5	13	13	8120	1.32	5.2/6.3

Note: (a) Ethylene – propylene – ethylene copolymers with 11 and 14% of ethylene, respectively

wt% LLDPE-A were studied. The 50:50 blends containing 0 or 10% EP-1 or EP-2 were labeled: BL, BL-1 and BL-2, respectively.

The capillary flow. The interlayer slip, apparent in uncompatibilized blend, BL, disappeared upon addition of EP-1 or EP-2. As shown in Fig. 21 the slip affected the entrance-exit pressure drop, P_e, ("Bagley correction") more than viscosity.

The second interesting observation based on ICR and RMS results is related to the need for pressure correction in capillary flow. Already in Fig. 19 an agreement between the dynamic viscosity, η', and corrected for pressure effect capillary shear viscosity, η, was shown. In Fig. 22 five different measures of viscosity are shown for LLDPE-A (four for the other samples): steady state elongational viscosity, $\eta_E/3$, complex and dynamic viscosity, η^* and η', as well as the steady state capillary viscosity corrected and uncorrected for the pressure effects, η_{corr} and $\eta(ICR)$, respectively. There is a double equivalence of data points: $\eta_{corr} \simeq \eta'$ and $\eta(ICR) \simeq \eta^*$. The latter equivalence has a form of an apparent Cox-Mertz rule. Evidently the true steady state viscosity for LLDPE with the narrow molecular weight distribution requires the pressure correction. When the data are properly corrected the equivalence between the steady state shear viscosity and the loss-part of complex viscosity is found. On the other hand, neglecting the pressure corrections results in coincidental equivalence between $\eta(ICR)$ and η^*.

In agreement with the polydispersity data in Table IV for PP the pressure correction was significantly smaller than that determined for LLDPE-A. For the blends the correction was reduced even further, suggesting that molecular weight distribution in the phase adjacent to the capillary wall is broader than that of either polymeric component, indicating that LLDPE-A and PP may be partially miscible.

The Dynamic flow. The flow curves, η' vs. ω in Fig. 22, were not corrected for the apparent yield stress. For PP and LLDPE-A the curves nearly reached the Newtonian plateau and the Cole-Cole plots were found to be semi-circular indicating that $\sigma_y \simeq 0$. However, for blends the situation is less clear. Judging by the flow curves for BL, BL-1 and BL-2 at low deformation rates, the Newtonian plateau seems to be far away. This may indicate the incipient yield stress. To clarify this point η'' vs. η' was plotted in Fig. 23. An onset of the second relaxation mechanism is visible. The long relaxation times in BL may only originate in the interphase interactions. These usually lead to the presence of the apparent yield stress.

The high frequency cross point coordinates (G_x, ω_x) were used for calculation of the Maxwellian zero-shear viscosity, η_{0M}, from Equation 38. For the neat polymers η_{0M} was in agreement with η_0 calculated from the shear and extensional responses. For BL, BL-1 and especially BL-2 $\eta_{0M} < \eta_0$. The presence of an apparent yield stress is the most likely explanation for the discrepancy.

The extensional flow. The stress growth functions in shear (η^+, labelled RMS) and extension, η_E^+, are shown in Fig. 24. The dependencies for all five materials are similar. There is an excellent agreement between η_E^+ and $3\eta^+$. Note the absence of strain hardening, SH, clearly visible (see Figs. 13 and 14) in blends of

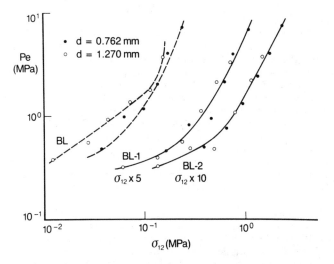

Figure 21.   Bagley pressure correction term, Pe, versus shear stress, σ_{12}, for BL, BL-1, and BL-2.   Data for BL-1 and BL-2 have been displaced horizontally by a factor of 5 and 10 respectively.

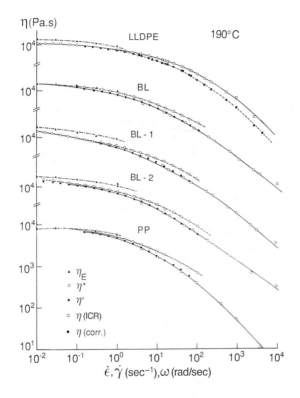

Figure 22.   Elongational, η_E, complex, η^*, dynamic, η', and shear, η, viscosities versus strain rate, $\dot{\varepsilon}$, frequency, ω, or shear rate, $\dot{\gamma}$, for (from the top): LLDPE, BL, BL-1, BL-2 and PP at 190°C.   The data corrected for pressure effects are shown as full squares.   For clarity the traces are displaced vertically each by one decade.

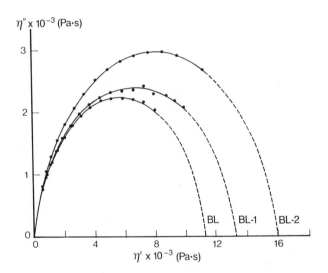

Figure 23.   Imaginary part, η'', versus real part, η', of the complex viscosity for BL, BL-1, and BL-2.

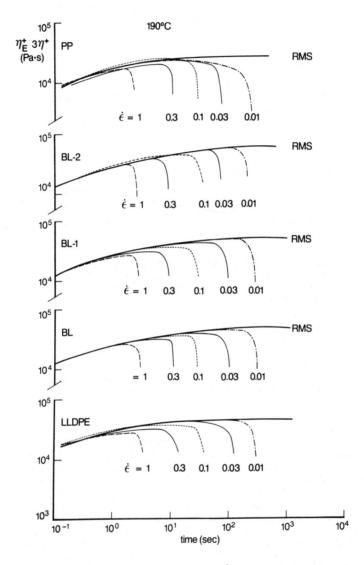

Figure 24. Comparison of shear, $3\eta^+$, and elongational, η_E^+ time dependent viscosities for (from top) PP, BL-2, BL-1, and LLDPE at 190°C.

LLDPE similar to LLDPE-A polymer, the LPX-30. For PP a strain thinning effect may be suspected. In Fig. 22 the steady state viscosities for PP and BL $\eta_E(\dot{\varepsilon}) \simeq 3\eta(\omega)$ were reported. For the other samples $\eta_E > 3\eta$ was observed, but the departure from equality was small.

The initial slope S of the relation shown in Fig. 24 was calculated according to Equation 39. For the two homopolymers S increased with M_z/M_n in accord with data presented in Fig. 15. For blends, S was above average indicating partial miscibility.

Summary. The following observations are worth noting: (i) The blends were prepared from relatively low molecular weight, nearly isoviscous polymers. They were immiscible, although the rheological data indicated broadening of the relaxation time spectrum, consistent with partial miscibility of the lowest molecular weight fractions. (ii) The partial miscibility was not sufficient to prevent the interlayer slip in capillary flow; for this purpose a compatibilizing block copolymer had to be added. (iii) The capillary flow data for polymers with narrow molecular weight distribution should be corrected for the pressure effects; after correction the equivalence $\eta \simeq \eta'$ was found. (iv) For the deformation rates ω, $\dot{\gamma} > 10^{-2}$ sec the apparent yield stress was unimportant. (v) The strain hardening in extension was absent for PP-1, LLDPE-A and their blends.

Detailed Studies of LLDPE/PP Blend Flow (40, 41, 69)

Polypropylene PP-2 was blended either with LLDPE-B (System-1) or LLDPE-C (System-2). The homopolymer characteristics are given in Table IV. In each system seven blends containing 0, 5, 25, 50, 75, 95 and 100 wt% PP was prepared.

Neat polymers and their blends were studied in dynamic shear field (using RMS) and in constant shear stress field using Rheometric Stress Rheometer, RSR. The molecular parameters of polymers and blends were determined by Size Exclusion Chromatography in trichlorobenzene at 140°C. The morphology of freeze-fractured specimens was characterized in Scanning Electron Microscope, SEM, Jeol JSM-35CF.

Molecular weights. The number, weight and average molecular weights, M_n, M_w and M_z, respectively are shown in Fig. 25. The broken lines represent the dependencies predicted by Equations 15 to 17. There is good agreement between the experiment and predictions for samples continuing less than 75% PP; for PP-rich samples the experimental data indicate degradation. It is worth pointing out that within the experimental error (indicated by a bar in Fig. 25) the molecular weights of neat PP before and after compounding/extrusion were the same. Thus PP degraded only in the presence of LLDPE. The extent of this effect was larger for higher molecular weight LLDPE-B than LLDPE-C. It primarily affected the highest molecular weight fractions, reducing M_w and M_z but leaving M_n virtually intact. The responsible mechanism seems to be shear degradation of PP catalyzed by LLDPE or rather by same its impurities.

Morphology. The SEM of freeze-fractured System-1 specimens indicates dispersed droplet morphology in blends containing 5, 75 and 95% PP

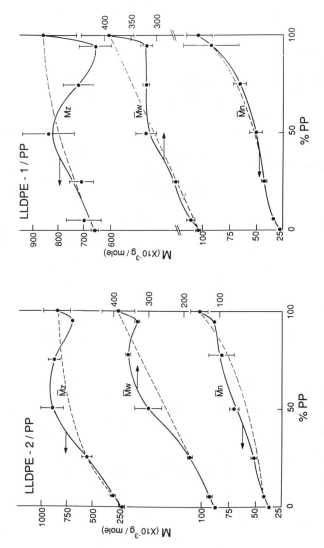

Figure 25. Compositional variation of average molecular weights for System 2 (left) and 1. Points – experimental, broken lines – calculated from Equations 15 to 17.

and co-continuous structure for the remaining blends containing 25 and 50% PP. In System-2 co-continuous morphology for 50% PP blends was observed.

The apparent yield stress. The complex viscosity η^* vs. ω for PP blends with LLDPE-B and LLDPE-C is shown in Fig. 26. The plot clearly indicates possible yield stress behavior especially for blends containing 50% PP. The apparent yield stress in dynamic flow data was calculated using Equation 23, with F = G' or F = G". The yield stress values as well as the assumed matrix material for calculating F_m are listed in Table V. For both systems the maximum value of the apparent yield stress occurred at 50% PP. In fact, there is a direct correlation — in a given system the yielding is primarily observed in blends having a co-continuous structure. As before (53) $G_y' > G_y''$

In the last column of Table V the directly measured shear stress values at yield are listed. These measurements were carried out in RSR, setting the desired level of stress and recording the strain after 200 s. The results are shown in Fig. 27. There is good agreement between G_y' and σ_y for System-1 but not for system-2. The two sets of experiments are carried out on a different time scale. Since, according to Equation 24, the apparent yield stress in blends depends on time one should not expect the same results from the dynamic and creep measurements. Furthermore, the differences in less viscous System-2 are expected to be larger than those for System-1.

The frequency dependence of data [corrected for the yield stress by means of Equation 25] was found to follow Equation 19. Having established values of the four parameters: η_o, τ_1, m_1 and m_2 the relaxation spectra were computed from Equation 29. These are shown in Fig. 28. For verification of the procedure $G'_{calc}(\omega)$ was computed from Equation 35. In Fig. 29 its values are compared with experimental G'(ω) data uncorrected for the yield stress. Furthermore, knowing η', the loss modulus, G''_{calc}, as well as the yield values for G' and G" it was possible to compute $\eta^* = \eta^*(\omega)$ dependence. These values are shown as lines intersecting the experimental data points in Fig. 26.

The concentration dependence of η_0 computed from Equation 20 is shown in Fig. 30, where the solid points represent the experimental data and the open points their values corrected for the effects of PP degradation. For System-1 there is strong negative deviation (NDB) from the log additivity rule, viz. Equation 1, but for System-2 NDB is visible at low PP content, converting to positive deviation (PDB) at high. It is worth recalling that η_0 was computed from corrected for the yield stress values of η'. The NDB behavior, indicative of interlayer slip, reflects poor miscibility in System-1 and that at low concentration of PP in System-2. The emulsion-like behavior of System-2 at high PP content reflects a better interphase interaction.

The Cole-Cole plot for System-2 is presented in Fig. 31. A "classical" semi-circular dependence was obtained for blends containing 0, 5, 25, 95 and 100% PP. For the remaining two blends

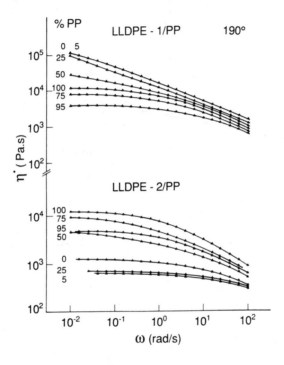

Figure 26. Complex viscosity vs. frequency, η^* vs. ω, for System 1 (top) and 2. Points are experimental, lines computed from the frequency relaxation spectrum.

Table V.   Apparent yield stress for PP blended with
LLDPE-B   or LLDPE-C

No.	Blend	PP (wt%)	Material	G'_y (Pa)	G''_y (Pa)	σ_y (Pa)
1.	System-1	5	PE	0	0	–
2.		25	PE	40	0	–
3.		50	PP	58	1.0	36
4.		75	PP	0.8	0	–
5.		95	PP	0	0	–
6.	System-2	5	PE	0	0	–
7.		25	PE	0.5	0.2	–
8.		50	PP	5	0	52
9.		75	PP	2	1.4	–
10.		95	PP	0	0	–

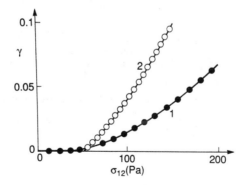

Figure 27.   Shear strain, γ, as a function of shear stress, σ_{12}, for blends containing 50% of each polymer.   Curve 1: PP/LLDPE-1, curve 2: PP/LLDPE-2

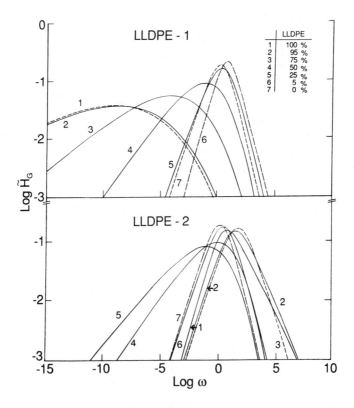

Figure 28.   Reduced frequency relaxation spectrum for LLDPE/PP
System 1 (top) and System 2.

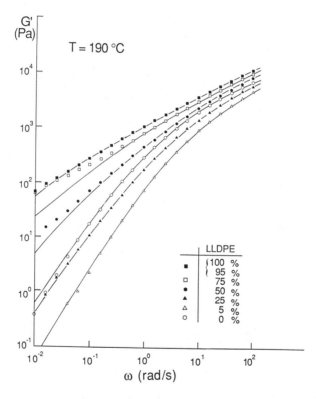

Figure 29.   Storage shear modulus vs. frequency for System 1.
The points are experimental, uncorrected for the apparent yield
stress, the lines computed from the relaxation spectrum.

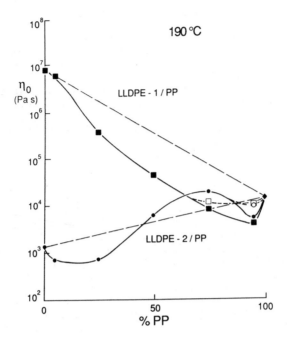

Figure 30.   Zero shear viscosity, η_0, vs. PP content in Systems 1 and 2; open points — data corrected for PP-degradation.

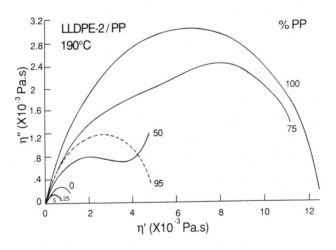

Figure 31.   Cole-Cole plot, η'' vs. η', for LLDPE/PP System 2 at 190°C.

with 50 to 75% PP there is an indication of a bimodal relaxation
spectrum. For this system the bimodality cannot be generated
applying a mixing rule to the homopolymer spectra. Clearly an
influence of the interphase is responsible for the apparent yield
stress. The dependence for System-1 shows a departure from
semi-circular behavior for blends containing 5, 25 and 50% PP but
none for those with 0, 75, 95 and 100% PP.
   If the appearance of bimodality of the relaxations in Cole-Cole
plot is equated with formation of co-continuous structure, then in
System-1 these occur at 5 to 40 wt% PP whereas in System-2 at 50 and
75 wt% PP. Paul and Barlow proposed the following dependence (<u>70</u>):

$$\phi_i / (1-\phi_i) = \eta_1/\eta_2 \tag{42}$$

where ϕ_i is the volume fraction of the dispersed phase inverting
into the continuous and η_1, η_2 are viscosities of the dispersed and
continuous phases. Assuming that the shear stress in the compounding
extruder was about 10 kPa, then Equation 42 predicts inversion at 8
to 79% of PP in System-1 and -2, respectively. The direct SEM
observations indicate dispersed structures at 5% but co-continuous
ones at 25 and 50% PP in System-1 as well as at 50 and 75% PP in
System-2. It is apparent that phase inversion, takes place via
co-continuous structure. Equation 42 seems to predict the inversion
concentration fairly well. However, the co-continuous structure is
observed not only near the predicted ϕ_i but also at 50:50 loading
where the volume fractions of the two polymers are nearly equal. The
broadening of the inversion region from a single ϕ_i to a wide range
of composition (including equi-volumetric) most likely is due to a
non-equilibrium nature of morphology.

The cross-point coordinates $G_x \equiv G' = G''$ at $\omega = \omega_x$ were
determined for all compositions then the Maxwellian viscosity η_{0M}
was calculated from Equation 38. The results are shown in Fig. 32 as
η_{0M} vs. η_0. In agreement with the previously discussed data for
LLDPE blends (see Fig. 12) initially $\eta_0 \simeq \eta_{0M}$; only for higher
values: $\eta_0 > 100$ kPas, there is a deviation from equality.

In summary, the PP/LLDPE blends are immiscible and the degree of
incompatibility increases with molecular weight of the homopolymers.
An increase of molecular weight also shifts the appearance of the
apparent yield stress to higher frequencies accessible in standard
test equipment. The yield stress seems to be associated with flow
and deformability of co-continuous blend structure. The
co-continuous morphology extended from the concentration predicted by
Paul and Barlow's empirical equation up to equi-volume polymer
content. The flow curves, corrected for the yield stress, could be
described by the four-parameter equation, which in turn lead to the
spectrum of relaxation times and derivation of other linear
viscoelastic functions in good agreement with directly measured
values.
   There are several other observations important for those
interested in PP/LLDPE rheology, but their generality is uncertain.

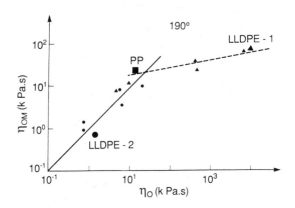

Figure 32. Maxwellian viscosity vs. zero-shear viscosity for LLDPE/PP Systems 1 and 2. Larger symbols indicate homopolymer values.

Table VI. Summary of rheological beavior of LLDPE-X blends with LLDPE-Y, LDPE and PP

No.	PROPERTY	LLDPE-Y	LDPE	PP
1.	Miscibility with LLDPE-X in the melt	depends on LLDPE-Y type	immiscible	partially miscible
2.	Need for pressure correction in capillary flow	yes	depends on polydispersity	
3.	η(capillary) = η'	yes	yes	yes
4.	Extrudate swell	depends on LLDPE-Y type	very large	large
5.	Interlayer slip	no	no	yes
6.	Presence of apparent yield stress, σ_y	no	yes	yes
7.	Frequency dependence	after correcting for σ_y it can be described by a four-parameter equation, then frequency relaxation spectrum and linear viscoelastic functions can be calculated		
8.	Concentration dependence of η_0	depending on type of LLDPE-Y: additive or sigmoidal	goes through a maximum	sigmoidal or goes through a minimum
9.	Cole-Cole plot	depends on LLDPE-Y type	bimodal	semicircular for low MW; then bimodal

(Continued on next page.)

Table VI. Summary of rheological beavior of LLDPE-X blends with LLDPE-Y, LDPE and PP (cont.)

No.	PROPERTY	LLDPE-Y	LDPE	PP
10.	Cross-point coordinates	$\eta_0 \simeq \eta_{0M}$ for η_0	100 kPas; above this limit η_0	η_{0M}
11.	Strain hardening in extensional flow	present	strong	absent
12.	Initial slope of η_E^+, S	the value depends on polydispersity		
13.	Maximum strain at break, ε_b	$\lesseqgtr 3.2$	> 3.2	$\lesseqgtr 3.2$ depending on $\dot\varepsilon$ and concentration
14.	The ratio $\eta_E(\dot\varepsilon)/3\eta(\omega)$ at $\dot\varepsilon = \omega$	> 1	$\gg 1$	$\simeq 1$
15.	Shear degradation	yes	no	yes of PP
16.	The activation energy of flow η (kJ/mole) at constant stress	30 ± 2	42 ± 2	45 ± 3

An example of these is the strain softening for PP or LLDPE-induced degradation of PP. The behavior is complex but self-evident.

PART IV. GENERAL CONCLUSIONS

The similarities and differences between LLDPE blends with LLDPE, LDPE and PP are summarized in Table VI. The LLDPE was found to be: immiscible with LDPE (even of low MW), miscible or not with another LLDPE resin (miscibility seems to be limited to the same polymerization products) and compatible with PP. Miscibility broadens MWD what affects the rheological behavior in a predictable way. On the other hand the immiscibility especially at lower loading leads to dispersed morphology; upon imposed strain this may turn into a co-continuous structure. In most blends the log-additivity rule is disobeyed; when the interfacial interactions are attractive a positive deviation from the rule is observed, when they are repulsive an interlayer slip leads to negative deviation. Immiscibility is also responsible for the apparent yield stress and usually large extrudate length reduction. There is little difference in extensional behavior between miscible and immiscible blends although the latter system show reduction of the maximum strain at break and a slower increase of the transient viscosity, η_E^+, with straining time (MWD effect). Working for years with these "simple" polymers one is continuously surprised how diverse their behavior can be.

NOTATION

$\underline{a}$	= coefficient in Equation 23
a_i, b_i	= coefficients in Equations 11 and 12
B	= extrudate swell
C_i	= polynomial coefficients in Equation 9
C_n	= n-paraffins
d	= capillary diameter
D	= extrudate diameter
EPR, EPDM	= binary and ternary copolymers of ethylene and propylene
$E_\sigma, E_{\dot\gamma}$	= activation energy of flow either at constant stress or constant rate of shear
F, F_a, F_y, F_m	= rheological function, its apparent, yield and in the matrix value; Equations 23 and 24
F_y^∞	= high frequency yield stress in Equation 24
G', G'', G^*	= storage, loss and complex shear modulus (Pa)

$G_x \equiv G' = G''$ = cross-point shear modulus (Pa)

HDPE   = high density PE

H_G, $\tilde{H}_G$ = frequency relaxation spectrum and H_G/η_0; Equation 29

$\tilde{H}_{G,max}$ = maximum value of $\tilde{H}_G$

i = $\sqrt{-1}$

Im   = imaginary part of a complex function

J_e = steady state compliance

L   = capillary length

LCB   = long chain branching

LDPE   = low density polyethylene

LLDPE   = linear LDPE

L_0, L_∞ = specimen length at $t = 0$ and $t = \infty$

m_i = parameters of Equation 20

M_n, M_w, M_z, M_η = number, weight, z- and η-average MW

MW   = molecular weight

MWD   = molecular weight distribution

n = power-law exponent

N_1 = first normal stress difference

NDB   = negative deviating blends

P   = total pressure loss in capillary flow (Pa)

PDB, PNDB   = positive, positive-negative deviating blends

P*   = pressure reducing factor in Equation 10

P_e = capillary end-effects pressure loss (Pa)

PE   = polyethylene

PP   = polypropylene

Q   = extruder output

r^2 = correlation coefficient squared

r_i	= function defined in Equation 29
Re	= real part of a complex function
R_T, $R_{T,0}$	= Trouton ratio at finite and zero rate of deformation, respectively
s	= variable in Equation 32
S	= initial slope defined in Equation 39
SCB	= short chain branching
SEC	= size exclusion chromatography
SH	= strain hardening
t	= time
T	= temperature (°C)
T*	= temperature reducing factor (K)
u	= exponent in Equation 24
ULDPE	= ultra low density PE
VLDPE	= very low density PE
w_i	= weight fraction of polymer
α, β	= parameters in Equation 21
β_1, β_{12}	= slip coefficient and its intrinsic value
γ	= shear strain in dynamic tests
$\dot{\gamma}$	= shear rate (1/s)
ε, ε_b	= Hencky strain, strain at break
$\dot{\varepsilon}$	= strain rate in extension (1/s)
η	= steady state shear viscosity (Pa·s)
η'	= dynamic shear viscosity
η_0, η_{0M}	= zero-shear viscosity and η_0 for Maxewellian liquid
η^+_{calc}	= compted from H stress growth function in shear

η_E , $\eta_{E,0}$ = uniaxial extensional viscosity at finite and zero deformation rate (Pa·s)

$\eta_{E,lin}^+$ = linear viscoelastic component of the stress growth function in uniaxial extension

η_E^+ , η^+ = stress growth function in uniaxial extension and in shear

η^* = complex viscosity

λ = viscosity ratio

θ_i = functions defined in Equation 29

ϕ_i = volume fraction

ρ = density (kg/m^3)

σ_b = extensional stress at break (Pa)

σ_{11} = extensional stress (Pa)

σ_{12} = shear stress (Pa)

τ_i = primary relaxation time (s)

τ_y = relaxation time of a network; Equation 24

ω = angular frequency (rad/s)

ω_x = cross-point frequency

ω_{max} = frequency for maximum of $\tilde{H}_G$

LITERATURE CITED

1. Anon., _Chem. Market Reporter_ 1987, pg 3, 23.
2. Ruof, M.; Fritz, H.-G.; Geiger, K. _Kunststoffe_, 1987, _77_, 480.
3. Romanini, D. _Polym.-Plast. Technol. Eng._ 1982, _19_, 201.
4. Nowlin, T.E. _Prog. Polym. Sci._ 1985, _11_, 29.
5. Keii, T.; Soga, K., Eds. _Catalytic Polymerization of Olefins_, Kodansha, Tokyo, 1986.
6. "Modern Plastics Encyclopaedia - 1988", McGraw-Hill, New York, 1987.
7. Plochocki, A.P. _Trans. Soc. Rheology_ 1976, _20_, 287.
8. Plochocki, A.P. In _Polymer Blends_; Paul, D.R; Newman, S., Eds.; Academic Press, New York, 1978.
9. Plochocki, A.P. _Polym. Eng. Sci._ 1982, _22_, 1153.
10. Utracki, L.A. _Adv. Plast. Technol._ 1985, _5_, 41: _J. Elastom. Plast._ 1986, _18_, 177.

11. Utracki, L.A.   <u>Polymer Alloys and Blends</u>; Hanser V., München, 1989.
12. Dobrescu, V.   In <u>Rheology</u>; Astarita, G.; Marrucci, G. and Nicolais, L., Plenum Press, New York, 1980.
13. Chuang, H.-K.; Han, C.D.   <u>J. Appl. Polym. Sci.</u> 1984, <u>29</u>, 2205.
14. Utracki, L.A.; Schlund, B.   <u>Polym. Eng. Sci.</u> 1987, <u>27</u>, 1512.
15. Schlund, B.; Utracki, L.A.   <u>Polym. Eng. Sci.</u> 1987, <u>27</u>, 1523.
16. Dumoulin, M.M.; Utracki, L.A.; Lara, J.   <u>Polym. Eng. Sci</u> 1984, <u>24</u>, 117.
17. Vadhar, P.; Kyu, T.   <u>Polym. Eng. Sci.</u> 1987, <u>27</u>, 202.
18. Stehling, F.C.; Wignall, G.D.   <u>A.C.S. Polym. Prepr.</u> 1983, <u>24</u>, 211.
19. Bates, F.S.; Wignall, G.D.   <u>Macromolecules</u> 1986, <u>19</u>, 932, 1938.
20. Ree, M.; Kyu, T.; Stein, R.S.   <u>J. Polym. Sci., B. Polym. Phys.</u> 1987, <u>25</u>, 105.
21. Kammer, H.W.; Socher, M.   <u>Acta Polym.</u> 1982, <u>33</u>, 658.
22. Curto, D.; La Mantia, F.P.; Acierno, D.   <u>Rheol. Acta.</u> 1983, <u>22</u>, 197.
23. La Mantia, F.P.; Curto, D.; Acierno, D.   <u>Acta. Polym.</u> 1984, <u>35</u>, 71.
24. Santamaria, A.; White,J.L.   <u>J. Appl. Polym. Sci.</u> 1986, <u>31</u>, 209.
25. Speed, C.S.   <u>Plast. Eng.</u> 1982, <u>38</u>, 39.
26. Nancekivell, J.   <u>Canad. Plast.</u> 1985, <u>43(1)</u>, 28; 1985, <u>43(9)</u>, 27.
27. Acierno, D.; Curto, D.; La Mantia, F.P.; Valenza, A.   <u>Polym. Eng. Sci.</u> 1986, <u>26</u>, 28.
28. La Mantia, F.P.; Valenza, A.; Acierno, D.   <u>Polym. Bull.</u>, 1986, <u>15</u>, 381.
29. Ghijsels, A.; Ente, J.J.S.M; Raadsen, J.   <u>Rolduc Meeting 2</u>, 1987.
30. Deanin, R.D; D'Isidoro, G.E.   <u>A.C.S. Org. Coat. Plast. Div. Preprints</u> 1980, <u>43</u>, 19.
31. Bartlett, D.W.; Barlow, J.W; Paul, D.R.   <u>J. Appl. Polym. Sci.</u> 1982, <u>27</u>, 2351.
32. Alle, N.; Lyngaae-Jørgensen, J.   <u>Rheol. Acta.</u> 1980, <u>19</u>, 94, 104.
33. Alle, N.; Andersen, F.E; Lyngaae-Jørgensen, J.   <u>Rheol. Acta.</u> 1981, <u>20</u>, 222.
34. Lin, C.-C.   <u>Polym. J.</u> 1979, <u>11</u>, 185.
35. Valenza, A.; La Mantia, F.P.; Acierno, D.   <u>Eur. Polym. J.</u> 1984, <u>20</u>, 727.
36. Kubicka, H; Woźniak, T.   <u>Prace Inst. Włókiennictwa (Łódź)</u> 1977, <u>27</u>, 57.
37. Plochocki, A.P.   <u>Adv. Polym. Technol.</u> 1983, <u>3</u>, 405.
38. Santamaria, A.; Munoz, M.E.; Pena, J.J.; Remiro, P.   <u>Angew. Makromol. Chem.</u> 1985, <u>134</u>, 63.
39. Dumoulin, M.M.; Farha, C.; Utracki, L.A.   <u>Polym. Eng. Sci.</u> 1984, <u>24</u>, 1319.
40. Dumoulin, M.M.   PhD thesis, Ecole Polytechnique, Montreal, Canada, 1988.
41. Dumoulin, M.M.; Utracki, L.A.   <u>CRG/MSD Symposium</u>, Hamilton Ont., June, 1984.
42. Utracki, L.A.   In <u>Current Topics in Polymer Science</u>; Ottenbrite, R.M.; Utracki, L.A.; Inoue, S., Eds., Hanser V., München, 1987.
43. Utracki, L.A.   In <u>Rheological Measurements</u>; Collyer, A.A.; Clegg, D.W., Eds.; Elsevier Appl. Sci. Publ., London, 1988.

44. Schlund, B; Utracki, L.A. Polym. Eng. Sci. 1987, 27 359, 380.
45. Utracki, L.A.; Schlund, B. Polym. Eng. Sci. 1987, 27 367.
46. Utracki, L.A.; Sammut, P. unpublished.
47. Utracki, L.A. Effect of Polydispersity on the Pressure Gradient of Shear Viscosity, 36th Canadian Chem. Eng. Conf., Sarnia, Ont., Canada Oct. 5-8, 1986.
48. Simha, R.; Somcynsky, T. Macromolecules 1969, 2, 342.
49. Tremblay, B.; Utracki, L.A. 4th Annual Polymer Processing Society meeting, Orlando FA, May 8-11, 1988.
50. Havriliak, S.; Nagami, S. Polymer 1967, 8, 161.
51. Utracki, L.A. Proceed. IUPAC 28th Macromol. Symp., Amherst MA, July 12-16, 1982.
52. Utracki, L.A. S.P.E. ANTEC Tech. Papers 1988, 34, 1192; Polym. Eng. Sci. 1988, 28, 1401.
53. Utracki, L.A.; Sammut, P. NRCC/IMRI Symposium "Polyblends-'88", Boucherville, Que. Canada, Apr. 5 and 6, 1988; Polym. Eng. Sci., 1988, 28, 1405.
54. Utracki, L.A. Proceeding CSChE 37th Conference, Montreal, Que. Canada, 18-22 May 1987.
55. Watanabe, H.; Sakamoto, T.; Kotaka, T. Macromolecules 1985, 18, 1008.
56. Watanabe, H.; Kotaka, T. Macromolecules 1987, 20, 530, 535.
57. Rubinstein, M.; Helfand, E.; Pearson, D.S. Macromolecules 1987, 20,822.
58. Ferry, J.D. Viscoelastic Properties of Polymers, third edition, J. Wiley & Sons, New York, 1980.
59. Marin, G. PhD Thesis, Université de Pau, France, 1977.
60. Diogo, A.C.; Marin, G.; Monge, Ph. J. Non-Newtonian Fluid Mech. 1987, 23, 435.
61. Ajji, A.; Choplin, L.; Prud'homme, R.E. NRCC/IMRI Symposium "Polyblends - '88", Boucherville, Que. Canada, Apr. 5 and 6, 1988: J. Polym. Sci., Polym. Phys. Ed., 1988, 26, 2279.
62. Zeichner, G.R.; Patel, P.D. 2nd World Congress Chem. Eng. Montreal, Canada, Oct. 1981; Proceed. 1981, 6, 333.
63. Zeichner, G.R.; Macosko, C.W. Proceed. IUPAC Macromolecular Symposium, Oxford, UK, 1982, P. 861; SPE ANTEC Tech. Papers 1982, 28, 79.
64. Rouse, P.E. Jr. J. Chem. Phys. 1953, 21, 1272.
65. Dumoulin, M.M.; Utracki, L.A.; Carreau, P.J. 36-th Canadian Chem. Eng. Conference, Sarnia, Ont. Canada, Oct. 1986.
66. Utracki, L.A.; Kamal, M.R.; Al-Bastaki, N.M. SPE ANTEC Tech. Papers 1984, 30, 417.
67. Gleissle, W. In Rheology, G. Astarita, Marruci, G.; Nicolais, L., Eds.; Plenum Press, New York, 2, 457, 1980.
68. Graessley, W.W. Adv. Polym. Sci. 1974, 16, 1.
69. Dumoulin, M.M; Utracki L.A. and Carreau, P.J. Rheol. Acta Suppl. 1988, 26, 215; In "Two Phase Polymeric Systems", L.A. Utracki, Ed., Hanser V., München, 1989.
70. Paul, D.R. and Barlow J.W. J. Macromol. Sci., Rev. 1980, C18, 109.

RECEIVED April 27, 1989

Chapter 8

Dynamic Melt Rheology of Polyethylene–Ionomer Blends

Gary R. Fairley and Robert E. Prud'homme

Centre de Recherche en Sciences et Ingénierie des Macromolécules, Chemistry Department, Laval University, Québec G1K 7P4, Canada

The addition of a copolymer has been shown to be a method of improving the mechanical properties of polyethylene/polyamide blends. One copolymer which has had particular success is poly(ethylene-co-methacrylic acid) (EMA) where the acid groups are partially neutralized by metal ions (EMA-salt). Dynamic melt rheology studies were carried out on PE/EMA and PE/EMA-salt in order to better understand the role of EMA-salt as a compatibilizer in the PE/EMA/PA system. The time-temperature superposition principle was applicable in all cases for G. Also, G' super master curves were constructed for blends of PE/EMA and PE/EMA-salt when the EMA and EMA-salt are derived from the same parent polymer. Superposition of G'' was possible for all blends containing EMA in the free acid form, but not for those in the salt form, with the extent of deviation from superposability being a function of EMA-salt concentration.

Poly(ethylene-co-methacrylic acid) (EMA), where the acid groups are partially on fully neutralized (EMA-salt) by metal ions, are interesting materials because of their unique properties as homopolymers (1-2) and their ability to compatibilize certain incompatible blends (3-14).

EMA-salts are light weight transparent materials possessing low temperature impact and flex toughness, good abrasion and solvent resistance (1-2).

EMA (5) and EMA-salts (15) have been used in binary blends in order to increase the impact strength and tensile strength of polyamides (PA). EMA and EMA-salts have also recently been used to improve the toughness of poly(ethylene terephthalate) while maintaining low permeability to hydrocarbons and other organic solvents (16).

0097–6156/89/0395–0211$06.00/0
© 1989 American Chemical Society

EMA and EMA-salts can also be used to improve the mechanical properties of incompatible blends (3-14). The role of the EMA or EMA-salt in these systems is not fully understood. In a recent publication, we have studied the polyethylene/EMA-salt/PA system by looking at the two binary systems from which it is made: PE/EMA-salt and EMA-salt/PA (17). Concerning the EMA-salt/PA system, it has been suggested that an amidation reaction can occur between the NH_2 terminal groups of the PA and the COOH groups of the EMA, in addition to possible hydrogen bonding (5). However, very little work has been carried out in order to detect interactions in the PE/EMA-salt system.

In our previous study (17), PE/EMA-salt blends were shown to be compatible (not in the thermodynamic sense) for a given mixing technique and parameters. It was also pointed out that there is separate crystallization of PE and EMA-salt in PE/EMA-salt blends, without any perturbation of the crystal structure or degree of crystallinity of the other component. Since the interactions leading to the compatibilization of the mechanical properties in PE/EMA-salt blends do not occur in the crystalline phase, they must be present in the amorphous phase.

Thus, it is the object of the present study to investigate the dynamic melt properties of PE/EMA and PE/EMA-salt blends in order to better understand the morphology of these systems, as well as to shed some light on the existence of ionic domains in PE/EMA and PE/EMA-salt blends.

EXPERIMENTAL

Materials

The LDPE's were obtained from Monsanto (LDPE-M8011) and Dow Chemical (LDPE-493c); the poly(ethylene-co-methacrylic acid))(Nucrel^{r}-1214) and ionomers (Surlyn^{r}-8660 and Surlyn^{r}-9950) were graciously provided by the Dupont Chemical Company.

Surlyn-9950 was refluxed in p-xylene at 130°C for approximately 1 h, after which the polymer gel was placed in a solution of 2:1 (v/v) 2N HCl/tetrahydrofuran (THF). The mixture was refluxed for 24 h with vigorous stirring in order to convert all the neutralized acid groups to their free acid form. This EMA-Zn-0 sample was then refluxed in a 1:2(v/v) solution of saturated zinc sulfate/THF with vigorous stirring for a period of 5 days to obtain an EMA-Zn salt containing more salt than the original EMA-Zn-20 sample. The percent ionization, which was verified by IR spectroscopy (18), was found to be 40%.

Some characteristics of the homopolymers and copolymers used in this study are shown in Table I: percent acid content, percent neutralization, number-average molecular weight (Mn), polydispersity index (Mw/Mn) and the nomenclature used.

All pure components and blends were prepared by mixing in a Mini Max molder, model CS-183 (19). The mixing of all samples was carried out at 150°C for a period of 10 min. The samples were subsequently injected into a cylindrical mold cavity with a diameter of 15.8 mm and a thickness of 3.0 mm.

All blends were prepared with the LDPE-493c sample, except those with EMA-Na and EMA which were prepared with the LDPE-M8011 sample.

Measurements

Dynamic mechanical properties were measured on a Rheometrics System-4 rheometer with parrallel plate geometry. The cylindrical samples were placed between the parrallel plates and melted. The gap between the plates was subsequently reduced to 1.2 mm and, after relaxation of the sample, the appropriate measurements carried out.

Dynamic mechanical properties of all pure components and blends were measured as a function of percent strain and indicated a linear viscoelastic region up to approximately 30-35 percent. Therefore, all rheological experiments were conducted at a strain rate of 20 percent. In cases where thermal degradation occurred (as seen in time sweep), the heating chamber was continuously purged with liquid nitrogen. Frequency sweeps, and in some cases frequency-temperature sweeps, were performed on all pure components and blends.

RESULTS

Homopolymers and Copolymers

Figure 1 shows the frequency dependence of the absolute value of the complex viscosity η^* for EMA-Zn-0 and EMA-Zn-40 as compared to EMA-Zn-20. It can be seen in this figure that the zero shear viscosity η_o is asymptotically approached for all the EMA-salt and EMA samples. This result is consistent with the observation of Earnest, Macknight et al. (18,20-21) who have shown that the zero-shear viscosity is asymptotically approached for EMA and its methyl ester; however, it is not for its 70% neutralized sodium salt, which is not inconsistent with the fact that the largest salt content of Fig. 1 is 40%.

Figure 2 shows the master curve of G' and the pseudo-master curve of G" for pure EMA-Zn-20, for a temperature range of 120 to 250°C, using a reference temperature of 150°C. As can be seen, there is good superposition of the G' data onto a single curve. Using the same shift factors a_t and reference temperature as used in G', superposition of G" was attempted. There is a clear breakdown of time-temperature superposition, particularly at high frequencies. This thermorheological complexity is characteristic of EMA-salts which exhibit microphase separation of ionic domains (18, 20-21).

For the same reference temperature and temperature range, thermorheological complexity was also displayed in the EMA-Na and EMA-Zn-40 samples.

In contrast, EMA shows good superposition of both G' and G" over a temperature range of 120°C to 250°C, using a reference temperature of 150°C and the same shift factors (Figure 3). Identical results were obtained for the superposability of G' and G" for EMA-Zn-0.

Table I : Characterization of the Polymers Used

POLYMER	CODE	% ACID	ION	% ION	Mn(kg/mol)	M_w/M_n
EMA-Zn-0	_____	15	—	—	16	2.9
EMA-Zn-20	Surlyn 9950	15	Zn	20	16	2.9
EMA-Zn-40	_____	15	Zn	40	16	2.9
EMA-Na	Surlyn 8660	9	Na	50	13	3.1
EMA	Nucrel 1214	12	—	—	18	3.9
LDPE	Dow 493c	0	—	—	139	6.0
LDPE	Monsanto 8011	0	—	—	—	—

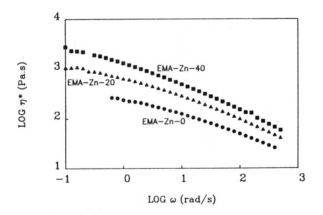

Figure 1. Complex viscosity η^* versus angular frequency ω for EMA-Zn-40, EMA-Zn-20 and EMA-Zn-0.

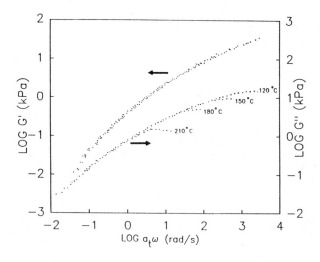

Figure 2.   G' master curve and G" pseudo-master curve for pure EMA-Zn-20.

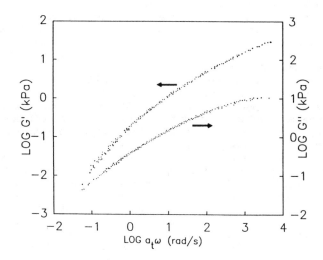

Figure 3.   G' and G" master curves for pure EMA.

Blends

The following five binary systems were studied: PE/EMA, PE/EMA-Na, PE/EMA-Zn-40, PE/EMA-Zn-20 and PE/EMA-Zn-0. These systems were investigated for compositions of 20, 40, 60 and 80 percent by weight PE over a temperature range of 120 to 250°C.

Master curves of G' were constructed for all compositions of the five different binary systems using a reference temperature of 150°C. A typical result of a G' master curve is given in Figure 4, which is a 40/60 blend of PE/EMA-Zn-20. As seen, satisfactory superposition is attained for G' in these blends. Using the same shift factors as for G', superposition was attempted for G". As can be seen, there is a definite breakdown of the time-temperature superposition principle at high frequencies (Fig. 4). Identical results were obtained for PE/EMA-Zn-40 and PE/EMA-Na blends. For these PE/EMA-salt systems, as the amount of ionomer present in the blend decreases, the frequency range over which superposition of G" is possible increases. As seen in a different perspective, the decrease in G" as a function of temperature at high frequencies is reduced upon the addition of PE to PE/EMA-salt binary systems.

Master curves can be constructed for both G' and G" over a temperature range of 120 to 250°C, using a reference temperature of 150°C, for both PE/EMA-Zn-0 and PE/EMA binary systems. A typical result is given in Figure 5, which shows master curves of G' and G" for a 60/40 PE/EMA-Zn-0 blend.

To summarize these results, we find superposability of both G' and G" for PE/EMA systems (EMA in the free acid form) and superposability of G' but not of G" for PE/EMA-salt systems. Furthermore, the degree of non-superposibility of G" in PE/EMA-salt systems increases with an increase in ionomer content. These results indicate that the non-superposibility of G" in PE/EMA-salt systems is due solely to the presence of ionic domains which occur as a result of the neutralization of the free acid groups by metal ions in the pure EMA copolymer.

However, the fact that superposition applies to G' and G" does not necessarily imply that the systems behave as single phase systems since it has been shown (22-23) that two-phase blends can act as thermo-rheologically simple materials for polymer blends consisting of polymers of high polydispersity, although we believe that this is a relatively rare case.

For polymers, it has been suggested that the Cole-Cole representation of the imaginary part (η") of the complex viscosity versus its real part (η') shows certain important differences for homogeneous systems as compared to heterogeneous systems (22-27). According to this proposal, for homogeneous systems, a unique circular arc is given which passes through the origin whereas for a heterogeneous system, a series of interpenetrating arcs is given.

Figure 6 shows Cole-Cole plots for pure EMA-Zn-20 at 150°C and 180°C. Unfortunately due to the high poly-dispersity of the pure polymers studied, the circular arc passing through the origin is too far from the real axis, not permitting the unequivocal determination of the existence or absence of more than one circular arc. This is a characteristic of all pure components studied herein as well as their binary blends.

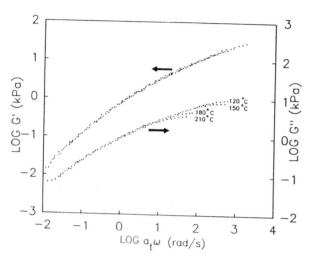

Figure 4.   G′ master curve and G″ pseudo-master curve for a 40/60 blend of PE/EMA-Zn-20.

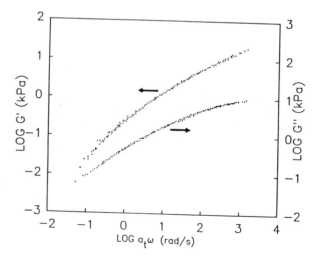

Figure 5.   G′ and G″ master curves for a 60/40 blend of PE/EMA-Zn-O.

Han (28-34) has found that the temperature independence of G' when plotted against G" to be a universal feature of all homogeneous visco-elastic fluids. This author has also postulated the general rule that, for a miscible blend, the G'-G" plot is invariant to changes in blend concentration, while for heterogeneous blends such invariance is not observed. However, Roland (35) has shown an example of a miscible blend exhibiting a variation in the G'-G" plot as a function of composition, which is at variance with the above-mentioned proposal.

In this study, G'-G" plots of the pure polymers show a temperature invariance for those polymers not containing metal ions. For example, Figure 7 shows G'-G" plots for pure EMA-Zn-20 at three different temperatures. As can be seen and as is the case for EMA-Na and EMA-Zn-40 samples, there is a deviation in the G'-G" plot at high frequencies as the temperature is increased.

Binary blends of PE/EMA-Zn-0 and PE/EMA show no temperature dependence of the G'-G" plots as a function of temperature. In the case of blends of PE/EMA-Zn-20, PE/EMA-Zn-40 and PE/EMA-Na, there is an uprise in the G'-G" plot at high frequencies. A typical result is given in Figure 8, which is for a 60/40 PE/EMA-Zn-20 blend at three different temperatures. In blends of PE/EMA-salt, the amount of deviation in G'-G" plots as a function of temperature is reduced as the concentration of PE in the blend is increased. The deviation of G'-G" as a function of temperature, in blends containing EMA-salt, can be simply related to the non-superbility of G" for pure EMA-salts.

Since plots of G'-G" are temperature dependent in some of the systems studied, their composition dependence was investigated at a constant temperature. As seen in Figure 9 for the PE/EMA-Zn-20 system, there is a strong composition dependence of G'-G" plots as is the case for all the systems studied. Furthermore, the composition dependence becomes more important as the second component in the blend becomes more different than pure PE (increase in percent acid content and increase in percent neutralization). It is well known (36-39) that in a two phase polymer system where there is a sharp interface and no interactions between the phases that interlayer slippage frequently occurs. This interlayer slippage gives rise to a reduction in blend viscosity or a negative deviation from additivity if the viscosity is plotted as a function of composition at a given temperature and frequency.

Figure 10 shows the absolute value of the complex viscosity for the PE/EMA-Zn-20 system at 150°C for 0.1, 1, 10, 100 and 500 Hz as a function of composition. These results indicate that the complex viscosity varies linearly as a function of composition. This is a typical result for PE/EMA and PE/EMA-salt blends, as the PE/EMA-Zn-0, PE/EMA, PE/EMA-Zn-40 and PE/EMA-Na show the same linear relationship between the complex viscosity and the composition at 150°C. The scatter of the data becomes more important at lower frequencies because the precision of these values is lower.

Figure 11 shows the absolute value of the complex viscosity as a function of composition for the PE/EMA-Zn-20 system at 210°C at 1, 10, 100, and 500 Hz. In this case, the system shows positive deviations from the additivity rule (straight line).

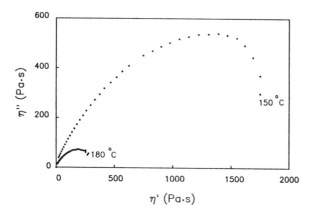

Figure 6.   A Cole-Cole representation of the imaginary part of the complex viscosity (η'') as a function of its real part (η') for pure EMA-Zn-20 at 150 and 180°C.

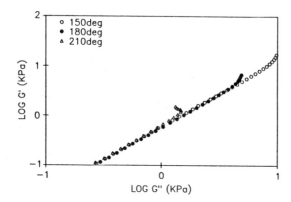

Figure 7.   G' vs. G" for pure EMA-Zn-20 at 150, 180 and 210°C.

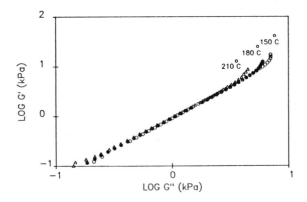

Figure 8. G' vs. G" for a 60/40 PE/EMA-Zn-20 sample at 150, 180 and 210°C.

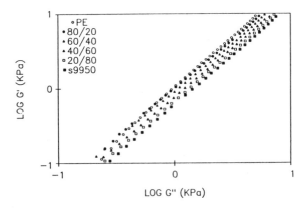

Figure 9. G' vs. G" for pure PE, pure EMA-Zn-20, and 80/20, 60/40, 40/60, 20/80 PE/EMA-Zn-20 blends.

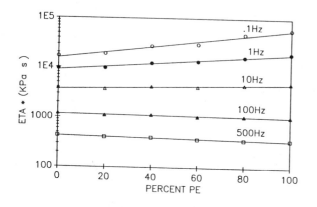

Figure 10. Complex viscosity ($\eta*$) versus composition for PE/EMA-Zn-20 at 150 ^{0}C and 0.1, 1, 10, 100 and 500 Hz.

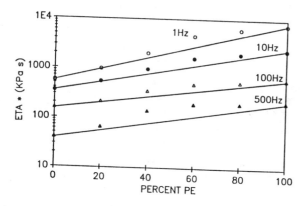

Figure 11. Complex viscosity ($\eta*$) versus composition for PE/EMA-Zn-20 at 210 ^{0}C and 1, 10, 100 and 500 Hz.

In general, for PE/EMA or PE/EMA-salt systems, a plot of complex viscosity as a function of composition follows the additivity rule or shows positive deviations from additivity, depending on the temperature and frequency.

Super-Master Curves

Earnest and MacKnight (18) have found that super-master curves, or composite-master curves, can be constructed for G' of EMA, EMA-salt and EMA-ester, all derived from the same parent polymer. In this work, we have studied three systems containing the same polyethylene, blended with three different components derived from the same parent polymer. These systems are PE/EMA-Zn-0, PE/EMA-Zn-20 and PE/EMA-Zn-40. Not only for the pure copolymers was it found that super-master curves could be constructed for G', but also for all blends containing an identical PE content. Figure 12 shows these super-master curves as a function of composition of the blends; as can be seen, there is quite a regular variation of G' versus frequency as a function of composition. The value of G', at a given low frequency, increases as the PE content is increased and the amount of increase diminishes with the frequency until we have convergence of all super-master curves at high frequencies. If we use EMA-Zn-0 or the corresponding PE/EMA-Zn-0 blend as a reference, we can define a frequency shift factor a'_T as the frequency needed to superpose pure EMA-Zn-20, or pure EMA-Zn-40, or the corresponding blend of PE/EMA-Zn-20 or PE/EMA-Zn-40, onto the corresponding reference curve. Figure 13 shows the composition dependence of the a'_T's for both PE/EMA-Zn-40 and the PE/EMA-Zn-20 system. As can be seen, there exists a linear relationship between the frequency shift factor a'_T and the composition of the blend.

These a'_T's can be translated into temperature shift factors ΔT's according to the method of Shohamy and Eisenberg (40). These ΔT's are given as a function of composition in Figure 14. As can be seen, there exists a linear relationship between the temperature shift factors and the composition in both the PE/EMA-Zn-40 and PE/EMA-Zn-20 systems.

DISCUSSION

The neutralization of the acid groups in EMA is known to influence their rheological and other physical properties (20). It has been established (18) that, for the percent ion contents of the EMA-salts used in this study, ionic clusters exist giving rise to ionic microdomains which remain intact well above the crystalline melting temperature. These ionic microdomains have an average radius of 8-10 nm and contain approximately 70 ions surrounded by a shell of hydrocarbon chains (41). These ionic microdomains act as thermoreversible cross-links; an increase in the percent neutralization results in an increase in the concentration of thermoreversible cross-links and a corresponding increase in the melt viscosity.

The superposability of G' and the non-superposibility of G" in pure EMA-salts has been reported by Earnest and MacKnight (18) for an EMA-salt above the so-called critical cluster concentration,

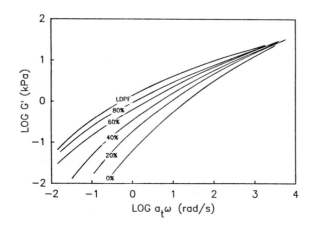

Figure 12.   G' super-master curves for pure PE, pure EMA-Zn as well as PE/EMA-Zn blends containing 20, 40, 60 and 80% PE.

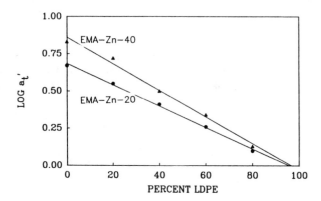

Figure 13.   Log a_T' as a function of blend composition for PE/EMA-Zn-40, PE/EMA-Zn-20 and PE/EMA-Zn-0 systems.

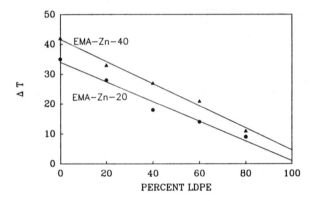

Figure 14. ΔT as a function of blend composition for PE/EMA-Zn-40, PE/EMA-Zn-20 and PE/EMA-Zn-O systems.

but for samples containing a much lower acid concentration than the ones used in the present study. Furthermore, pure EMA has been shown to be a thermorheologically simple material. In the case of EMA and EMA-salts, the fact that G' is superposable is due to the mechanism of energy storage: G' is a function of the elastic nature of the polymer chain and not the nature or form of crosslinks. The effect of the ionic domains, acting as quasi-crosslinks, is simply to increase the storage modulus without the introduction of additionnal energy storage mechanisms. In pure EMA, there are no ionic domains and subsequently no thermoreversible crosslinks. Thus, using the same shift factors as in G', G" can also be made to superpose.

For blends of LDPE with EMA-salts, G' is superposable but there is a clear breakdown of the time-temperature superposition principle at high frequencies for G". Furthermore, the frequency range over which G" is superposable decreases with increasing ionomer (EMA-salt) content. In other words, the inability of EMA-salt to increase G" at high frequencies, as a function of temperature, increases as the content of EMA-salt in the blend is increased. These results lead to the conclusion that above the crystalline melting temperature of the two components, the breakdown of the time-temperature superposition principle in G" is due solely to the presence of ionic domains in PE/EMA-salt blends.

The linear relationship (or positive deviations from additivity) between the complex viscosity and the composition of PE/EMA or PE/EMA-salt blends eliminates the case of a two phase system with no interactions exhibiting interlayer slippage.

The fact that G' super-master curves can be constructed for EMA and its corresponding salts is a general characteristic of these materials. In fact, Earnest and Macknight (18) have shown that a G' super-master curve can be constructed for EMA, Ema-salt and the corresponding ester derivative, derived from an EMA containing 4% acid. Since a super-master curve can be constructed for these three different materials, it can be concluded that the mechanisms for energy storage via G' depend only on the nature of the polymer main chain. Furthermore, the only differences between the EMA, EMA-salt and EMA-ester studied by Earnest and Macknight are in the nature and form of the crosslinks, the main chains being identical. Likewise in our case, EMA-Zn-40 and EMA-Zn-20 can be made to superpose with EMA-Zn-0 for G'. Furthermore, super-master curves can be constructed for blends of equivalent PE content of PE/EMA-Zn-40, PE/EMA-Zn-20 and PE/EMA-Zn-0. Using the PE/EMA-Zn-0 system as a reference, the frequency shift factors needed to construct master curves for PE/EMA-Zn-40 blends and PE/EMA-Zn-20 blends both show a linear variation as a function of PE content.

These frequency shift factors can be easily translated into temperature shift factors and show a linear relationship for both PE/EMA-Zn-40 and PE/EMA-Zn-20 as a function of PE content. Earnest and Macknight (18) have indicated that the ΔT's reflect a difference in the enthalpies for the two types of interactions involved (H-bonding versus thermoreversible crosslinking). These authors state that, as the percent ionization is increased in an EMA-salt, the ΔT for (EMA-salt-EMA-ester) should increase accordingly. Following this argument, but chosing EMA as the reference, the ΔT for (EMA-salt-EMA) is expected to increase as the

percent neutralization in the EMA-salt increases.   Earnest and
Macknight have argued that this is due to the increase in size and
perfection of the ionic domains.   However, it is difficult to
understand how the increase in ΔT upon neutralization could be
solely due to an increase in cluster size and pertection and not
also due to an increase in the number of clusters.   However, in our
case, we do see an increase in the ΔT as the percent neutralization
increases which we believe to be due to either an increase in size
or number of clusters or a combination of the two.   The linear
decrease in the ΔT for (PE/EMA-salt-PE/EMA) blends could be
attributed to a decrease in the number of clusters and not due to
an increase in their size or perfection.   It is believed that the
size of these ionic domains (at constant neutralization) does not
change significantly with temperature.   Thus it would be difficult
to imagine how the addition of PE, which has no specific
interactions with an EMA-salt, could pull out ion pairs from an
ionic domain which is much more viscous than the surrounding matrix
in a PE/EMA-salt-system.   If the addition of PE to EMA-salt has any
effect on the ionic clusters, it would probably be to increase the
hydrocarbon layer surrounding the ionic domains.

CONCLUSIONS

In this study, we have shown that if a 15% (molar) acid content
EMA-salt is neutralized to 20% or more, there is no superposition
of G".   With the EMA in the free acid form, superposition is
observed with both G' and G" at least within the temperature and
frequency range used herein.
    In blends of PE and EMA-salt, where the salt concentration is
20 percent of more, there is lack of superposability of G", and
superposability of G', over the temperature and frequency ranges
used.   Blends of PE with EMA show superposability of both G' and
G".
    In the case of EMA-salts and blends of EMA-salts with PE, the
results shown are in agreement with the presence of inhomogeneities
in the melt in the form of ionic microdomains.   That is the lack of
superposition of G" has been attributed to the presence of ionic
domains (18) and the deviation from superposition increases as the
amount of EMA-salt and, consequently, the number of ionic domains
is increased.   In samples containing no salt and hence no ionic
domains, there is superposition of G".
    However, a very important question remains unanswered: do
PE/EMA and PE/EMA-salt blends exhibit one single PE-EMA or PE-
EMA-salt phase or do they contain separate PE and EMA or EMA-salt
phases in the melt?   With respect to the melt rheology data, if
there are two separate phases in PE/EMA or PE/EMA-salt blends,
non-superposibility of both G' and G" would most likely occur.   As
previously mentioned, this is not the case, which suggests that in
the melt blends of PE and EMA or PE and EMA-salt form one phase
with ionic microphase separation occuring in blends containing EMA-
salt, when the salt concentration is above the critical salt
concentration.
    The complex viscosity as a function of composition curves
would not be inconsistent with the presence of one phase in the
melt.

The presence of one phase in the melt for PE/EMA or PE/EMA-salt blends is not inconsistent with the thermal analysis results presented previously (17). It has been determined that there is separate crystallization of PE and EMA or EMA-salt in PE/EMA of PE/EMA-salt blends. In fact, even for two polymers which form one phase in the melt, co-crystallization is quite difficult to induce. Kyu et al. (42) have shown that blends of linear and branched polyethylenes do not have the ability to cocrystallize and Prud'homme (43) has shown that it suffices to have a small difference in molecular weight between two poly(ethylene oxides) to have separate crystallization. Therefore, homogeneity in the melt does not hinder the separate crystallization of the two components of the mixture.

ACKNOWLEDGMENTS

The authors thank the National Sciences and Engineering Research Council of Canada and the Department of Education of the Province of Quebec (F.C.A.R. and Action structurante programs) for the grants (R.E.P.) and scholarship (G.R.F.) that supported this study.

REFERENCES

1. Purgett, M.D.; MacKnight, W.J.; Vogl, O. Polym. Eng. Sci., 1987, 27, 1461.
2. Rees, R.W.; Vaughn, D.J. Polymer Preprints, 1965, 6, 296.
3. Iwami, I.; Kawasaki, H.; Kodama, A. German Pat. No. 2,613, 968, Oct. 14, 1976. (C.A. 86: 6190j).
4. Japanese Pat. No. 58 108 251, Jun. 28, 1983 to Toray Ind. Inc. (C.A. 100: 86635d).
5. MacKnight, W.J.; Lenz, R.W.; Musto, P.V.; Somani, R.J. Polym. Eng. Sci., 1985, 25, 1124.
6. Netherland Pat. No. 6,504,219, Dec. 23, 1965 to Continental Can Co. Inc. (C.A. 64: 19916b).
7. Starkweather Jr., H.W.; Mutz, M.J. U.S. Pat. No. 4,078,014, March 07, 1978 (C.A. 88: 192150k).
8. Armstrong, R.G. U.S. Pat. No. 3,373,222, March 12, 1967 to Continental Can Co. Ltd. (C.A. 68: 87957t).
9. Murch, L.E. U.S. Pat. No. 3,845,163, Oct. 29, 1974 to duPont de Nemours E.I. and Co. (C.A. 82: 99239n).
10. Preto, R.J.; Scheckman, E.Z. U.S. Pat. No. 3,873,667, March 25, 1975, (C.A. 83: 60919c).
11. Schuster, R.A. British Pat. No. 2,035,938, June 25, 1980, (C.A. 93: 240862w).
12. Japanese Pat. No. 57,133,130, Aug. 17, 1982 to Mitsubishi Petrochemical Co. Ltd., (C.A. 98: 35666x).
13. Subramanian, P.M. U.S. Patent No. 4,410,482, Oct. 18, 1983.
14. Subramanian, P.M. U.S. Patent No. 4,444,817, April 24, 1984.
15. Deanin, R.D.; Jherwar, J. SPE RETEC, 1985, 11, 24.
16. Subramanian, P.M. Polym. Eng. Sci., 1987, 27, 1574.
17. Fairley, G.; Prud'homme, R.E. Polym. Eng. Sci. 1987, 27, 1495.
18. Earnest Jr., T.R.; MacKnight, W.J. J. Polym. Sci., Polym. Phys. Ed., 1978, 16, 143.
19. Maxwell, B. SPE Journal, 1972, 28, 24.

20. Sakamoto, K.; MacKnight, W.J.; Porter, R.S. J. Polym. Sci.,
     Part A-2, 1970, 8, 277.
21. Earnest Jr., T.R. Ph.D. Thesis, University of Massachusetts,
     May 1978.
22. Wisniewski, C.; Marin, G.; Monge, Ph. Eur. Polym. J., 1985, 21,
     479.
23. Marin, G.; Labaig, J.J.; Monge, Ph. Polymer 1975, 16, 223.
24. Vernag, V.; Michel, A. Rheol. Acta., 1985, 24, 627.
25. Cole, K.S.; Cole, R.H. J. Chem. Phys., 1941, 9, 341.
26. Ishida, Y.; Amano, O.; Takaganagi, M, Kolloid J., Band 172,
     1960, 126.
27. Ajji, A.;l Choplin, L.; Prud'homme, R.E. J. Polym. Sci., Polym.
     Phys. Ed., 1988, 26, 2279.
28. Han, C.D.; Chuang, H-K. J. Appl. Polym. Sci., 1985, 30, 4431.
29. Han, C.D.; Chuang, H-K. J. Appl. Polym. Sci., 1985, 30, 2431.
30. Han, C.D.; Kim, Y.W. J. Appl. Polym. Sci., 1975, 19, 2831.
31. Chuang, H-K.; Han, C.D. J. Appl. Polym. Sci., 1985, 30, 165.
32. Man, C.D.; Yu, T.C. J. Appl. Polym. Sci., 1971, 15, 1163.
33. Chuang, H-K.; Han, C.D. J. Appl. Polym. Sci., 1985, 30, 2457.
34. Chuang, H-K.; Han, C.D. J. Appl. Polym. Sci., 1984, 29, 2205.
35. Roland, C.M. J. Polym. Sci., Polym. Phys. Ed., 1988, 26, 839.
36. Willis, J.M.; Favis, B.D. private communication.
37. Garcia-Rejon, A.; Alvarez, C. Polymer Eng. Sci., 1987, 27, 640.
38. Lin, C.-C. Polymer Journal, 1979, 11, 185.
39. Utracki, L.A. in "Current Topics in Polymer Science", R.M.
     Ottenbrite, L.A. Utracki and S. Inoue, Eds.), 1987, II. Rheology
     and Polymer Processing/Multiphase Systems, Chap. 5.1, page 7.
40. Shohamy, E.; Eisenberg, A. J. Polym. Sci., Polym. Phys. Ed.,
     1976, 14, 1211.
41. MacKnight, W.J.; Taggart, W.P.; Stein, R.S. J. Polym. Sci., Part
     C., 1974, 45, 113.
42. Kyu, T.; Hu, S.R.; Stein, R.S. J. Polym. Sci., Polym. Phys.
     Ed., 1987, 25, 89.
43. Prud'homme, R.E. J. Polym. Sci. Polym. Phys. Ed., 1982. 20,
     307.

RECEIVED April 14, 1989

INTERPENETRATING   NETWORKS

Chapter 9

Ternary Phase Diagrams for Interpenetrating Polymer Networks Determined During Polymerization of Monomer II

L. H. Sperling, C. S. Heck, and J. H. An

Center for Polymer Science and Engineering, Department of Chemical Engineering, Materials Research Center, Whitaker Laboratory #5, Lehigh University, Bethlehem, PA 18015

Important aspects of polymer I/monomer II/polymer II ternary phase diagrams were determined for the system cross-polybutadiene-inter-cross-polystyrene as monomer II, styrene, is polymerized. Information on the mechanisms of phase separation suggest first nucleation and growth, followed by a modified spinodal decomposition. Studies on the same system by small-angle neutron scattering and light-scattering both yield negative diffusion coefficients, but different numerical values of both the diffusion coefficients and the domain sizes.

As a class, multicomponent polymer materials encompass polymer blends, grafts, blocks, AB-crosslinked copolymers, and interpenetrating polymer networks, IPN's. Each of these represents a distinct way of joining two or more polymers by a variety of methods, plain or fancy. Together, they constitute one of the fastest growing fields within polymer science, because with an increasing understanding of the interrelationships among synthesis, morphology, and mechanical behavior. Thus, we can fabricate materials which are tough impact-resistant plastics, reinforced elastomers, semi-permeable membranes, biomedical materials, sound and vibration dampers, or a host of other materials (1-4). Each of these materials is also a combination of two

Interpenetrating polymer networks are defined as a combination of two polymers, each in network form. From a practical point of view, an IPN is comprised of two polymers which cannot be separated chemically, do not dissolve or flow, and are not bonded together. Like most other multicomponent polymer materials, IPN's usually phase separate due to their very small entropy of mixing. However, the presence of the crosslinks tends to reduce the resulting domain size, hence yielding a unique method of controlling the final morphology.

DEVELOPMENT OF THE IPN CONCEPT

There are two general methods of synthesizing IPN's, see Figure 1. For sequential IPN's, polymer network I is synthesized, and monomer II plus crosslinker and activator is swelled in and polymerized. For simultaneous interpenetrating polymer networks, SIN's, both monomers and their respective crosslinkers and activators are mixed together and polymerized, usually by separate and non-interfering kinetic methods such as stepwise and chain polymerizations. Of course, there are many

0097–6156/89/0395–0230$06.00/0

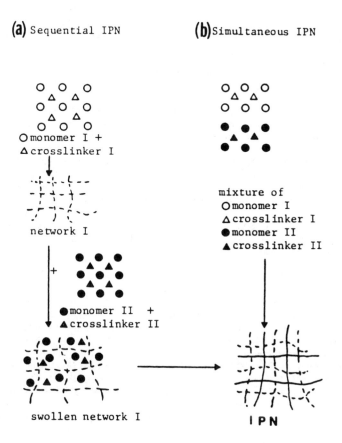

Figure 1.   Schematic of IPN synthetic methods.

intermediate methods. Combined with choices of bulk, emulsion, suspension, and solution polymerization, very many possibilities exist.

Since this paper will be restricted to sequential IPN's based on cross-polybutadiene-inter-cross-polystyrene, PB/PS, it is valuable to examine the range of possible compositions, see Figure 2 (5). The PB/PS IPN polymer pair models high-impact polystyrene, and in fact, many of the combinations made are actually more impact resistant than the commercial materials. In general, with the addition of crosslinks, especially in network I, the phase domains become smaller. The impact resistance of high-impact polystyrene, upper left, is about 80 J/m. In the same experiment, the semi-I IPN, middle left is about 160 J/m, and the full IPN, lower left, is about 265 J/m (6). Since the commercial material had perhaps dozens of man-years of development, and the IPN composition was made simply for doctoral research with substantially no optimization, it was obvious that these materials warranted further study.

Shortly thereafter, Yeo, et al. (7,8) showed that the domain size depended quantitatively on the crosslink density, the interfacial surface tension, and the temperature. For many compositions involving non-polar polymer pairs and moderate crosslinking levels, domain sizes of the order of 50-100 nm were predicted. While the theory was developed for spheres for simplicity, it was already known that both phases tended to be cocontinuous, especially for midrange compostions (5,6).

In TEM studies by Fernandez, et al. (9) on thin-sliced materials, it was shown that early in the polymerization of the styrene in PB/PS IPN's the domains tended to be spherical, while later in the polymerization the domains tended to be ellipsoidal in nature. The latter were modeled as irregularly shaped cylinders, which resemble ellipsoidal structures on thin sectioning. In more recent experiments involving small-angle neutron scattering, SANS, it was concluded that the phase separation involved a mixture of nucleation and growth, and spinodal decomposition kinetics (10).

THEORY

From a thermodynamic point of view, phase diagrams may be constructed by changing the temperature (11), pressure (12), or composition of a material. The present experiments are concerned with changes in composition at constant temperature and pressure, leading to a ternary phase diagram with polymer network I at one corner, monomer II at the second corner, and polymer network II at the third corner. According to classical concepts, at first there should be a mutual solution of monomer II in network I, followed by the binodal (nucleation and growth kinetics) and finally the spinodal (spinodal decomposition kinetics).

It must be emphasized that the two kinetic schemes are quite different, Figure 3 (10). Nucleation and growth kinetics occur frequently in the precipitation of salts from saturated solutions. This is the kinetic model usually taught in undergraduate classes. Spinodal decomposition, much less understood, was first studied by Cahn (13) and Cahn and Hilliard (14). It involves the formation of compositional waves, the amplitudes of which grow with time. While the wave length may be initially constant, frequently a coarsening effect is noted with time. Often, dual phase continuity in the form of cylinders within a matrix is observed. The more polymer physicists study the kinetics of phase separation, the greater the importance of spinodal decomposition in modern polymer science.

A powerful instrument for the study of polymers in the bulk state, including during polymerization, is small-angle neutron scattering, SANS. Studies involving SANS utilize differences in scattering length, rather than differences in refractive index, as for visible light. Usually, differences in scattering length are brought about by using deuterated molecules as a portion of the sample. For many irregularly

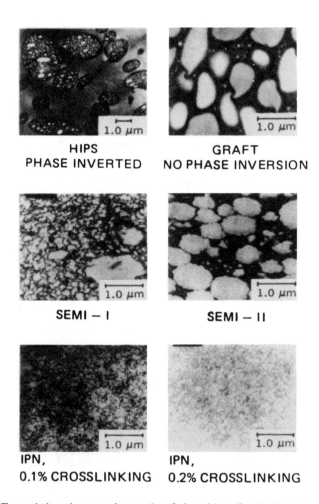

Figure 2. Transmission electron micrographs of six polybutadiene/polystyrene sequential IPN s and related materials, the polybutadiene portion stained with osmium tetroxide. Upper left: high-impact polystyrene, commercial. Upper right: a similar composition made quiescently. Middle left: semi-I IPN, PB (only) crosslinked. Middle right: semi-II IPN, PS (only) crosslinked. Lower left: full IPN, both polymers crosslinked. Lower right: full IPN, PB with higher crosslink level. (Reproduced from ref. 5. Copyright 1976 American Chemical Society.)

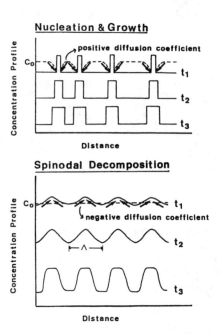

Figure 3.   Kinetics of phase separation in multicomponent polymer materials. (Reproduced from ref. 10. Copyright 1988 American Chemical Society.)

shaped domain structures, the theory of Debye et al. (15) provides an excellent approximation. Debye defined a correlation length, a, as the average distance from a random position and in a random direction in the sample to the first domain interface. From the correlation length, several other quantities can be calculated.

The correlation length, ℓ, for phases 1 and 2, is defined as the average distance across a random domain by a randomly placed line:

$$\ell_1 = a/(1-\phi_1) \tag{1a}$$

and

$$\ell_2 = a/(1-\phi_2) \tag{1b}$$

where the quantity ϕ represents the volume fraction of the phase in question.

The specific surface area, S_{sp}, is defined as the surface area per unit volume between domains of the two phases:

$$S_{sp} = 4\phi_1\phi_2/a \tag{2}$$

The wavelength, Λ, of a system, assuming spinodal decomposition kinetics (16), is equated to the maximum wave number, β_m,

$$\Lambda = 2\pi/\beta_m \tag{3}$$

Assuming that a shoulder or maximum is found in a scattering pattern, the wavelength is determined directly. Otherwise, algebraic techniques may be employed (10).

An important quantity that can be calculated is the diffusion coefficient, D. As illustrated in Figure 3, nucleation and growth results in a positive diffusion coefficient, because diffusion of the separating component is measured as movement from the original concentration to the depleted zone ahead of the growing new phase domain.   By contrast, spinodal decomposition diffusion is measured as the spontaneous movement from the original composition to a new, more concentrated phase.   In terms of growth rate of the phase domain amplitude, $R(\beta)$ and the diffusion mobility M,

$$R(\beta) = -D\beta^2 - 2M\gamma\beta^4 \tag{4}$$

and γ is gradient energy coefficient divided by the second derivative of the Gibbs free energy with respect to composition.   The scattered intensity (either light or neutrons) at an arbitrary angle over time yields the quantity R, and hence D.

SANS PATTERNS DURING A POLYMERIZATION

Figure 4 (10) shows the SANS scattering patterns at various stages of a PB/PS IPN polymerization.   At 3% polystyrene, the low scattering intensity and small angular dependence suggest that phase separation has not yet taken place. The next three data curves show a shoulder, which can be interpreted according to equation (3) as domains of the order of 60 nm.   Then the shoulder disappears, suggesting greater disorder.   The latter is bourn out by electron microscope studies, see Figure 5 (17). At low conversions, spherical domains are formed, followed by what appear to be ellipsoidal structures.   These can be modeled as truncated irregularly shaped cylinders.   At latter conversions, where the shoulders in the scattering patterns vanish, the structures appear more blurred.   It is thought that this blurring may

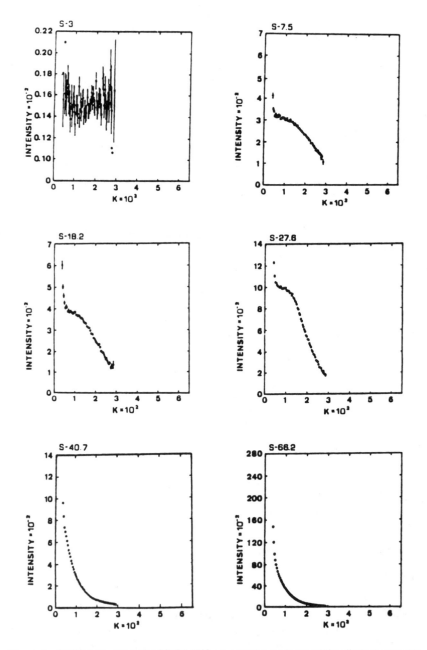

Figure 4. SANS intensities of a PB/PS IPN, on samples made by adding limited amounts of styrene monomer, and polymerizing to completion. Number following the S represents the weight-fraction of polystyrene. (Reproduced from ref. 10. Copyright 1988 American Chemical Society.)

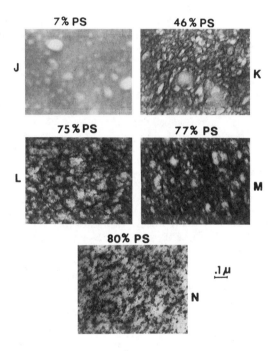

Figure 5.   Collage of PB/PS photomicrographs produced by evaporating the unreacted monomer before TEM.

arise from the increased internal viscosity of the material on high conversion, and subsequent decrease in a diffusional control of phase separation by any mechanism.

Following equation (1), the polystyrene transverse lengths were found to increase with weight fraction of polystyrene, Figure 6 (<u>18</u>). This suggests an increase in the average polystyrene domain diameter with increase in conversion. On the other hand, the specific surface area, Figure 7 (<u>18</u>), is seen to go through a broad maximum near midrange compositions. The apparent decrease in S_{sp} during the latter stages of conversion might be indicative of the blurred order observed via TEM. Thus, the two experiments, highly complementary, are in agreement.

LIGHT-SCATTERING STUDIES AS A FUNCTION OF TIME

A series of PB/PS IPN's were photopolymerized by ultraviolet light in a special glass sandwich cell. The sample could be removed periodically from the photopolymerization box, and the scattering intensity, I, recorded while still in the sandwich cell. Then the sample was returned to the photopolymerization box. Time of polymerization, t, was recorded as the time the sample spent in front of the UV light. The light-scattering instrument was a Brice-Phoenix 2000 photometer, using the mercury blue line, 435 nm. The angle used below, after correction for refraction, was approximately 20°.

For nucleation and growth, assuming that the polymerization of monomer II forms spheres of constantly increasing volume yields (<u>19</u>)

$$I = kt^2 \qquad (5)$$

The basis for equation (5) is that, for spheres small in comparison to the wavelength of the radiation, the scattering intensity increases as the square of the volume. A plot of scattered intensity vs. time squared should be linear if nucleation and growth kinetics are followed. As shown in Figure 8, three portions of the scattering pattern are distinguishable. At very short times, the scattering intensity is low, and increasing rapidly. This is taken as the onset of phase separation. At times longer than about $4x10^5$ minutes squared, the curve drifts upward, suggesting that nucleation and growth kinetics may not be strictly followed.

For spinodal decomposition, Lipatov, et al. (<u>19</u>) assumed that I depends on the square of the difference in refractive index of the original phase and the new phase, but that the phase dimensions are constant. Then,

$$\ell nI(\beta,t) = \ell nI(\beta,O) + 2R(\beta)t \qquad (6)$$

This suggests a plot of ℓnI vs. time. As shown in Figure 9, a straight line is obtained for times above 600 minutes, corresponding to the region of upward drift in Figure 8. However, the data are also compatible with spinodal decomposition kinetics throughout the polymerization, see dashed line.

Thus, the experiment appears indecisive. The most likely answer is that both kinetic mechanisms are active through the larger portion of the polymerization, and/or the theory illustrated above in equations (5) and (6) does not accurately portray the physical situation. For example, even though the data in Figure 8 and 9 represent conversions up to about 40% polystyrene, the transverse lengths increase as shown in Figure 6.

The data in Figures 8 and 9 can be converted to weight percent conversion making use of earlier experiments by Fernandez at al. (<u>20</u>). The beginnings of a phase diagram can be determined making use of all of the data now available, see Figure 10. The reaction follows the line of arrows. The dashed line is the binodal, where nucleation and growth begins in the range of approximately 3-6% polystyrene.

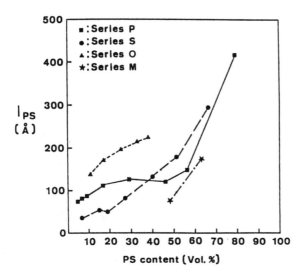

Figure 6. Increase in the polystyrene transverse lengths with polystyrene content. (Reproduced from ref. 18. Copyright 1988 American Chemical Society.)

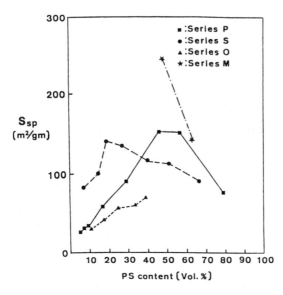

Figure 7. The specific interfacial surface area goes through a broad maximum as polystyrene content increases. (Reproduced from ref. 18. Copyright 1988 American Chemical Society.)

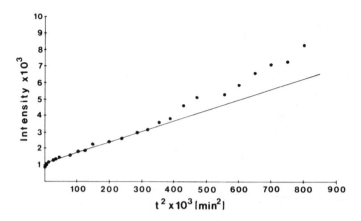

Figure 8.   Nucleation and growth kinetics for PB/PS IPN polymerization.

PB/PS IPN'S

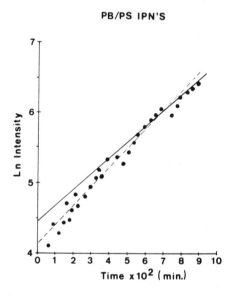

Figure 9.   Spinodal decomposition kinetics for PB/PS IPN polymerization. Same data as in Figure 8.   Dashed line is the best fit assuming spinodal decomposition throughout.

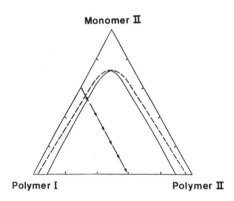

Figure 10.   Ternary phase diagram at constant temperature.

The solid line is on the spinodal, where spinodal decomposition begins at approximately 20-25% polystyrene. When the line joining Polymer I and Polymer II is reached, polymerization is completed, producing a sample of 80% polystyrene. While the phase diagram is only for one composition, perhaps it will be helpful in suggesting future experiments. Existing theory of phase separation for IPN's (21) predict domain wavelengths which are both larger and smaller than the typical distances between network crosslinks. In the latter case, they anticipate a coarsening of the structure until the domain size becomes comparable to the distances between crosslinks. This latter corresponds to the 60 nm domain size determined from the shoulder of the SANS patterns.

Lastly, it is interesting to look at the diffusion coefficients, see Table I. The values of D are all negative, supporting the idea that spinodal decompositon is important. However, the values found for light-scattering do not agree with those found by SANS, and are not in proportion to their respective wavelengths. The question arises, are the two experiments actually measuring different quantities? Other data in the literature, also shown in Table I, show similar differences.

Table I. Summary of Diffusion Coefficient and Wavelength Results
From Several Studies

Investigators	Method	Diffusion Coefficient $D(m^2/s)$	Wavelength (nm)	Ref.
Sperling, L. H., et al.	Light Scattering	-1.8×10^{-8}	1,600	(a)
An, J. H., et al.	SANS	-1.1×10^{-2}	54	(b)
Hill, R. G., et al.	SANS	-1.1×10^{-1}	30	(c)
Nojima, S., et al.	Light Scattering	-9.6×10^{-1}	--	(d)
Hashimoto, T., et al.	Light Scattering	-9.5×10^{-7}	1,970	(e)
Lipatov, Y. S., et al.	Light Scattering	-4.1×10^{-8}	1,500	(f)

(a) L. H. Sperling, C. S. Traubert, and J. H. An., Polym. Mat. Sci. Eng., 58, 889 (1988).
(b) J. H. An and L. H. Sperling, in "Cross-Linked Polymers: Chemistry, Properties, and Applications," R. A. Dickie, S. S. Labana, and R. S. Bauer, Eds., ACS Symp. Series No. 367, American Chemical Society, Washington, D.C., 1988.
(c) R. G. Hill, P. E. Tomlins, and J. S. Higgins, Polymer, 26, 1708 (1985).
(d) S. Nojima, Y. Ohyama, M. Yamaguchi, and T. Nose, Polym. J., 14, 907 (1982).
(e) T. Hashimoto,, J. Kumahi, and H. Kawai, Macromolecules, 16, 641 (1983).
(f) Y. S. Lipatov, O. P. Grigor'yeva, G. P. Kovernik, V. V. Shilov, and L. M. Sergeyeva, Makromol. Chem., 186, 1401 (1985).

THE PAST, PRESENT, AND FUTURE OF IPN RESEARCH

The first modern scientific paper on IPN's was written by Millar in 1960 (22). In 1979, there were a total of about 125 papers and about 75 patents in the field of IPN's, and three products. By 1985, there were several hundred papers and patents, and fifteen products. Today, the production of papers and patents is well over 100 per year. Also, the fraction of papers and patents that utilize IPN notation is increasing, showing that more scientists and engineers are cognizant of the growing body of literature surrounding the new field. (Many of the earlier papers spoke of graft copolymers that were crosslinked, etc.)

Interpenetrating polymer networks are important because their crosslinks offer a novel method of controlling domain size and shape; many mechanical properties such as impact strength depend on the size of the rubber domain. Thus, small, nearly uniform domains can be generated.

Dual phase continuity offers many advantages, because rubber/plastic compositions yield tough, leathery materials. Many of the compositions described above, for example, contain two continuous phases, with cylinders of polystyrene meandering within the polybutadiene matrix. Since all IPN's are crosslinked, it may be that their greatest advantage will lie in products which are leathery or rubbery, but can not be permitted to flow.

ACKNOWLEDGMENTS

The authors wish to thank the National Science Foundation for support through Grant No. DMR-8405053, Polymers Program. The SANS experiments were performed at NCSASR, funded by NSF Grant No. 7724458 through interagency agreement No. 40-367-77 with DOE under contract DE-AC05-84R-21400 with Martin Marietta Energy System, Inc. The authors also wish to thank G. D. Wignall for his time and excellent suggestions thoughout this project.

LITERATURE CITED

1. Sperling, L. H. Interpenetrating Polymer Networks and Related Materials; Plenum: New York, 1981.
2. Xiao, H. X.; Frisch, K. C.; Al-Khatib, S. In Cross-Linked Polymers: Chemistry, Properties, and Applications; Dickie, R. A.; Labana, S. S.; Bauer, R. S., Eds.; ACS Symposium Series No. 367; American Chemical Society: Washington, D.C., 1988.
3. Lipatov, Yu. S.; Karabanova, L.; Sergeeva, L.; Gorbach, L.; Skiba, S. Vysokomol. Soedin, B, Krat. Soobshch 1986, 29, 274.
4. Hourston, D. J.; Satgurunthan, R.; Varma, H. C. J. Appl. Polym. Sci. 1987, 33, 215.
5. Donatelli, A. A.; Sperling, L. H.; Thomas, D. A. Macromolecules 1976, 9, 671.
6. Donatelli, A. A.; Sperling, L. H.; Thomas, D. A. Macromolecules 1976, 9, 676.
7. Yeo, J. K.; Sperling, L. H.; Thomas, D. A. Polymer 1983, 24, 307.
8. Yeo, J. K.; Sperling, L. H.; Thomas, D. A. Polym. Eng. Sci. 1982, 22, 190.
9. Fernandez, A. M.; Wignall, G. D.; Sperling, L. H. In Multicomponent Polymer Materials; Paul, D. R.; Sperling, L. H., Eds.; ACS Adv. in Chem. No. 211; American Chemical Society: Washington, D.C., Ch. 10.
10. An, J. H.; Sperling, L. H. In Cross-Linked Polymers: Chemistry, Properties, and Applications; Dickie, R. A.; Labana, S. S.; Bauer, R. S., Eds.; ACS Symposium Series No. 367; American Chemical Society: Washington, D.C., 1988; Ch. 19.
11. Bauer, B. J.; Briber, R. M.; Han, C. C. Polym. Prepr. 1987, 28(2), 169.
12. Lee, D. S.; Kim, S. C. Macromolecules 1984, 17, 268.
13. Cahn, J. W. J. Chem. Phys. 1965, 42, 93.
14. Cahn, J. W.; Hilliard, J. E. J. Chem. Phys. 1958, 28, 258.
15. Debye, P.; Anderson, H. R.; Brumberger, H. J. Appl. Phys. 1957, 28, 679.
16. Olabisi, O.; Robeson, L. M.; Shaw, M. T. Polymer-Polymer Miscibility; Academic Press: New York, 1979.
17. Fernandez, A. M. Ph.D. Thesis, Lehigh University, Pennsylvania, 1984.

18. An, J. H.; Sperling, L. H. In Cross-Linked Polymers: Chemistry, Properties and Applications; Dickie, R. A.; Labana, S. S.; Bauer, R. S., Eds.; ACS Symposium Series No. 367; American Chemical Society: Washington, D.C., 1988.

19. Lipatov, Y. S.; Grigor'yeva, O. P.; Kovernik, G. P.; Shilov, V. V.; Sergeyeva, L. M. Makromol. Chem. 1985, 186, 1401.

20. Fernandez, A. M.; Widmaier, J. M.; Sperling, L. H. Polymer 1984, 25, 1718.

21. Binder, K.; Frisch, H. L. J. Chem. Phys. 1984, 81, 2126.

22. Millar, J. R. J. Chem. Soc. 1960, 1311.

RECEIVED April 27, 1989

Chapter 10

Synthetic Sequence Effects on Cross-Linked Polymer Mixtures

R. B. Fox, D. J. Moonay, J. P. Armistead, and C. M. Roland

Polymeric Materials Branch, Chemistry Division, Naval Research Laboratory, Washington, DC 20375

Essentially immiscible mixtures of a polyol-based polyurethane (PU) and poly(n-butyl methacrylate) (PBMA) having both, one, or neither component crosslinked were prepared by photopolymerization of the PBMA precursors in the presence of slowly reacting PU precursors. Within the 30-70% PU range, a morphology of PBMA particles in a continuous PU matrix was found. Particle size decreased as the delay time between the PU and the PBMA initiation increased. Crosslinking the PBMA resulted in irregularly-shaped particles, while spherical particles formed in the absence of crosslinker. Initial moduli and dynamic mechanical properties varied continuously with increasing delay time, with a minimum midway during the PU formation. The effect is more pronounced in the series with irregular particles. Loss tangent peak separation decreased in all of the mixture types as particle size decreased, with a greater effect when both components were crosslinked. Creep and extraction experiments in the PU(x)/PBMA(1) system indicate, respectively, that apparent crosslink density decreases, and grafting, while extensive, decreases with increased delay time. Precursor diluent effects are suggested as the origin of many of the property and structure changes resulting from altering the sequence of component formation.

In recent years, much attention has been paid to multicomponent polymer mixtures in which one or both components are crosslinked and in which the potential exists for mutual entanglement or interpenetration of the chains of the components. A few miscible pairs have been reported, but immiscibility is by far the most common case. Combinations of polyurethanes and acrylics or methacrylics have long been attractive because the components can, in principle, be formed by independent and non-interfering polymerization reactions.

There are at least four general types of combinations of
crosslinked (x) and linear (l) polymers in a two-component system:
both components crosslinked (xx), one or the other component
crosslinked (lx or xl), and both components linear (ll). Where at
least one of the components has been polymerized in the presence of
the other, the xx forms have often been called interpenetrating
polymer networks (IPN), the lx and the xl forms termed "semi-IPNs",
and the last, linear or in situ blends. There are also a number of
ways in which the components can be formed and assembled into a
multicomponent system. Sequential IPNs are prepared by swelling one
network polymer with the precursors of the second and polymerizing.
Simultaneous IPNs are formed from a mixture of the precursors of both
components; polymerization to form each component by independent
reactions is carried out in the presence of the other precursors or
products. Usually, the simultaneous IPNs that have been reported are
extremes in the component formation sequence: the first component is
formed before the second polymerization is begun. Sequential IPNs and
simultaneous IPNs of the same composition do not necessarily have the
same morphology and properties.

In a simultaneous system, the sequence and rates of formation
and phase separation of the components in each other's presence can
strongly affect the morphology and properties of the resulting
materials. A few attempts have been made to coordinate the reaction
rates and thereby achieve true simultaneity. For example, in an
epoxy/acrylic system, the epoxy network can be formed thermally and
the acrylic network photolytically; the proper combination of heat
and light is capable of providing simultaneous reactions (1). In a
similar system (2-3), kinetic control was achieved through set
prereaction times for the epoxy, followed by the addition of the
acrylic precursors; the acrylic initiator and crosslinker
concentrations were also controlled. Greatest molecular mixing and
smallest domain size in these two-phased systems resulted when the
rates of the polymerization reactions were closest to simultaneous;
these conditions did not, however, provide the best mechanical
properties in the products.

A polyurethane (x)/poly(methyl methacrylate) (l) system in which
the methyl methacrylate was polymerized at various intervals after the
gelation of the polyurethane was studied by Allen and coworkers (4).
They found that for polymerizations at 50°C, increasing the post-
gelation time prior to acrylic formation decreased the size of the
spherical poly(methyl methacrylate) particles that formed in the
polyurethane matrix but had little effect on the shear moduli.

Other synthetic approaches to the kinetic problem have been
taken. Variations in catalyst concentration for the formation of each
component network from linear polyurethanes and acrylic copolymers
have been used along with a rough measure of gelation time (5) to
confirm the earlier (2-3) results. Kim and coworkers have investigated
IPNs formed from a polyurethane and poly(methyl methacrylate) (6) or
polystyrene (7) by simultaneous thermal polymerization under varied
pressure; increasing pressure resulted in greater interpenetration and
changes in phase continuity. In a polyurethane-polystyrene system in
which the polyurethane was thermally polymerized followed by
photopolymerization of the polystyrene at temperatures from 0° to
40°C, it was found (8) that as the temperature decreased, the phase-

separated polystyrene particles decreased in size from about 0.2 to 0.03 μm. Based on the shape and the inward shift of the peaks of the tan δ curves of materials made at 0°, two families were observed: the xx and xl had a greater inward T_g shift than the lx and ll combinations.

Kinetics and mechanism of component formation can also be affected. In a polyurethane-polyester IPN system, it has been shown (9, 10) that the polyurethane polymerization in the presence of polyester resembles that in solution before gelation and a bulk polymerization after gelation; it was also found that the polyester reaction mechanism was affected by the presence of the polyurethane phase. Similarly, a study (11) of the formation of poly(methyl methacrylate) (x) in the presence of a polyurethane (x) showed that the latter acted as a diluent to maintain the T_g of the mixture below that of the former, and that it produced a high initiation rate and early gelation time for the acrylic polymer.

A polyurethane (PU)/poly(n-butyl methacrylate) (PBMA) system has been selected for an investigation of the process of phase separation in immiscible polymer mixtures. Within this system, studies are made of the xx, lx, xl, and the ll forms. In recognition of the incompatibility of PBMA with even the oligomeric soft segment precursor of the PU, no attempt was made to equalize the rates of formation of the component linear and network polymers. Rather, a slow PU formation process is conducted at room temperature in the presence of the PBMA precursors. At suitable times, a relatively rapid photopolymerization of the PBMA precursors is carried out in the medium of the slowly polymerizing PU. The expected result is a series of polymer mixtures essentially identical in component composition and differing experimentally only in the time between the onset of PU formation and the photoinitiation of the acrylic. This report focuses on the dynamic mechanical properties of these materials and the morphologies seen by electron microscopy.

EXPERIMENTAL

Materials. The polyurethane precursor materials were Adiprene L-100 (Uniroyal, Inc.), a poly(oxytetramethylene glycol) capped with toluene diisocyanate, eq. mol. wt. 1030; 1,4-butanediol (BD) and 1,1,1-trimethylolpropane (TMP); and, as catalyst, dibutyltin dilaurate (DBTDL). Acrylic precursors included n-butyl methacrylate (BMA), washed with 10% aq. NaOH to remove inhibitor; tetramethylene glycol dimethacrylate (TMGDM) crosslinker; and benzoin sec-butyl ether (BBE) as a photosensitizer. These materials were dried appropriately but not otherwise purified.

Sample Preparation. For each of the xx, lx, and xl mixtures, three PU:PBMA compositions, 3:7, 5:5, and 7:3, were made; with the ll mixture, only the 5:5 composition was prepared. An equivalent weight ratio of 0.95 for OH:NCO was maintained for the PU components; a 3.5:1 (by weight) BD:TMP mixture was used for the PU(x) samples. Samples containing PBMA(x) were made with a 60:1 (mole ratio) BMA:TMGDM mixture. These ratios yield a calculated M_c of 1700 and 4200 g/mol for the PU(x) and PBMA(x), respectively. In each mixture, 0.05 wt.%

DBTDL, based on the total sample weight, and 0.1 % BBE, based on the weight of the BMA, were used as catalysts.

For a given composition, a mixture of appropriate weights of all of the precursors except DBTDL was stirred until it was homogeneous. The DBTDL was added and the time, t_o, noted. Following degassing, the mixture was transferred to a series of cells composed of thin glass plates faced by Teflon shim stock, separated by a 1 mm gasket, and held together by spring clamps. After a suitable interval, $t_1 - t_o$, each cell was suspended for 2 h in a Rayonet Photoreactor fitted with UV sources nominally rated at 300 nm. The temperature within the photoreactor was about 30°C; all other operations were conducted at room temperature, 23°C. About 24 h after t_o, each cell was disassembled and the sheet product heated at 90°C in vacuum for 48 h Small weight losses at any step were assumed to be BMA and were taken into account in calculating the composition of the sample.

Reference polymers and networks were prepared by analogous procedures. Two PU networks were made, one from the precursors alone and the other in the presence of 50 wt.% of BMA, which was subsequently removed by evaporation.

Characterization. Opacity of a sample was determined from its absorption at 700 nm. Dynamic mechanical characterization was carried out with an automated Rheovibron DDV-IIC (IMASS) in the tensile mode with a heating rate of 1.5°/min; data taken at 11 Hz are reported here. The same sample was used for the entire temperature range of -100° to 150°C. Because of the magnitude of the load cell compliance, properties of our samples in the glassy region below about -40°C were not viewed in any quantitative sense.

Initial moduli at room temperature were obtained with an Instron Model 4206 at a strain rate of 2/min; ASTM D638 type V specimens were used. The Instron was also used in the creep experiments, in which deformation under a 1 MPa tensile load was continuously monitored for 10^5 sec, followed by measurement of the recovered length 48 h after load removal. Strain dependence of the elastic modulus was determined by deforming specimens to successively larger tensile strains and, at each strain level, measuring the stress after relaxation after it had become invariant for 30 min.

Thin sections for transmission electron microscopy were dry cut on a Reichert Ultracut ultramicrotome with a FC4D cryostatic unit; temperatures were -100°C for the specimen and -80°C for the diamond knife. Sections stained with RuO_4 vapor as well as unstained sections were viewed by means of a Zeiss EM10 electron microscope.

RESULTS

For each composition a series of five to seven samples were produced with increasing delay time between the beginning of the PU formation (t_o) and the initiation (t_1) of the PBMA formation. All series showed a smooth decrease in opacity and in PBMA domain size with increased delay time. The data for the 5:5 compositions are shown in Table I. In this and later Tables, designations such as "11-4.5" refer to the 11 series, sample 4, nominal composition containing a PU weight fraction of 0.5. For similar delay times the 3:7 compositions were

Table I. 5:5 PU-PBMA Mixtures

Sample	xBMA	t(i)-t(o) min	k(700) cm^{-1}	Domain Size, µm
lx-1.5	.48	16	9.3	.92
lx-2.5	.49	137	5.2	.4
lx-3.5	.49	258	2.4	.3
lx-4.5	.48	389	1.5	.15
lx-5.5	.49	674	.7	.1
lx-6.5	.48	1284	.7	.057
xl-1.5	.49	13	12.9	1.5
xl-2.5	.49	133	6.4	1
xl-3.5	.49	548	1.4	.25
xl-4.5	.49	1247	.55	.12
xl-5.5	.49	1539	.43	.09
xx-1.5	.48	15	8.9	1.1
xx-2.5	.49	76	4.5	.65
xx-3.5	.49	197	2.6	.48
xx-4.5	.49	586	.75	.24
xx-5.5	.49	1286	.37	.03
xx-6.5	.48	1508	.42	.12
ll-1.5	.47	22	9.8	1.3
ll-2.5	.47	145	3.6	.45
ll-3.5	.47	266	1.8	.2
ll-4.5	.47	925	.74	.07
ll-5.5	.47	1466	.81	.057

more opaque and had larger domains, while the 7:3 compositions were
less opaque and had smaller domains.

Electron Microscopy. All of the samples produced in this work exhibit
a two-phase morphology of PBMA particles in a PU matrix. There is no
indication of continuity in the PBMA phase. Micrographs of RuO₄-
stained sections of the lx-n.5 and ll-n.5 series are displayed in
Figure 1, in which the light areas are the acrylic phase. The
micrographs clearly show the decrease in domain sizes with increasing
tᵢ in both series. Qualitatively similar morphologies were observed
in the lx and xx series, in which irregular particles formed, at least
at the shorter delay times. Likewise, the xl and ll series had similar
morphologies with spherical particles. The major difference between
the two families at shorter delay times appears to be that of particle
shape and therefore surface area. Plots of subjectively-estimated
domain sizes against opacity and irradiation delay time for the 5:5
series in Table I show that, within experimental error, there is
little difference among the series.

The coarsest and finest morphologies are amplified in Figures 2
and 3, in which the extremes of the 5:5 series are shown at higher
magnification. Again, the lx and xx series and the xl and ll series
form two morphological families at short delay times, where the
domains are large (Figure 2). In both cases, the particle size
distribution appears to be bimodal, with fine particles of about the
same size and shape as those seen at long delay times, as in Figure
3. This Figure also indicates the slight tendency toward smaller
particle size in the lx series relative to the ll series when
irradiation delay times are similar. There is a generally smooth
transition in both particle size and shape with increasing delay times
in each series.

Dynamic Mechanical Spectroscopy. Typical spectra, exemplified by the
5:5 compositions at the shortest delay times, are shown in Figure 4.
It is evident that these are strongly phase-separated materials, with
the temperatures of the tan δ maxima near those of the corresponding
homopolymers and networks (Table II). The two morphological groups
seen in the electron micrographs have their counterparts in Figure 4:
the PU transitions of the samples with spherical particles are
suppressed relative to the PU transitions in xx-1.5 and lx-1.5, with
irregular particles.

TABLE II

REFERENCE POLYMERS

	PU(x)	PU(l)	PBMA(x)	PBMA(l)
Tₘₐₓ, tan δ	-22	-24	74	70
Tan δₘₐₓ	.98	.98	1.25	1.53
E'(20°), MPa	2.76	2.85	428	324

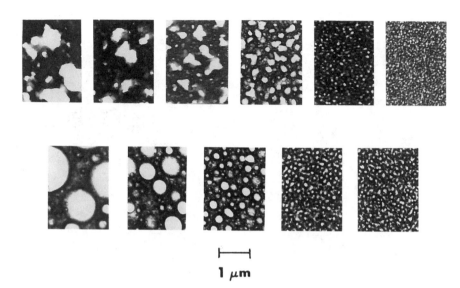

1 μm

Figure 1. Electron micrographs of the 1x-n.5 series (upper, delay times, l. to r., 16, 137, 258, 389, 674, 1284 min) and the 11-n.5 series (lower, delay times, l. to r., 22, 145, 266, 925, 1466 min).

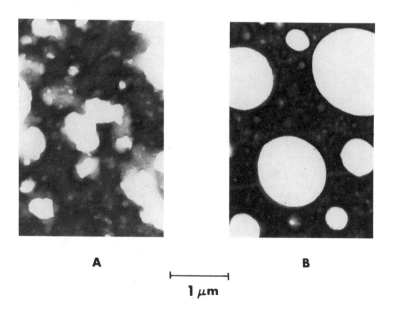

A   B

1 μm

Figure 2. Electron micrographs of (A) 1x-1.5 and (B) 11-1.5.

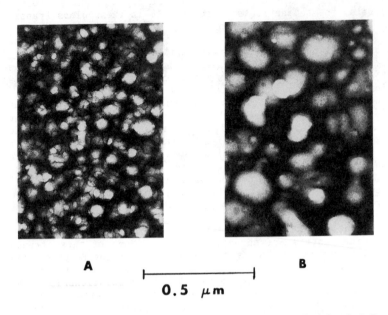

Figure 3. Electron micrographs of (A) 1x-6.5 and (B) x1-4.5.

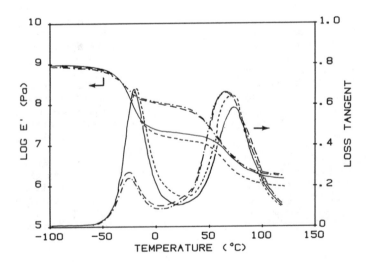

Figure 4. Dynamic mechanical spectra at 11 Hz: 1x-1.5 (_·_); x1-1.5 (----); xx-1.5 (— —); 11-1.5 (———).

Specific data from the spectra of the reference polymers and all of the mixtures are given in Tables II-V. As lightly crosslinked materials, the homonetworks differ little in their glass transitions from their linear counterparts.

Storage moduli (E') have been used frequently in the qualitative evaluation of phase continuity through the use of mixing rules. Such rules cannot be applied here, since inspection of the Tables shows that there is considerable variation in E' at 20°C in the plateau region between the major relaxations, with a minimum as one goes through each of the series. This is especially true for the lx and xx series. The minimum value of E'(20°C) as well as the irradiation delay time corresponding to that point increases as the weight fraction of PU decreases in these two series. In the xl compositions, the delay time corresponding to the minimum does not shift with increasing PU weight fraction.

Family relationships are emphasized by a comparison of E'(20°C) values for the four mixture types at one composition, as shown in Figure 5. Again, the families lx, xx and xl, ll relate well to the irregular and spherical particle morphologies seen in the electron micrographs. Damping peak behavior also follows the family pattern. Peak height ratios are shown in Figure 6. Most of these trends are the result of changes in tan δ peak heights for the PU phase. Stress-strain curves for these materials give initial moduli (Figure 7) that also show the same family patterns. The family relationships have thus been established by three independent characterization methods.

Although this is a strongly phase-separated system, small shifts in T_g seen in T_{max} for tan δ (Tables III-V) are consistent with interactions between the phases. Most of the shifts result from reductions in the high temperature T_g corresponding to the PBMA phase. Reductions in the peak temperature are accompanied by small (up to 10°C) increases in the half-width and a slight asymmetry, suggesting an incorporation of PU into a PBMA-rich phase. Differences in temperature between the two tan δ peaks for the 5:5 compositions are plotted against rough estimates of domain size in Figure 8. The family relationships noted above are absent in these plots. Shifts are greatest in the xx series. At the smallest domain size, corresponding to the longest irradiation delay times, there is some reduction in the T_g difference in all the series. Samples with the largest domain sizes have T_g differences less than those of the parent materials (Table II), which were prepared in bulk rather than in solution.

Creep, Swelling, and Extraction Studies. Additional indications of component interaction were found in the results of creep experiments with the xl series shown in Tables VI and VII. Creep in the xl series is fully recoverable, i. e. there is no permanent set, consistent with a PUx continuous phase. The increase in recoverable compliance, however, indicates a reduction in apparent crosslink density with increasing delay time before irradiation. This result is reinforced by the data in Table VII; C_1 and C_2 are the material constants in the Mooney-Rivlin equation. The rubbery plateau modulus and the crosslink density of PUx prepared in BMA, which mimics xl formation, is less than that of PU prepared neat.

Table III. Dynamic Mechanical Properties of the xx and 11 Series

Sample	t-t(o) min	E'(20°C) MPa	Low Tan δ T °C	max	High Tan δ T °C	T_g(high) max -T_g(low)	
xx-1.3	10	204	-25	.132	65	1.003	90
xx-2.3	139	210	-23	.13	67	.983	90
xx-3.3	262	206	-23	.129	68	.952	91
xx-4.3	563	198	-23	.142	68	.927	91
xx-5.3	1868	102	-21	.265	68	.816	89
xx-6.3	2134	122	-22	.237	65	.814	87
xx-7.3	2784	144	-20	.188	70	.826	90
xx-1.5	15	85.5	-25	.271	67	.66	92
xx-2.5	76	78.8	-23	.28	68	.63	91
xx-3.5	197	55.3	-23	.324	63	.612	86
xx-4.5	586	27.4	-21	.482	66	.515	87
xx-5.5	1286	59	-22	.272	59	.6	81
xx-6.5	1508	61	-22	.275	58	.597	80
xx-1.7	18	23.9	-21	.473	66	.451	87
xx-2.7	139	9.06	-19	.71	70	.3	89
xx-3.7	259	6.56	-20	.74	70	.306	90
xx-4.7	669	11.3	-20	.546	58	.339	78
xx-5.7	1272	14.3	-19	.491	55	.384	74
xx-6.7	1512	17.2	-21	.48	55	.43	76
11-1.5	22	19.9	-20	.674	71	.587	91
11-2.5	145	17.8	-20	.67	71	.554	91
11-3.5	266	14	-20	.74	71	.54	91
11-4.5	925	15	-20	.528	71	.56	91
11-5.5	1466	14.5	-21	.363	66	.59	86

Table IV. Dynamic Mechanical Properties of the 1x Series

Sample	t-t(o) min	E'(20°C) MPa	Low Tan δ T °C	max	High Tan δ T °C	max	Tg(high) -Tg(low)
1x-1.3	2	208	-25	.132	70	.934	95
1x-2.3	124	181	-23	.137	70	.933	93
1x-3.3	252	96	-23	.194	70	.843	93
1x-4.3	371	123	-25	.171	67	.827	92
1x-5.3	677	98	-20	.256	74	.823	94
1x-6.3	1528	141	-22	.186	71	.925	93
1x-1.5	16	91.6	-24	.244	67	.663	91
1x-2.5	137	74.5	-24	.285	69	.63	93
1x-3.5	258	53.2	-22	.377	70	.58	92
1x-4.5	389	31.4	-20	.473	72	.525	92
1x-5.5	674	39	-20	.432	70	.54	90
1x-6.5	1284	95.4	-24	.224	62	.67	86
1x-1.7	1	22.2	-23	.517	67	.417	90
1x-2.7	129	6.1	-20	.787	72	.37	92
1x-3.7	261	6.72	-19	.83	73	.34	92
1x-4.7	385	6.74	-19	.799	70	.33	89
1x-5.7	710	12	-21	.607	61	.35	82
1x-6.7	1309	18	-21	.492	59	.43	80
1x-7.7	1601	21.7	-21	.486	60	.424	81

Table V. Dynamic Mechanical Properties of the xl Series

Sample	t-t(o) min	E'(20°C) MPa	Low Tan δ T °C	max	High Tan δ T °C	max	Tg(high) -Tg(low)
xl-1.3	1	46.3	-20	.423	71	.885	91
xl-2.3	123	43.8	-18	.422	71	.87	89
xl-3.3	245	40.6	-18	.425	70	.834	88
xl-4.3	647	38.6	-17	.408	69	.827	86
xl-5.3	1338	45.9	-16	.434	69	.84	85
xl-6.3	1630	49	-17	.402	68	.858	85
xl-1.5	13	14.5	-19	.655	69	.62	88
xl-2.5	133	14.8	-19	.683	70	.58	89
xl-3.5	548	11.8	-17	.703	72	.56	89
xl-4.5	1247	16.3	-18	.597	70	.59	88
xl-5.5	1539	21.2	-17	.545	70	.61	87
xl-1.7	1	6.37	-20	.812	70	.37	90
xl-2.7	134	5.57	-19	.837	71	.34	90
xl-3.7	286	6.06	-19	.828	70	.328	89
xl-4.7	599	5.45	-19	.838	70	.33	89
xl-5.7	1308	7.43	-19	.728	67	.34	86
xl-6.7	1575	8.59	-19	.665	62	.388	81

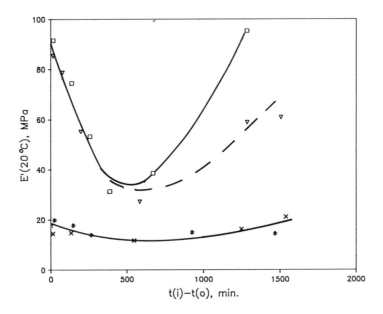

Figure 5. Storage moduli at 20°C vs. delay time prior to irradiation: □ 1x-n.5; x x1-n.5; ∇ xx-n.5; # 11-n.5.

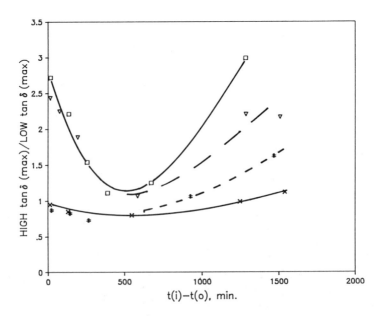

Figure 6. Ratios of damping peak heights vs. delay time prior to irradiation: □ 1x-n.5; x x1-n.5; ∇ xx-n.5; # 11-n.5.

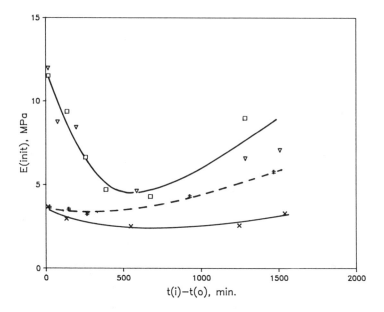

Figure 7. Initial moduli at 23°C vs. delay time prior to irradiation: □ lx-n.5; x xl-n.5; ∇ xx-n.5; # ll-n.5.

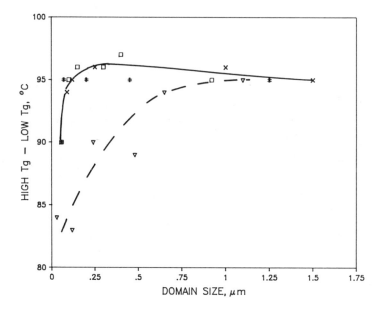

Figure 8. T_g difference vs. estimated particle size, normalized to 95° for largest sizes: □ lx-n.5; x xl-n.5; ∇ xx-n.5; # ll-n.5.

Table VI. Creep and Recovery in the 5:5 xl Series

	Recoverable Compliance, MPa^{-1}		Creep Viscosity, GPa.s	
	23°C	80°C	23°C	80°C
xl-1.5	0.22	2.0	∞	>10
xl-2.5	0.42	2.1	∞	>10
xl-3.5	0.58	1.9	∞	>10
xl-4.5	0.99	5.5	∞	>10
xl-5.5	1.21	7.3	90	70

Table VII. Dry PU(x) Properties

	Crosslinked Neat	Crosslinked in 50% BMA Solution
Soluble Fraction	3%	21%
Swelling Ratio	5	13
C_1 (MPa)	0.48	0.05
C_2/C_1	1.9	4.8

Direct evidence for chemical interaction, such as grafting, was found through initial extraction experiments with this series. Exhaustive Soxhlet extraction of xl-1.5, 3.5, and 5.5 with chloroform or tetrahydrofuran removed only 7, 9, and 18%, respectively, of the possible PBMA in these samples.

DISCUSSION

Particle Size. First impressions of the sequential effects are visual: the longer the delay before irradiation, the more transparent the final material. There were no indications of homogeneous phase formation in the micrographs. Transparency can be associated with phase-separated particle sizes smaller than the wavelengths of visible light. At some time, the PU matrix will reach a viscosity reflecting chain entanglement and the onset of gelation that markedly reduces PBMA precursor diffusion. We associate the minima in property plots such as Figures 5-7 with this viscosity elevation, which we term the "critical" viscosity. In a fluid medium, given the incompatibility of PBMA and Adiprene L-100, PBMA precursor diffusion is likely to produce large PBMA particles. With increasing medium viscosity, PBMA precursor motion is reduced, and the particle sizes become smaller. After a semi-solid matrix has formed, PBMA is being generated from precursors in an essentially immobile PU(1) solution or a swollen gel that is still undergoing reaction, a situation studied by Allen and coworkers (4). Completeness of PU formation will thus generally dictate PBMA particle size under our conditions.

The above simplistic view, while consistent with our data, ignores the possibilities, for example, of phase interaction, grafting, hydrogen-bonding, or the completeness of PBMA formation during the irradiation period. With respect to the latter, the bimodal distribution of particle sizes seen in Figure 2 for samples irradiated before the "critical" viscosity point is reached supports the formation of additional particles of a size and shape obtained during irradiations after the "critical" point. A similar finding has been reported (<u>3</u>) for an epoxy-butyl acrylate system.

Particle Shape. The delineation of two families of materials on the basis of particle shape is very clear from the electron microscope evidence. That the families are xx, 1x and x1, 11 points directly to the presence or absence of crosslinking in the PBMA as a source of particle shape differentiation. Overall concentration of the PBMA precursors is not a strong factor, since the irregular particles were observed in all three of the xx compositions, but an examination of the effect of crosslinker concentration alone was not carried out. In related work on an epoxy-butyl acrylate system in which component polymerization rates and the simultaneity of the reactions were matched, it was reported (<u>3</u>) that, prior to gelation of the matrix, irregular particles of crosslinked acrylate were formed but spherical particles were found in the absence of crosslinker. Together with the observation of an apparent bimodal size distribution, our results are similar, even though our system and conditions are markedly different from those in the earlier study.

While no explanation was offered for the irregular domain formation in the epoxy-butyl acrylate work (<u>3</u>), we have observed, as have others (<u>12, 13</u>), the propensity of BMA to undergo proliferous ("popcorn") polymerization under conditions that allow both crosslinking and chain scission. These are precisely the conditions we have with photoinitiation in our xx and 1x series. It is suggested that small centers of tight acrylic networks form in the early stages of irradiation, and that these, in combination, may form nuclei for irregular particle formation. Such nuclei may form more readily in the relatively fluid PU precursor medium existing early in the PU formation process than in a later, more viscous, matrix medium.

Size and shape do not seem to be closely related, but after the "critical" viscosity point is reached, there is less differentiation among the two "shape families", as seen in Figure 3. In part, this may be the result of a longer time for equilibration of the PBMA precursors with the medium as its viscosity increases.

Interphase Interaction. Phase separation in this system is fairly complete, as demonstrated by both electron microscopy and the dynamic mechanical spectra. Nonetheless, interaction is consistent with the extent of separation of the tan δ peaks, taken as a measure of T_g separation. Inspection of the data in the Tables shows that the greatest reduction in T_g separation occurs in the xx series, that there is some reduction in all series at the longest irradiation delay times, and that there is greater reduction with increasing proportions of PBMA in the mixtures. The changes with increasing delay time are fairly smooth. From the electron micrographs, total particle surface area and the proportion of fine particles increases with increased delay time in all series.

Particle <u>shape</u> may play the strongest role in interactive processes between matrix and particle. Chains or segments of PU close

to a relatively glassy particle will tend to have restricted motion
at the temperatures for normal PU relaxation. For a given particle
volume, a sphere (with minimal surface area), such as is observed in
the xl and ll series, is likely to affect fewer PU chains or segments
than is an irregularly-shaped particle. A larger proportion of the PU
in the xx and lx series, with irregular particles, will undergo
relaxation near the temperatures for PBMA relaxation. This is
reflected in the xx and lx series in the tan δ for PU, which is
strongly suppressed in peak height relative to the xl and ll series
but little changed in shape or location on the temperature scale. It
is also seen in the trends in the shape and position of the high
temperature tan δ peaks. Thus we conclude that restricted motion of
matrix chains or segments in the close vicinity of the phase-separated
domains of PBMA is a major source of apparent interphase interaction.

Mechanical Properties. Phase-separated particle shape and size,
controllable by the sequential timing of phase formation, gives a
measure of control over transition region properties, as seen above.
Within experimental error, areas under the tan δ curves for a given
composition are constant, even though the temperature ranges for
effective damping change markedly.

Other properties from the dynamic mechanical data, such as the
values of E'(20°C) (Figure 5), the ratios of the tan δ peak heights
(Figure 6) or initial moduli from stress-strain tests (Figure 7) show
similar trends, i. e. there are two families of data, the lx, xx and
xl, ll series, with the latter showing less variation with delay time
than the former. These curves, as noted above, also show minima that
reflect attainment of our assumed "critical" matrix viscosity. The
"tightness" of the network, as well as viscosity, determine the course
of curves such as that in Figure 5. Matrix PU in the short delay time
samples is largely formed in the absence of solvents and will have a
higher E'(20°C) than the PU formed in the presence of PBMA precursors
in the long delay time samples. Allen and coworkers (4) reported that
the initial modulus of their PUx (neat) was about twice that of dried
PUx formed in a 1:1 mixture with methyl methacrylate. This effect
alone would be sufficient to account for the E'(20°C) decreases below
the delay time corresponding to the critical viscosity. Countering
this is the influence of particle size, which steadily decreases with
increasing delay time. Smaller particles will restrict the motion of
a greater proportion of the PU, resulting in a larger effective
particle volume fraction and an increase in E'(20°C). Particle shape
amplifies these phenomena and accounts for the differences between the
families.

Based on the electron micrograph evidence, we have assumed a
simple morphology of PBMA particles in a PU matrix with invariant
composition through any series. However, the creep and extraction
studies suggest otherwise, at least for the xl series. While the
evidence for a continuous PU matrix is clear, the apparent crosslink
density in the system decreases with increasing delay time to PBMA
formation. Preliminary extraction studies on the xl series indicate
the possibility of grafting, the degree of which decreases with
increasing delay time before initiation of the acrylic. This may be
a surprising result in the light of reports (14, 15) that no grafting
occurs in polyurethane-poly(methyl methacrylate) systems in which the

PU was fully formed before initiation of the acrylic. In the present system, polyol concentration decreases as delay time increases. It is well-established (16) that organotin compounds are transesterification catalysts. Thus, it is suggested that, in the presence of the tin catalyst and during irradiation, transesterification between BMA and one or more of the OH groups in the polyols may lead to the formation of PU chains terminated with methacrylate moieties to provide grafting sites for BMA polymerization.

The crosslinking and grafting results in the xl series may have their counterparts in the other series as well. It is possible that such basic changes in structure will account for some of the trends in properties with changes in formation sequence of the components in these polymer systems. We intend to pursue the issues raised in this broad survey in future publications.

SUMMARY AND CONCLUSIONS

This PU-PBMA system is strongly phase-separated in each of the ll, xl, lx, and xx forms; all show a morphology of PBMA particles in a PU matrix over the range of 30-70% PU. Particle size is readily controlled by the timing of the relatively rapid formation of PBMA in a slowly-forming PU matrix. With an increasing delay before initiation of the PBMA precursors, the PU viscosity has increased, and particle size is reduced; an increasing proportion of the PU is formed in the presence of the PBMA precursors as solvent. The combined effects produce a minimum in the values of certain mechanical properties as delay time increases. Interaction between phases reflected in a decreasing T_g in the PBMA phase is largely a function of decreasing phase size; it is greatest in the xx series but takes place to some extent in the lx and xl series as well. In the xl series, a diluent effect tends to reduce the apparent matrix crosslink density as the delay before PBMA formation increases; significant grafting also appears to take place.

In this system, particle shape is dictated by crosslinking in the PBMA phase; irregular particles are formed in the lx and xx series, and spherical particles are formed in the xl and ll series. Particle shape exerts a strong influence on the mechanical properties through the extent to which matrix adjacent to the relatively glassy particles is prevented from participating in normal matrix relaxation processes. Irregular particles have the greater effect in this respect, resulting in a suppression of the PU relaxation and an increase in the moduli in the temperature region between the relaxations of the two phases. Regions for efficient damping can thus be controlled through both crosslinking and the sequence of phase formation.

LITERATURE CITED

1. Sperling, L. H.; Arnts, R. R. *J. Appl. Polym. Sci.* 1971, *15*, 2317.
2. Touhsaent, R. E.; Thomas, D. A.; Sperling, L. H. *J. Polym. Sci. Sympos.* 1974, *46*, 175.
3. Touhsaent, R. E.; Thomas, D. A.; Sperling, L. H. In *Toughness and Brittleness of Plastics*; Deanin, R. D.; Crugnola, A. M., Eds.; Advances in Chemistry Series No. 154; American Chemical Society: Washington, DC, 1974; p. 206.

4. Allen, G.; Bowden, M. G.; Blundell, D. J.; Hutchinson, F. G.;
   Jeffs, G. M.; Vyvoda, J. Polymer 1973, 14, 597; Allen, G.; Bowden,
   M. J.; Blundell, D. J.; Jeffs, G. M.; Vyvoda, J.; White, T.
   Polymer 1973, 14, 604.
5. Klempner, D., Yoon, H. K., Frisch, K. C., Frisch, H. L. In
   Chemistry and Properties of Crosslinked Polymers; Labana, S. S.,
   Ed.; Academic Press: New York, 1977; p. 243; and in Polymer
   Alloys II; Klempner, D.; Frisch, K. C., Eds.; Plenum Press: New
   York, 1979; p. 185.
6. Lee, D. S.; Kim, S. C. Macromolecules 1984, 17, 268.
7. Lee, J. H.; Kim, S. C. Macromolecules 1986, 19, 644.
8. Kim, B. S.; Lee, D. S.; Kim, S. C. Macromolecules 1986, 19, 2589.
9. Yang, Y. S.; Lee, L. J. Macromolecules 1987, 20, 1490.
10. Lee, Y. M.; Yang, Y. S.; Lee, L. J. Polym. Eng. Sci. 1987, 27,
    716.
11. Jin, S. R.; Widmaier, J. M..; Meyer, G. C.; Polymer 1988, 29, 346.
12. Breitenbach, J. W. Brit. Polym. J. 1974, 6, 119.
13. Breitenbach, J. W.; Goldenberg, H. Monatsh. Chem. 1973, 104, 500.
14. Djomo, H.; Morin, A.; Damyanidu, M.; Meyer, G. Polymer, 1983, 24,
    65.
15. Allen, G.; Bowden, M. J.; Lewis, G.; Blundell, D. J.; Jeffs, G.
    M. Polymer, 1974, 15, 13.
16. Poller, R. C.; Retout, S. P. J. Organomet. Chem. 1979, 173, C7.

RECEIVED April 27, 1989

Chapter 11

Energy-Absorbing Multicomponent Interpenetrating Polymer Network Elastomers and Foams

D. Klempner, B. Muni, and M. Okoroafor

Polymer Technologies, Inc., University of Detroit, Detroit, MI 48221

Two- and three-component interpenetrating polymer network (IPN) elastomers composed of polyurethanes (PU), epoxies (E), and unsaturated polyester (UPE) resins were prepared by the simultaneous technique. Fillers and plasticizers were incorporated by random batch mixing. The PU/E and PU/E/UPE ratios were varied. Enhanced energy absorbing abilities were demonstrated by dynamic mechanical spectroscopy. This was reflected in broad and high tan δ values as a function of temperature. The effects of fillers and plasticizers on the tan δ values varied from type to type.
Three-component IPNs prepared from polyurethane, epoxy, and unsaturated polyester resin resulted in even broader tan δ values when compared to two component (PU/E) IPN elastomers. Furthermore, the tan δ values for the three component IPN systems were still high after the transitions were apparently complete, which is of enormous significance in sound energy absorption applications. IPN foams prepared by using PU/E (two-component) showed excellent energy absorbing abilities. This was reflected in rebound, hysteresis, and sound absorption studies.

Interpenetrating polymer networks (IPNs) are relatively novel types of polymer alloys consisting of two or more crosslinked polymers held together by permanent entanglements with only accidental covalent bonds between the polymers, i.e. they are polymeric catenanes (1-6). IPNs possess several interesting characteristics in comparison to normal polyblends. Formation of IPNs is the only way of intimately combining crosslinked polymers, the resulting mixture exhibiting (at worst) only limited phase separation. Normal blending or mixing of polymers results in a multi-phase morphology due to the well known thermodynamic incompatibility of polymers. However, if mixing is accomplished simultaneously with crosslinking, phase separation may be kinetically controlled by permanent interlocking of entangled chains.

0097–6156/89/0395–0263$12.50/0
© 1989 American Chemical Society

Depending upon the miscibility of the polymers, IPNs exhibit varying degrees of phase separation (7-9). With highly immiscible polymers, only small gains in phase mixing occur. In cases where the polymers are more compatible, phase separation can be almost completely circumvented. Complete compatibility is not necessary to achieve complete phase mixing, since the permanent entanglements (catenation) can effectively prevent phase separation.

Incompatible polymer alloys (10-12) display separated T_gs and therefore exhibit several damping ranges corresponding to the glass transition temperature of the components. Homogeneous polymer alloys (8) show only one damping (T_g) which may be slightly broader than that of the individual components. Semi-miscible polymer blends (13), where the mixing between the polymers is extensive, have a very broad thermal damping range which leads to a wider frequency range.

Polymer systems with broad transitions (over a wide range of temperature and frequency) will be the most effective acoustical/absorbing materials. When polymers are at their glass transition, the time required to complete an average coordinated movement of the chain segments approximates the length of time of the measurement. If dynamic or cyclical mechanical motions are involved such as vibrational or acoustical energy, the time required to complete one cycle, or its inverse, the frequency becomes the time unit of interest. At the glass transition conditions, which involve both temperature and frequency effects, the conversion or degradation of mechanical or acoustical energy to heat reaches its maximum value. Thus in this study, IPNs of low and high T_g polymers along with fillers and plasticizers were prepared in an effort to obtain this "semi-miscible" behavior, to thus achieve the desired broad range acoustical energy absorption (14). These IPNs are based on polyurethane, epoxy, and unsaturated polyester resin networks prepared via the one-shot simultaneous polymerization technique (6). These polymers were studied because our previous studies of such IPNs showed a great potential for good sound and mechanical energy absorbing materials (15,16).

Thus, this research focused on the synthesis of IPNs which have a high and broad tan δ. Two-component IPN systems that proved the most promising were used as the foundation for foams. Three-component IPN foams have yet to be prepared.

In principle, noise contacts the foam structure in the form of sound pressure waves. The pressure wave within the foam structure is partially converted to heat energy and is dissipated. The common types of damping materials are foams (17,18) and homopolymers or copolymers, which exhibit efficient damping only in normal temperature (frequency) ranges corresponding to the glass transition of the polymer (19-21). In this paper, we present the synthesis of foams composed of IPNs of semi-miscible behavior, thus taking advantage of both the cellular structure of foams, as well as the broad damping behavior of the IPNs (22-24).

EXPERIMENTAL

Preparation

The materials used in this investigation are listed in Table
I.

Elastomers

Polyol Niax 31-28, epoxy DER-330, and chain extender Isonol-
100 were vacuum dried overnight at 80°C. All other materials
were used as received.
For preparation of polyurethanes (PU), ethylene oxide capped
poly(oxypropylene) glycerol containing 21 wt. % grafted
polyacrylonitrile (Niax 31-28, Union Carbide), MW=6000 was
blended with a carbodiimide-modified diphenylmethane diisocyanate
(Isonate-143L, Upjohn), N,N'-bis(2-hydroxypropyl)aniline (Isonol-
100, Upjohn), and dibutyltin dilaurate (T-12, M & T Chemical).
The IPN elastomers were synthesized by the mixing of two
components. In the case of two-component PU/E IPN elastomers,
one component contained Niax 31-28, Isonol-100, T-12 and epoxy
catalyst (BF₃-etherate, Eastman Chemical). The other component
contained epoxy resin (DER-330, Dow Chemical) and Isonate-143L.
Whereas in the case of three-component IPN elastomers composed of
PU/E/UPE, polyester resin catalyst (TBPB) and unsaturated
polyester resin were incorporated, respectively, in the former
and latter components.
The blend was mixed for sixty seconds at room temperature
(RT). It was poured into a hot mold (100°C) and cured for 30
minutes at 100°C on a platen press under a pressure of
approximately 2700 KPa. The elastomers were then post-cured in
an oven at 100°C for 16 hours and conditioned at 25°C and 50%
relative humidity for three days prior to testing.

Foams

PU foams were prepared by the one-shot, free-rise method.
A homogeneous liquid mixture consisting of Niax 31-28, Isonol-
100, silicone surfactant DC-193 (Dow Corning), T-12, Niax A-1
(Union Carbide), trichlorofluoromethane (Freon 11A, E.I. duPont
de Nemours & Co.), and water was mixed thoroughly with Isonate-
143L. It was put into a cold mold and allowed to free rise.
It was necessary to prepare IPN foams using a hot mold
method. One component contained Niax 31-28, DC-193 and three
different surfactants (L-540, L-5303 and L-5614) from Union
Carbide, T-12, 2,4,6-tris(dimethylaminomethyl) phenol (DMP-30,
Rohm & Haas), BCl₃-amine complex (XU-213, Ciba-Geigy), Freon
11A, and water. The other included Isonate-143L, DER-330, and
Freon 11A. The two components were mixed and poured into a hot
mold and cured at 90°C for 2-3 hours. The foams were post-
cured at 90°C for 16 hours, and conditioned at 25°C and 50%
relative humidity for three days.

Table I. Materials

Designation	Description	Supplier
Isonate-143L	Carbodiimide modified diphenyl-methane diisocyanate	Dow Chem. Co.
Niax 31-28	Graft copolymer of poly (oxypropylene)(oxyethylene) adduct of glycerol	Union Carbide
Isonol-100	N,N'-Bis(2-hydroxypropyl) aniline	Dow Chem. Co.
DER-330	Bisphenol A-epichlorohydrin epoxy resin	Dow Chem. Co.
UPE	Unsaturated polyester with 33% styrene	Budd Chem.
T-12	Dibutyltin dilaurate	M & T Chem. Inc.
Niax A-1	70% Bis(2-dimethylaminoethyl)ether in dipropylene glycol	Union Carbide
$BF_3(OC_2H_5)_2$	Boron trifluoride etherate	Eastman Chem.
TBPB	t-Butyl perbenzoate	Budd Chem.
DMP-30	2,4,6-Tris(dimethylaminomethyl) phenol	Rohm & Haas
XU-213	BCl_3-Amine complex	Ciba-Geigy
Freon 11A	Trichlorofluoromethane	E.I. duPont de Nemours & Co.
DC-193	Silicone copolymer surfactant	Dow Corning
L-540	Silicone copolymer surfactant	Union Carbide
L-5303	Silicone copolymer surfactant	Union Carbide
L-548	Silicone copolymer surfactant	Union Carbide
1,4-BD	1,4-Butanediol	BASF
Santicizer 141	2-Ethylhexyldiphenyl phosphate	Monsanto
Santicizer 148	Isodecyldiphenyl phosphate	Monsanto
Santicizer 160	Butylbenzyl phthalate	Monsanto
Stan-Flex LV	Aromatic processing oil	Harwick Chem.
Benzoflex 9-88	Dipropylene glycol dibenzoate	Velsicol Chem.
TCP	Tricresyl phosphate	C.P. Hall Co.
Graphite flake #1	Size: -20 Mesh + 80 Mesh	Asbury Graphite Mills Inc.
Graphite flake #2	Size: -50 Mesh + 200 Mesh	Asbury Graphite Mills Inc.
Graphite flake #2	Size: -80 Mesh or less	Asbury Graphite Mills Inc.
Dicaperl FP1010	Hollow glass bubble filler	Grefco Inc.

METHODS OF ANALYSIS AND TESTING

Dynamic Mechanical Spectroscopy

A Rheovibron (DDV-II, Toyo Measuring Instrument Co.) was used to measure the dynamic mechanical properties of the elastomers. The sample was a rectangular film with the dimensions of 24 x 2 x 1 mm. The measurements were carried out from -60°C to 180°C with a heating rate of 1 to 3°C/min at 110 Hz.

Density of Foams

The density was determined gravimetrically according to ASTM D-1622.

Sound Absorption of Foams (Impedance Tube)

The measurements were carried out using the impedance tube technique. It was run on the standard Bruel and Kjaer equipment (Standing Wave Apparatus Type 400Z, Beat Frequency Oscillator Type 1014, and Frequency Spectrometer Type 2112). Two measuring tubes with diameters of 30 and 100 mm were employed. The larger tube was used for frequencies from 100 to 2,000 Hz, the small one from 1,250 to 8,000 Hz. All samples were 50.8 mm in thickness.

Ultimate Mechanical Properties of the Elastomers

The tensile strength, elongation, and hardness were determined according to ASTM methods D-412 and D-2134, respectively.

Mechanical Properties of the Foams

The hysteresis (in compression) was determined on an Instron Universal Tester at a crosshead speed of 25.4 mm per minute. The area surrounded by the compression and retraction curves represents the mechanical energy absorbed by the sample. The specimens were 50.8 x 50.8 x 25.4 mm in dimensions. The ratio of percentage of energy absorbed to the total energy applied to the sample to compress the specimen 75% of its original thickness was calculated and reported as the hysteresis.
The stress at 50% compression, tensile properties, and Bashore rebound were measured according to ASTM D-3574, Section C, E, and H, respectively.

RESULTS AND DISCUSSION

IPN Elastomers

The ability of elastomers to absorb mechanical and

acoustical energy is indicated by their dynamic mechanical properties, as seen from Rheovibron studies. The presence of a broadened glass transition indicates the semi-miscible morphology in the elastomers (23). The formulations, properties and dynamic mechanical spectroscopy curves of PU/E and PU/E/UPE are shown in Tables II to VII and Figures 1 to 12.

Comparison of the Rheovibron spectra of 100% polyurethane (Fig. 1) with that of the IPNs (Figs. 2 to 12) shows the presence of "semi-miscible" morphology, as indicated by the broadening of the tan δ and E" peaks.

In general, for 100% polyurethane, it was observed that plasticizer has not brought any significant shift in T_g while it has resulted in increased magnitude and broadening of the tan δ peak.

Further analysis of the above results may be obtained by studying the solubility parameters of the polymers and plasticizers. Solubility parameters provide a measure of the extent of interaction possible between chemical species. To determine the solubility parameters, the structure of the smallest repeat unit was considered and the contribution of each atomic group to the total energy of vaporization and molar volume was summed over the molecular structure. The ratio is calculated as the cohesive energy. The calculated solubility parameters of the polymers and plasticizers are shown in Table II.

It may be seen that as the solubility parameter of the plasticizer approaches that of the polyurethane, the height of tan δ and the temperature range broadens. Santicizer -160 and tricresyl phosphate (TCP), which have similar high solubility parameters, and Benzoflex 9-88, which has an even higher value (close to that of PU), led to a broadening of the temperature range and high tan δ value to result in the greater area under the tan δ curve, as shown in Figures 1 and 2.

The effects of IPN formation with epoxy and UPE were to greatly broaden the tan δ peak and shift it to around room temperature. This is, of course, due to the interpenetrating effect, resulting in a semi-miscible morphology. The heights of tan δ peaks decreased. The overall area under the curve increased, indicating the enhancement in energy absorbing potential of these IPNs. No relation between the solubility parameters of the prepolymers and the plasticizers could be found regarding tan δ behavior.

As seen from Table IV and Figures 4 and 5, it is observed that fillers such as graphite and mica do not result in much enhancement of tan δ height or breadth. With plasticizers such as Benzoflex 9-88 and Santicizer-160, increased tan δ height accompanied by a narrower temperature range, was observed (Figs. 6 and 7). With Stan-Flux-LV, the lower level of plasticizer resulted in increased tan δ height and temperature range (Fig. 7). It is interesting to note that the solubility parameter of Stan Flux is the lowest, i.e. the farthest removed from the solubility parameters of the polymeric components.

The combination of filler and plasticizer (Benzoflex 9-88) resulted in reduction of tan δ and a broader temperature range

Table II.   Solubility Parameters of Various Components
and Plasticizers

Components	Solubility Parameter (Hildebrands) $(cal/cm^3)^{0.5}$
Isonate 143L (isocyanate)	10.69
Niax 31-28 (polyol)	11.23
Isonol 100 (chain extender)	11.64
DER 330 (epoxy resin)	10.61
Plasticizers	
TCP	9.0
Santicizer 141	8.6
Santicizer 148	8.56
Santicizer 160	9.28
Benzoflex 9-88	11.87
Stan Flux	8.0

Table III.　Formulations and Dynamic Mechanical Spectroscopy
Results for Polyurethane Elastomers

Fig. No.	1	1	1	2	2	2
Sample No.	1	2	3	4	5	6
POLYURETHANE						
Isonate 143L, g	6.48	6.48	6.48	6.48	6.48	6.48
Niax 31-28, g	62.24	62.24	62.24	62.24	62.24	62.48
Isonol-100, g	1.27	1.27	1.27	1.27	1.27	1.27
T-12, g	0.06	0.06	0.06	0.06	0.06	0.06
PLASTICIZERS						
Benzoflex 9-88, g	-	14	-	-	-	-
Santicizer-160, g	-	-	14	-	-	-
Stan Flux LV, g	-	-	-	14	-	-
TCP, g	-	-	-	-	14	35
DYNAMIC MECHANICAL PROPERTIES						
Temp. range (°C)[1]	-43to20	-50to44	-50to40	-46to40	-44to52	-48to28
	(63)	(94)	(90)	(86)	(96)	(76)
Height of tanδ	0.96	1.48	1.6	1.26	1.5	1.75
Temp. at tanδ max. (°C)	-24	-28	-26	-28	-28	-30
Area under the curve[2]	35.16	50.44	49.56	48.35	47.35	55.61

[1] Temperature range has been considered from the point of
inflection.
[2] Area under the curve is calculated by Simpson's rule method.

Table IV. Formulations and Dynamic Mechanical Spectroscopy
Results for PU/Epoxy IPN Elastomers

Fig. No.	3	4	4	5	5
Sample No.	7	8	9	10	11
POLYURETHANE					
Isonate 143L, g	3.89	3.89	3.89	3.89	3.89
Niax 31-28, g	37.35	37.35	37.35	37.35	37.35
Isonol-100, g	0.76	0.76	0.76	0.76	0.76
T-12, g	0.06	0.06	0.06	0.06	0.06
EPOXY					
DER-330, g	28	28	28	28	28
BF_3-Etherate, g	0.6	0.6	0.6	0.6	0.6
PU/Epoxy ratio	60/40	60/40	60/40	60/40	60/40
FILLERS					
#1 Graphite, g	-	7	-	-	-
#2 Graphite, g	-	-	7	-	-
#3 Graphite, g	-	-	-	7	-
Dicaperl FP 1010, g	-	-	-	-	7
DYNAMIC MECHANICAL PROPERTIES					
Temp. range (°C)	-20to108	-40to128	-40to120	-40to104	-20to100
	(128)	(168)	(160)	(144)	(120)
Height of tanδ	0.65	0.71	0.68	0.68	0.65
Temp. at tanδ max. (°C)	40	32	32	30	48
Area under the curve	43.90	49.77	45.62	47.08	41.57

Table V. Formulations and Dynamic Mechanical Spectroscopy
Results for PU/Epoxy IPN Elastomers

Fig. No.	6	6	6	7	7	7
Sample No.	12	13	14	15	16	17
POLYURETHANE						
Isonate 143L, g	3.89	3.89	3.89	3.89	3.89	3.89
Niax 31-28, g	37.35	37.35	37.35	37.35	37.35	37.35
Isonol-100, g	0.76	0.76	0.76	0.76	0.76	0.76
T-12, g	0.06	0.06	0.06	0.06	0.06	0.06
EPOXY						
DER-330, g	28	28	28	28	28	28
BF_3-Etherate, g	0.6	0.6	0.6	0.6	0.6	0.6
PU/Epoxy ratio	60/40	60/40	60/40	60/40	60/40	60/40
FILLERS						
#2 Graphite, g	-	-	3.5	-	-	-
PLASTICIZERS						
Benzoflex 9-88, g	7	35	35	-	-	-
Santicizer-160, g	-	-	-	35	-	-
Stan Flux LV, g	-	-	-	-	7	14
Sundex 748T, g	-	-	-	-	-	-
DYNAMIC MECHANICAL PROPERTIES						
Temp. range (°C)	-40to40 (80)	-36to>12	-40to>32	-40to>4	-26to>34	-40to>64
Height of tanδ	0.85	1.7	1.2	1.65	1.2	0.93
Temp. at tanδ max. (°C)	20-24	0	5	-7	8	16
Area under the curve	36.395	40.639	50.967			50.73

Table VI. Formulations and Dynamic Mechanical Spectroscopy
Results for PU/Epoxy IPN Elastomers

Fig. No.	8	9	10	9	10
Sample No.	18	19	20	21	22
POLYURETHANE					
Isonate 143L, g	3.89	3.24	3.24	2.59	2.59
Niax 31-28, g	37.35	31.12	31.12	24.90	24.90
Isonol-100, g	0.76	0.64	0.64	0.51	0.51
T-12, g	0.06	0.04	0.04	0.04	0.04
EPOXY					
DER-330, g	28	35.0	35.0	42.00	42.00
BF_3-Etherate, g	0.6	0.6	0.6	0.6	0.6
PU/Epoxy ratio	60/40	50/50	50/50	40/60	40/60
PLASTICIZERS					
Benzoflex 9-88, g	-	21.0	-	21	-
Santicizer-160, g	-	-	14	-	-
Stan Flux LV, g	-	-	-	-	14
Sundex 748T, g	7	-	-	-	-
DYNAMIC MECHANICAL PROPERTIES					
Temp. range (°C)	-40to>60	-12to60 (i.e. 72)	-28to44 (i.e. 72)	-39to56 (i.e. 96)	-12to71 (i.e. 83)
Height of tanδ	0.72	0.80	0.90	1.2	1.34
Temp. at tanδ max. (°C)	40	24-28	12	20	20
Area under the curve	-	34.253	31.1	42.71	48.03

Table VII. Formulations of Polyurethane/Epoxy/Unsaturated
Polyester Three-Component IPN Elastomers

Fig. No.	11	12	13	14
Sample No.	23	24	25	26
POLYURETHANE				
Isonate 143L, g	3.24	3.24	2.60	2.60
Niax 31-28, g	31.12	31.12	24.9	24.9
Isonol-100, g	0.63	0.63	0.50	0.50
T-12, g	0.065	0.065	0.052	0.052
EPOXY				
DER-330, g	14.0	17.50	14.0	7.0
BF_3-Etherate, g	0.30	0.40	0.30	0.15
UNSATURATED POLYESTER RESIN				
Polyester resin	21.0	17.50	28.0	35.0
TBPB	0.066	0.055	0.088	0.110
PU/E/UPe RATIO	50/20/30	50/25/25	40/20/40	40/10/50
DYNAMIC MECHANICAL PROPERTIES				
Temp. range (°C)	-32to>80	-36to>68	-32to>100	0to>120
Height of tanδ	0.75	0.81	0.72	1.02
Temp. at tanδ max. (°C)	>60	8	52-68	68-100

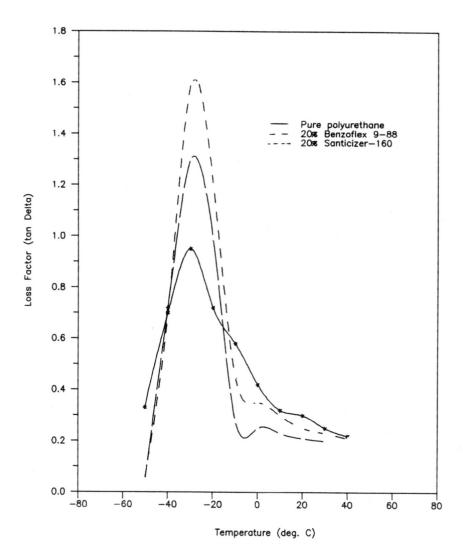

Figure 1. Temperature dependence of tan δ for elastomers
#1,2,3.

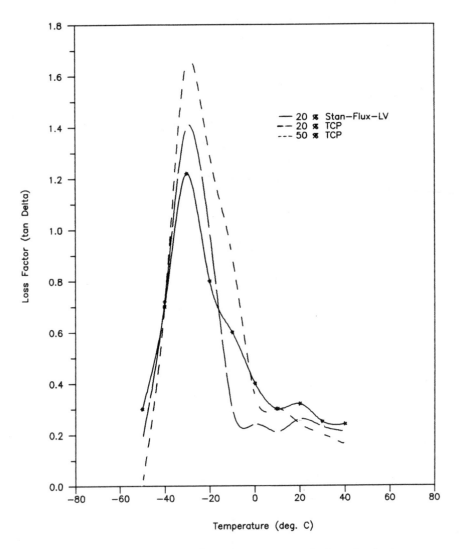

Figure 2.   Temperature dependence of tan δ for elastomers
         #4,5,6.

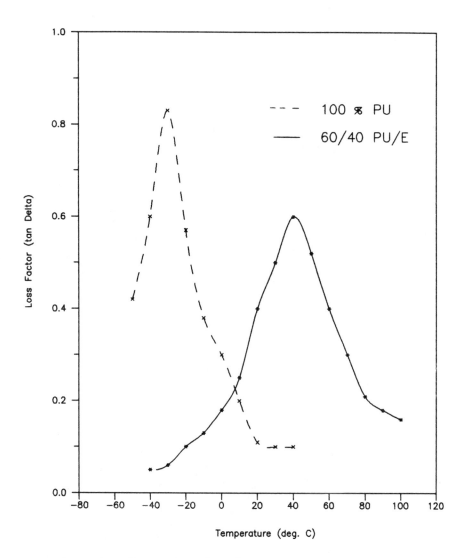

Figure 3.   Temperature dependence of tan δ for elastomers #1 and #7.

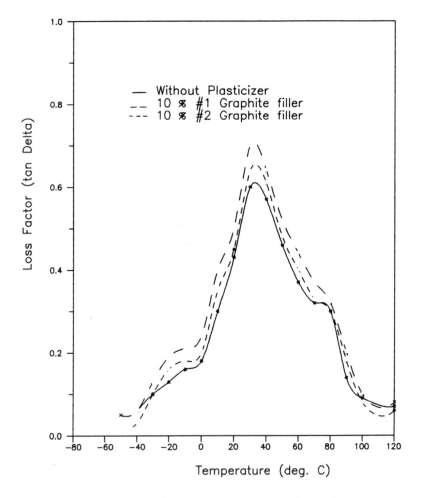

Figure 4.   Temperature dependence of tan δ for elastomers #8,9.

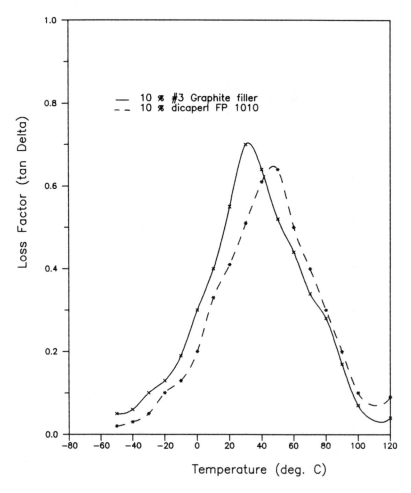

Figure 5. Temperature dependence of tan δ for elastomers #10,11.

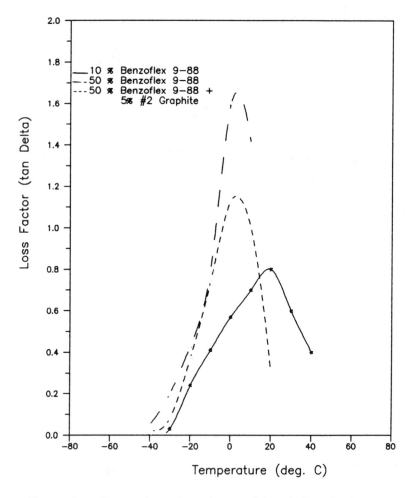

Figure 6.   Temperature dependence of tan δ for elastomers
             #12,13,14.

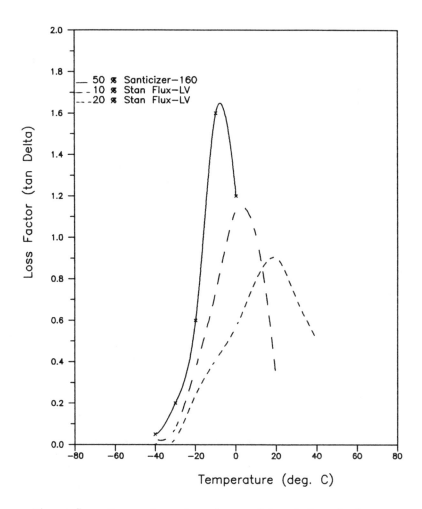

Figure 7.   Temperature dependence of tan δ for elastomers #15,16,17.

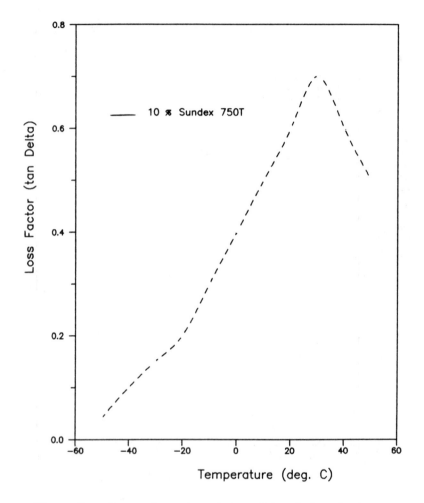

Figure 8.   Temperature dependence of tan δ for elastomer #13.

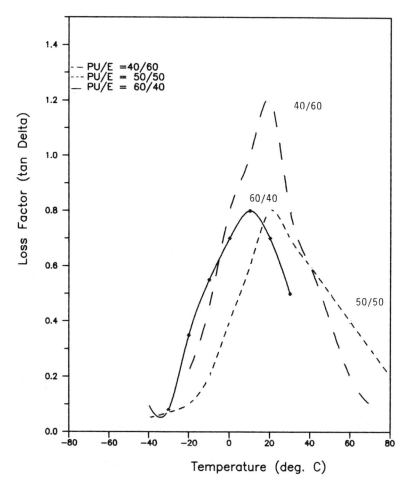

Figure 9.   Temperature dependence of tan δ for elastomers #19,21 (30% Benzoflex 9-88).

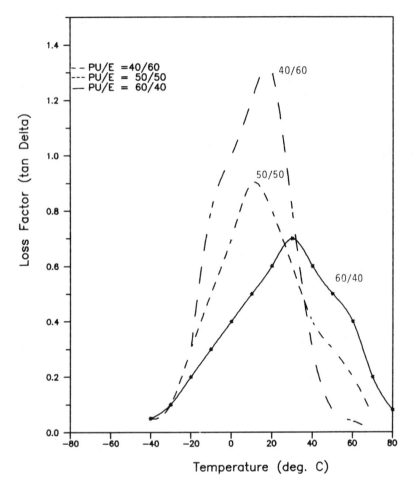

Figure 10.   Temperature dependence of tan δ for elastomers
#20,22 (20% Santicizer-160).

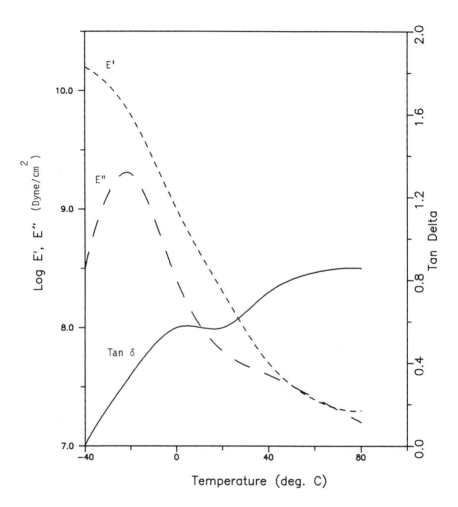

Figure 11. Dynamic Mechanical Spectroscopy of 50/20/30
PU/E/UPE IPN Elastomer with 2% Isonol-100
Post-cured 16 hrs.

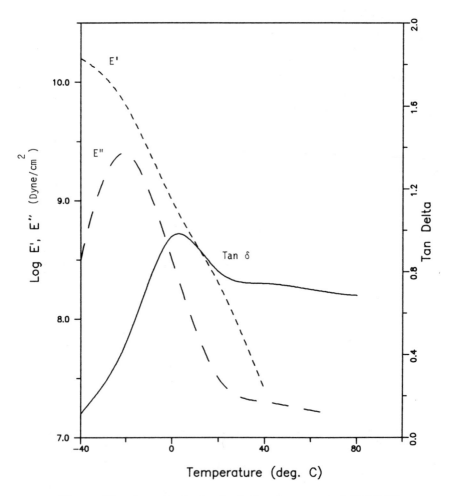

Figure 12.   Dynamic Mechanical Spectroscopy of 50/25/25
PU/E/UPE IPN Elastomer with 2% Isonol-100
Post-cured 16 hrs.

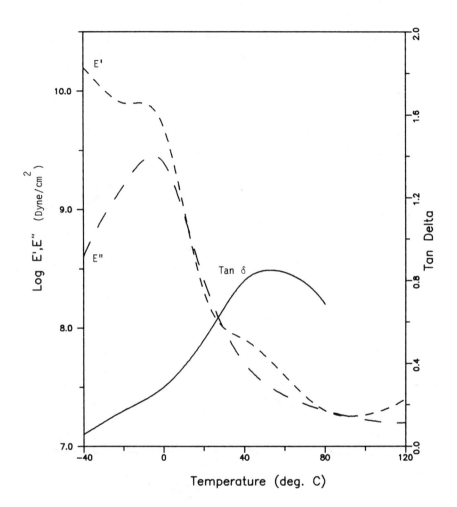

Figure 13. Dynamic Mechanical Spectroscopy of 40/20/40 PU/E/UPE IPN Elastomer with 2% Isonol-100 Post-cured 16 hrs.

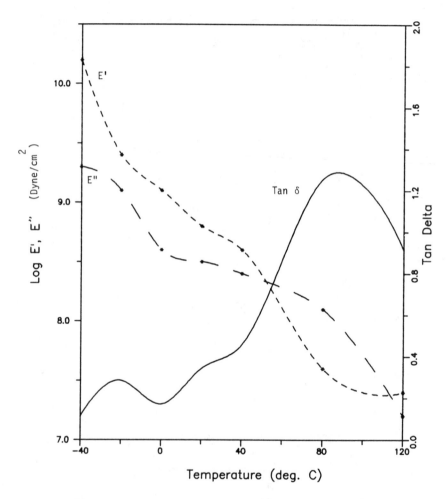

Figure 14. Dynamic Mechanical Spectroscopy of 40/20/50
PU/E/UPE IPN Elastomer with 2% Isonol-100
Post-cured 16 hrs.

(Fig. 6). With Sundex-750T, which is similar to Stan-Flux-LV, lower amounts resulted in broader tan δ peaks (Fig. 8).

As seen from Table IV and Figures 4 and 5, graphite filler does not result in much enchancement of tan δ height or breadth. With plasticizers such as Benzoflex 9-88 and Santicizer-160, increased tan δ height accompanied by a narrower temperature range were observed with Stan-Flux-LV. Further, it was observed that PU/E ratios of 40/60 and 50/50 yielded higher tan δ values and broader temperature range with plasticizer Benzoflex 9-88 and Santicizer-160, as compared to a PU/E ratio of 60/40 (Figs. 9 and 10).

In addition it was also observed that for a given IPN elastomer (#7), reduced post-curing time resulted in higher tan δ values and broader temperature ranges (Fig. 15). THF extraction showed that the epoxy had mostly gelled within the mold during the partial cure (low amounts extracted, see Table VIII). It was also observed (in general) that higher levels of chain extender (Isonol-100) resulted in increased broadening of the loss peaks as well as a decrease in magnitude and shift to slightly higher temperature (due to the increased amount of hard blocks in the segmented polyurethanes). This supports our earlier work (23).

In order to further improve this behavior, a third component (unsaturated polyester), which is virtually immiscible with PU and epoxy, was introduced. Significant effects were observed with the addition of unsaturated polyester. Keeping the amount of polyurethane constant and increasing the amount of unsaturated polyester (as opposed to epoxy) resulted in higher tan δ values and much broader temperature ranges. Further, it was also observed that the tan δ vs. temperature curves stayed high over a very broad temperature range (i.e. after the transition is apparently complete, tan δ still remains fairly high, which is of enormous importance in sound absorption applications (Figs. 11 to 14, Table VII). This behavior is due to the fact that the unsaturated polyester-styrene, with a lower solubility parameter than the polyurethane or epoxy, is immiscible with the latter two components. However, its transition apparently overlaps the PU/epoxy phase broad transition to result in tan δ maintaining a high value even after the PU/epoxy phase transition is complete.

IPN Foams

Two-component IPN foams consisting of polyurethane and epoxy were prepared by the one-shot, free-rise method. The effects of PU/E ratio on the sound absorption and mechanical energy attenuation characteristics were determined with varying levels of different fillers and plasticizers. The formulations (Table IX) were based on the best elastomer results. An average of over 90% absorption was obtained at high frequencies by the impedance tube method. However, this average drops dramatically at low frequencies. This reduction may be seen in Figs. 3 and 4 for 90/10 and 70/30 IPN foams with 20%

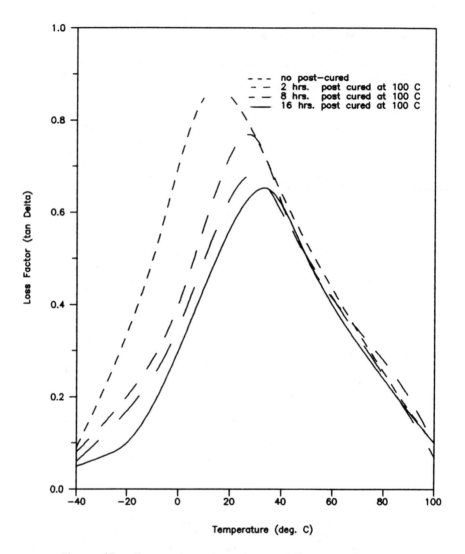

Figure 15.   Temperature dependence of loss tan δ for
              elastomers #7 for varying post-curing time.

Table VIII. Percent Weight Loss After THF Extraction

Sample No.	System	Wt. Loss in %
7	Uncured	14.10
7	2 Hrs. post cured	13.40
7	4 Hrs. post cured	12.10
7	8 Hrs. post cured	11.83
7	16 Hrs. post cured	8.80

Table IX. IPN Foam Formulations

Formulation Fig. No.	#1 16	#2 17	#3 18	#4 19	#5 20	#6 21	#7 22	#8 23	#9 24	#10 25
Isonate 143L	25	25	7.5	17.4	21.4	21.4	25	17.5	24	24
Niax 31-28	100	100	60	50	50	50	100	60	50	50
Water	1.05	1.05	0.77	0.76	0.95	0.95	1.05	0.77	0.90	0.90
1.4-BD	-	-	-	-	-	-	-	-	1.12	1.12
A-1	0.06	0.06	0.06	0.06	0.06	0.06	0.06	0.06	0.06	0.06
T-12	0.02	0.02	0.02	0.02	0.02	0.02	0.02	0.02	0.02	0.02
DC-193	0.5	0.5	0.1	0.1	0.5	0.5	0.5	0.1	0.2	0.2
L-540	0.5	0.5	0.3	0.30	0.5	0.5	0.5	0.2	0.3	0.3
DER-330	-	-	8.6	29.2	48.2	48.2	-	8.6	32.5	32.5
XU-213	-	-	0.25	0.4	0.72	0.72	-	0.25	0.49	0.49
DMP-30	-	-	0.12	0.22	0.36	0.36	-	0.12	0.24	0.24
Freon 11A	15	15	20	20	20	20	20	20	26	26
Plasticizer	-	25[1]	-	-	-	24[1]	12.5[2]	17.4[3]	22[4]	22[3]
Filler	-	-	-	-	-	60/40	100/0	-	-	-
PU/Epoxy	100/0	100/0	90/10	70/30	60/40	60/40	100/0	90/10	70/30	70/30
Filler, %	-	20	-	-	-	20	10	-	-	-
Plasticizer,%	-	-	-	-	-	-	-	20	20	20

[1]Graphite; [2]Rubber powder; [3]Santicizer 160; [4]Santicizer 148

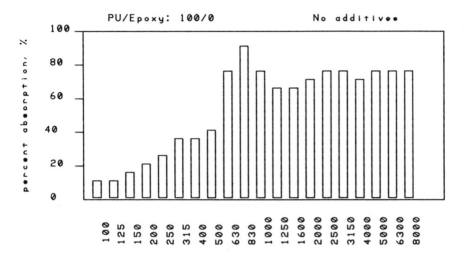

Figure 16.   Normal incidence sound absorption of foams
(Formulation 1).

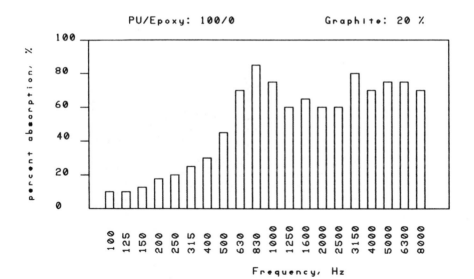

Figure 17.   Normal incidence sound absorption of foams
(Formulation 2).

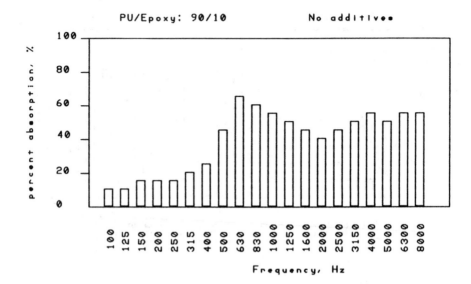

Figure 18. Normal incidence sound absorption of foams
            (Formulation 3).

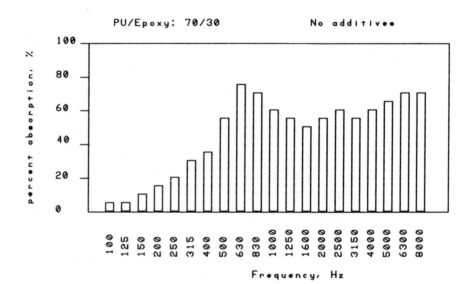

Figure 19. Normal incidence sound absorption of foams
            (Formulation 4).

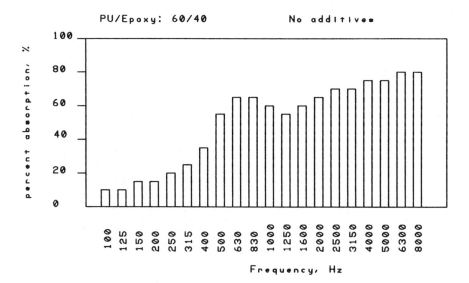

Figure 20. Normal incidence sound absorption of foams (Formulation 5).

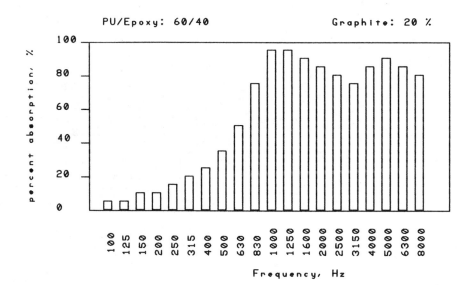

Figure 21. Normal incidence sound absorption of foams (Formulation 6).

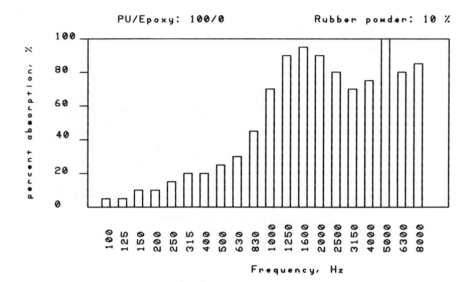

Figure 22. Normal incidence sound absorption of foams (Formulation 7).

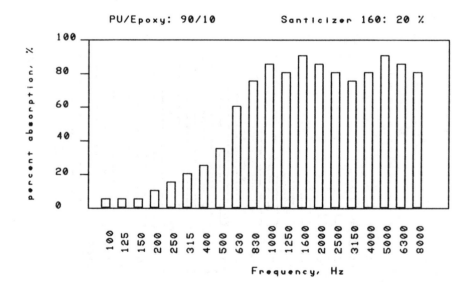

Figure 23. Normal incidence sound absorption of foams (Formulation 8).

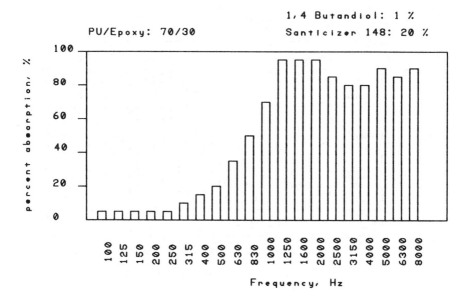

Figure 24.   Normal incidence sound absorption of foams
(Formulation 9).

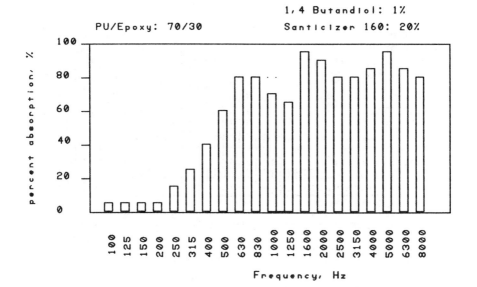

Figure 25.   Normal incidence sound absorption of foams
(Formulation 10).

plasticizer respectively. Addition of the epoxy to the
polyurethane system resulted in IPNs with higher sound
absorption (summary in Fig. 26). Figures 16 and 20
specifically show this effect on 60/40 IPN foam. Also,
addition of fillers such as graphite and rubber powder enhanced
the sound absorption. Figures 16 and 18, 20 and 21, and 18 and
22 and 24, show this enhancement for polyurethane, 60/40, and
90/10 IPN foams, respectively. This effect may also be
observed for 80/20 IPN foams with levels of graphite changing
from 0 to 20% (summary in Fig. 27). Introducing plasticizer
also increased the sound absorption. Figs. 18 and 23, and 19
and 25, indicate this increase for 90/10 and 70/30 IPN foams,
respectively.

Other properties of the PU foams were also affected by IPN
formation and addition of fillers and plasticizers to the
system (Table X). The tensile strength of the foams decreased
with increasing epoxy in the system (Fig. 28). The stress at
50% compression (Fig. 29) slightly decreased and then increased
with a steeper slope with increasing epoxy content. The
presence of fillers such as graphite did not affect the stress
at 50% compression. This may be seen in Fig. 30 for the PU
foam and 90/10 and 80/20 IPN foams. The tensile strength
decreased with increasing graphite content as shown in Fig. 31.

The energy absorbing characteristics of the foams,
hysteresis and rebound, were also affected by introducing epoxy
and fillers. The presence of epoxy dramatically enhanced the
hysteresis as shown in Fig. 32. The rebound decreased with
increasing epoxy in the system (see Fig. 33). The hysteresis
slightly increased with increasing graphite content as shown in
Fig. 34. The slight changes of the rebound with the presence
of graphite can be seen in Fig. 35.

The presence of plasticizers noticeably decreased the
stress at 50% compression and the tensile strength as well as
the hysteresis and the rebound as shown for 90/10 and 70/30 IPN
foams in Figs. 36-38.

CONCLUSIONS

Our study of pure polyurethane, PU/E, and PU/E/UPE IPN
elastomers, has shown that IPNs have very broad glass
transitions (broad tan δ vs. temperature peak) centered around
room temperature. The pure polyurethanes have a relatively
high and sharp T_g well below RT. Fillers such as mica and
graphite have not shown any significant effect on tan δ height
or temperature range.

IPNs of PU/E ratios of 40/60 and 50/50 have even larger
effects on tan δ height and breadth. PU/E/UPE IPN elastomers
with higher amounts of unsaturated polyester have even higher
and broader tan δ ranges.

Foams made from the two-component IPNs showed a
significant increase in energy absorbing ability as compared to
the 100% polyurethane foams. This is indicated by increased
sound absorption, increased hysteresis, and decreased rebound.

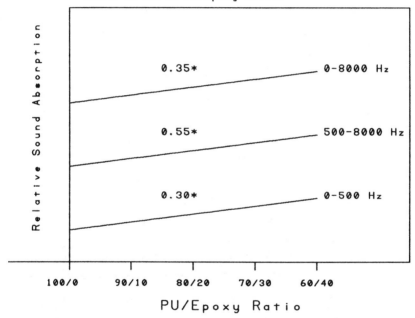

Figure 26. Effect of epoxy content on sound absorption.

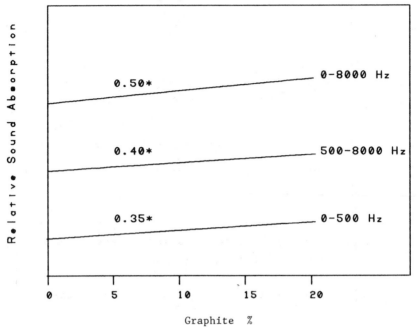

Figure 27.   Effect of graphite content on 80/20 IPN foam.

Table X.   Foam Properties

Formulation No.	Hyster- esis %	D (pcf)	Rebound %	Ts (KPa)	Elongation %	Stress (KPa) at 50% Compression
1	28	3.4	35	158	120	15
2	31	4.3	32	138	107	16
3	39	2.4	20	140	100	21
4	47	2.6	8	114	130	13
5	73	3.1	8	89	150	15
6	69	4.0	6	58	95	12
7	35	4.3	39	207	124	15
8	27	3.2	18	50	90	10
9	30	2.9	3	37	116	17
10	37	3.7	6	75	116	10

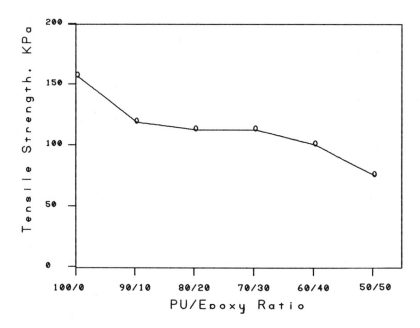

Figure 28.   Tensile strength of the foams v. epoxy content.

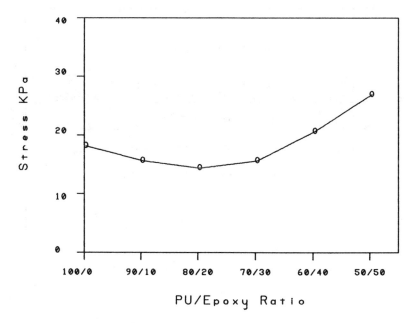

Figure 29.   Stress at 50% compression v. epoxy content.

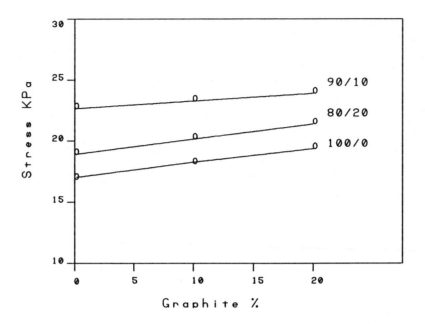

Figure 30.   Stress at 50% compression v. graphite content.

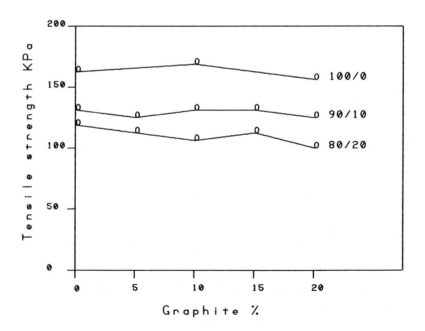

Figure 31.   Tensile strength v. graphite content.

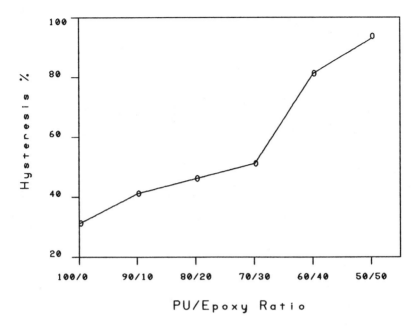

Figure 32.   Hysteresis v. epoxy content.

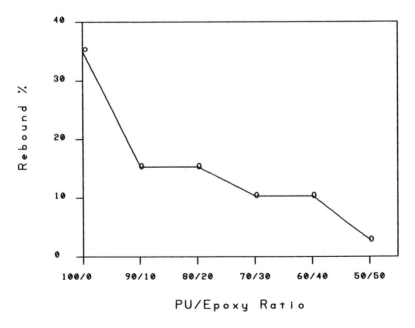

Figure 33.   Rebound of the foams v. epoxy content.

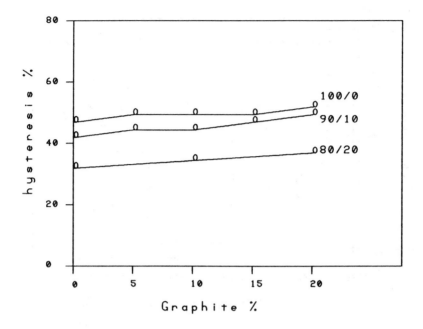

Figure 34. Hysteresis v. graphite content.

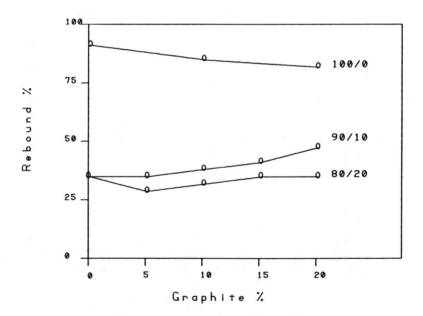

Figure 35. Rebound v. graphite content.

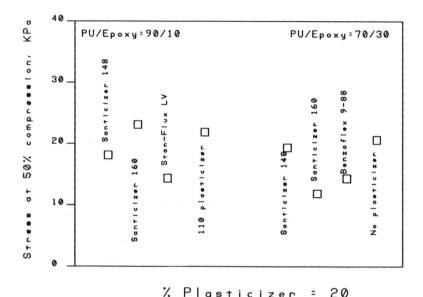

Figure 36.   Effect of plasticizer on stress at 50% compression.

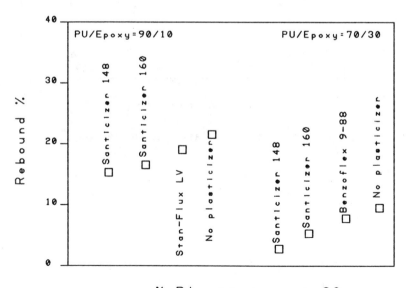

Figure 37.   Effect of plasticizer on tensile strength.

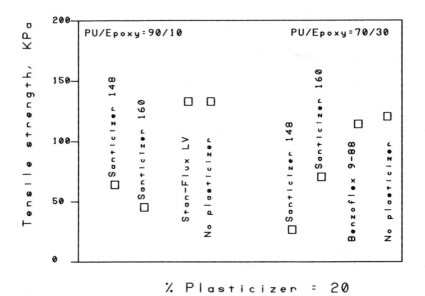

Figure 38. Effect of plasticizer on rebound.

Fillers, such as graphite and rubber powder, also increased the sould absorption while affecting only insignificantly the rebound and hysteresis.

ACKNOWLEDGMENT

The authors gratefully acknowledge the Army Research Office for their support of this work under contract DAAG 25-85-K0129. We would like to acknowledge Mr. Mark Brooks of H.L. Blachford Corp. for his assistance in the sound absorption measurement.

LITERATURE CITED

1. Lipatov, Yu and Sergeeva, L., "Vzaimopronikayushcie polimerovye setki" (Interpenetrating Polymer Networks), Naukova Dumka, Kiev, 1979, in Russian.
2. Millar, J. R., J. Chem. Soc., 1311, 1960; Shibayama, K. and Suzuki, Y., Kobunshi Kagaku, 23, 1966, 249; Shibayama, K., 12, 1962, 362; Shibayama, K., Kagaku, Kobunshi, 19, 1962, 219, ibid., 20, 1963, 221.
3. Frisch, H. L, Klempner, D. and Frisch, K. C., J. Polym. Sci., Polym. Letters, 7, 1969, 775; Frisch, H. L., Klempner, D. and Frisch, K. C., J. Polym. Sci., A-2, 8, 1970, 921; Matsuo, M., Kwei, T. K., Klempner, D. and Frisch, H. L., Polym. Eng. Sci., 10, 1970, 327; Klempner, D. and Frisch, H. L., J. Polym. Sci., B, 8, 1970, 525.
4. Frisch, H. L. and Klempner, D., Adv. Macromol. Chem., 2, 1979, 149.
5. Klempner, D., Angew. Chem., 17, 1978, 97.
6. Frisch, K. C., Frisch, H. L., Klempner, D. and Mukherjee, S. K., J. Appl. Polym. Sci., 18, 1974, 689.
7. Frisch, K. C., Klempner, D., Antczak, T. and Frisch, H. L., J. Appl. Polym. Sci., 18, 1974, 683.
8. Frisch, K. C., Klempner, D., Migdal, S., Frisch, H. L. and Ghiradella, H., Polymer Eng. Sci., 15, 1975, 339.
9. Matsuo, M., Kwei, T. K., Klempner, D. and Frisch, H. L., Polym. Eng. Sci., 19, 6, 1970, 327.
10. Sperling, L. H., Thomas, D. A., and Huelck, V., Macromolecules, 5, 1972, 340.
11. Klempner, D., Frisch, H. L. and Frisch, K. C., J. Polym. Sci., A2, 8, 1970, 921.
12. Manabe, S., Murackami, R., and Takayanagi, M., Mem. Fac. Eng. Kyushu University, 28, 1969, 295.
13. Grates, J. A., Thomas, D. A., Hickey, E. C. and Sperling, L. H., J. Appl. Polym. Sci., 19, 1975, 1731.
14. Frisch, H. L., Frisch, K. C. and Klempner, D., Polym. Eng. Sci., 14, 1974, 9, 646.
15. Frisch, H. L., Cifaratti, J., Palma, R., Schwartz, R., Foreman, R., Yoon, H., Klempner, D. and Frisch, K. C., "Polymer Alloys", edited by Klempner, D. and Frisch, K. C., 10, 1977, 97.

16. Kingsbury, H. B., Cho, K. H. and Powers, W. R., J. Cell.
     Plastics, No. 2, 1978, 113.
17. Frisch, K. C., in "Proceedings of the Workshop on Acoustic
     Attenuation Materials System", NMAB-339, National Acadeny
     of Science, Washington, D. C., 1, 1977.
18. Mizumachi, H., J. Adhes., 2, 1970, 292.
19. Ungar, E. E., in "Noise and Vibration Control", edited by
     Beranke, L. L., McGraw-Hill, New York, 1971.
20. Sperling, L. H., "Interpenetrating Polymer Networks and
     Related Materials", Plenum Press, New York, 1981.
21. Frisch, K. C., Frisch, H. L., Klempner, D. and Mukherjee,
     S. K., J. Appl. Polym. Sci., 18, 1974, 689.
22. Frisch, H. L., Frisch, K. C. and Klempner, D., Polym. Eng.
     Sci., 14, 1974, 646.
23. Klempner, D., Wang, C. L., Ashtiani, M. and Frisch, K. C.,
     J. Appl. Polym. Sci., 32, 1986, 4197-4208.
24. Muni, B. H., Klempner, D., and Okoroafor, M. O.,
     "Two- and Three-Component IPN Elastomers with Enhanced Sound
     Absorption Properties", A.C.S. 19th Central Regional Conf.,
     June 25, 1987.

RECEIVED April 10, 1989

Chapter 12

Fatigue Behavior of Interpenetrating Polymer Networks Based on Polyurethanes and Poly(methyl methacrylate)

T. Hur[1], J. A. Manson[†], R. W. Hertzberg[2], and L. H. Sperling

Center for Polymer Science and Engineering, Department of Chemical Engineering, Materials Research Center, Whitaker Laboratory #5, Lehigh University, Bethlehem, PA 18015

Although interpenetrating polymer networks (IPNs) are now beginning to be commercially exploited, little is known about many types of engineering behavior, such as fatigue. In this paper, energy-absorbing simultaneous interpenetrating networks (SINs) based on polyether-type polyurethanes (PU) and poly(methyl methacrylate) (PMMA) were prepared by both a one-shot procedure and a prepolymer procedure.

While the products have single and broad glass transitions, the SINs prepared by the prepolymer procedure show slightly broader transitions than those by the one-shot procedure. The percent energy absorption determined from dynamic properties and pendulum impact tests increases directly with PU content for both series. However, the values from the prepolymer procedure are always larger than those obtained from the one-shot procedure. These larger values from the prepolymer procedure, combined with a greater extent of phase separation, result in better fatigue resistance than for the one-shot procedure. The fatigue fracture surfaces of these SINs show that the dominant factor for the energy dissipating deformations is shear yielding rather than crazing.

Engineering plastics and composites are frequently subjected to cyclic loads during service, a condition which is inherently more damaging than the corresponding monotonic loading. For these reasons, an understanding of fatigue behavior is extremely important in many components, from molded articles to adhesive joints (1). Fatigue tests may be conducted using either notched or unnotched specimens. While fatigue data acquired from unnotched specimens do not usually distinguish between the initiation and propagation stages, the use of notched specimens allows concentration on the crack propagation stage. Fatigue crack propagation (FCP) studies with notched specimens constitute a conservative approach to the problems of failure, since the use of a notch implies that real materials do contain defects that may grow into catastrophic cracks. Because of the importance of FCP analysis, a

[†]Deceased
[1]Current address: Zettlemoyer Center for Surface Studies, Sinclair Laboratory #7, Lehigh University, Bethlehem, PA 18015
[2]Current address: Department of Materials Science and Engineering, Lehigh University, Bethlehem, PA 18015

0097–6156/89/0395–0309$06.00/0
© 1989 American Chemical Society

program was initiated by J. A. Manson and R. W. Hertzberg in the early 1970's to study the effects of polymer properties and loading conditions on the kinetics and energetics of FCP in a variety of plastics and related multicomponent systems (see Table I).

In the early works within this group, Skibo studied the effect of molecular weight on FCP rates in notched poly(vinyl chloride) (PVC) (2), while Kim studied poly(methyl methacrylate) (PMMA) (3). It has been suggested that resistance to chain disentanglement in cyclic deformation is favored by high molecular weight species. The specific nature of the FCP rate dependence on molecular weight was examined in terms of molecular weight and molecular weight distribution effects on fatigue behavior by Janiszewski et al. and Kim et al. (4,5). The addition of small amounts of high or medium molecular weight tails to the PMMA molecular weight distribution resulted in greater resistance to both fatigue and static loading, whereas the addition of low molecular weight tails had a deleterious effect. It is likely that during cycling, the entanglements in craze fibrils involving low molecular weight species are progressively broken down, but the longer molecules permit considerable strain hardening to occur in the fibrils. Rimnac et al. (6) showed that the overall craze stability improved dramatically with increasing molecular weight and was consistent with the markedly greater degree of FCP resistance observed with high molecular weight specimens.

Bretz et al. (7) examined the effects of molecular weight on FCP behavior in some semicrystalline polymers. Fatigue resistance in semicrystalline polymers such as high density polyethylene (HDPE) (8), poly(oxy methylene) (POM) resin (7), nylon 6,6 (7), and poly(ethylene terephthalate) (PET) (9) were typically higher at higher molecular weight. However, the sensitivity of crack growth rates (da/dN) to molecular weight was less than with an amorphous polymer, because effects of molecular weight reflect the amorphous regions, primarily. Meanwhile, Ramirez et al. (10) successfully differentiated the effects of crystalline content on FCP from effects of crystallite size and perfection in PET, and found that at a given crystallite size and perfection, FCP rates first increase slightly, then decrease as crystallization proceeds, and then increase catastrophically.

Attention has been given to detailed studies of the effects of such experimental variables as frequency and test temperature, since FCP behavior of most polymers is sensitive to these variables due to their viscoelastic nature. In general, FCP rate for polymers decreases with increasing cyclic frequency, although exceptions have been reported (1,11). According to Hahn et al. (12,13), this phenomenon has been attributed to crack tip blunting resulting from local hysteretic heating and to the inhibition of molecular motion due to a shortened deformation response time. Lang et al. (14) showed that an increase of the test frequency on hysteretic heating increased the crack-tip plastic zone size. Also FCP resistance was increased by increasing the material's strain to break. By contrast, the effect of test temperature on FCP rate of polymers has been studied much less extensively than cyclic frequency. Michel et al. (15,16) proposed a model which relates FCP rate to both mechanical and energetic driving forces as well as cyclic frequency and test temperature. Recently, Cheng et al. (17) reported that the FCP rate of PMMA increased with decreasing cyclic frequency and increasing test temperature, but conditions reversed near 100° C.

Increasing attention has been given to the modification of brittle or notch-sensitive polymers by the incorporation of rubbery components to improve the FCP resistance as well as the general toughness during static loading (18). Extensive studies of polymer blends in comparison with the matrix have revealed significant increases in FCP resistance in PVC modified with a methacrylate-butadiene-styrene (MBS) terpolymer (19) and rubber-toughened nylon-6,6 (13,20,21), provided the frequency is low enough to minimize extensive hysteretic heating. The second phase

Table I.   Summary of Fatigue Crack Propagation Ph.D. Dissertations at Lehigh
University, 1976-1988*

YEAR	NAME	TITLE
1976	M. D. Skibo	The Effect of Frequency, Temperature, and Materials Structure on FCP in Polymers
1978	J. Janiszewski	Molecular Weight Distribution Effects on FCP in PMMA
1980	P. E. Bretz	The FCP Behavior of Semi-Crystalline Polymers
1982	S. P. Qureshi	SINs based on Botanical Oils
	A. Ramirez	Effect of Thermal History on FCP in PET
	M. T. Hahn	The Effect of Test and Material Variables on FCP of Nylon 66 Blends
1983	S. M. Webler	Effect of Thermodynamic and Viscoelastic State on FCP in Polymers
	C. M. Rimnac	The FCP Response and Fracture Surface Micromorphology of Neat and Rubber-Modified PVC
1984	R. W. Lang	Applicability of Linear Elastic Fracture Mechanics to Fatigue in Polymers and Short-Fiber Composites
	J. C. Michel	A Unified Approach of Fracture Mechanics and Rate Process Theory
1988	W. M. Cheng	Chemistry and Mechanics Aspects of Fatigue in Polymers
	T. Hur	FCP Behavior in IPNs
1989	J. W. Hwang	FCP Behavior in Rubber-Modified Epoxy (expected)

*In memory of Dr. John A. Manson, 1929-1988.

appears to stimulate shear yielding or crazing in the matrix. Thus, the driving force required to maintain a given crack propagation rate was increased. More recently, Manson et al. (22) showed the improvement in FCP behavior in rubber-toughened polyoxymethylene.

However, little or no significant benefit of rubber modification has been seen in the few epoxies examined, even though the rubber improved the impact strength. The cause of this paradoxical behavior has not been established, but Manson et al. (23) proposed that the rubber-matrix interface may fail well ahead of the advancing crack, thus limiting the already low capacity for shear deformation in the matrix.

Because of the current trend toward the use of interpenetrating polymer networks (IPN) in engineering applications (24-27), consideration of the possible consequences of exposure to fatigue loading becomes extremely important. By combining elastomeric and glassy phases in the form of IPNs, it is possible to obtain improved properties such as FCP resistance, as well as tensile, impact strengths, and damping properties. Qureshi et al. (28) have reported that simultaneous interpenetrating networks (SINs) based on combinations of rubbery polyesters (derived from botanical oils) with crosslinked PS were significantly superior in fatigue resistance to the PS control. In this respect, the SINs are a new type of rubber-toughened plastics.

In view of these promising results, SINs based on polyurethane (PU) and PMMA were prepared by

(1) a one-shot procedure in which all the reactants were mixed and cured simultaneously (29), and

(2) a prepolymer procedure, in which PU was prepolymerized prior to mixing with PMMA prepolymer and cured simultaneously (30).

Another reason for interest is that certain polyurethanes are known to absorb significant energy and to attenuate shock loading effectively. Such energy-absorbing polymers have become important in orthopedic applications, and in sound and vibration damping. This paper describes the synthesis and characterization of a series of cross-polyurethane-inter-cross-poly(methyl methacrylate), PU/PMMA, SINs and an evaluation of their fatigue crack propagation behavior. Comparisons are made between the one-shot procedure and the prepolymer procedure.

EXPERIMENTAL

MATERIALS. Diphenyl methane 4,4' - diisocyanate (MDI), poly(oxypropylene glycol) (PPG) of molecular weight 2000, and trimethylol propane (TMP) were vacuum dried at 80° C for 5 hours. Methyl methacrylate (MMA) (laboratory reagent) was freed from inhibitor and water by passing through a neutral alumina column and molecular sieves. The crosslinking points of each network were introduced via trifunctional polyols (for the PU network) and tetraethylene glycol dimethacrylate (TEGDM) (for the PMMA network).

SNYTHESIS. The exact same formula was used for the two polyurethane syntheses described below. Based on 100 g, this was PPG; 54.5 g, TMP; 18.2 g, and MDI; 27.3 g. Two percent of TEGDM was used in all cases, based on MMA content. The PU catalyst was a mixture of triethylenediamine and a tin catalyst, dibutyl tindilaurate, 2:1, 0.1 wt-% based on PU, and AIBN, 0.4 wt-%, based on MMA.

ONE-SHOT PROCEDURE. All reactants were mixed together at room temperature and catalysts, and initiator were added to promote polyurethane and PMMA network formation. Prior to gelation, the solution was quickly degassed and poured into a mould. After the gelation, the casting was cured at 60° C for 48 hours.

PREPOLYMER PROCEDURE. The isocyanate-terminated PU prepolymer was prepared by reacting two equivalent weights of MDI with one equivalent weight of PPG at room temperature for an hour. Since the PPG may react with both groups on the MDI, some chain extension is expected, with oligomers of the prepolymer coexisting with the di-MDI-mono-PPG product, and some free MDI. Indeed, a viscosity increase was noted. It should be noted that any polymerization reaction, such as between MDI and PPG, will also raise the viscosity to some extent. The PU prepolymer was stored under vacuum (for not more than one day) because of its susceptibility to moisture. The mixture of MMA, TEGDM, and initiator was reacted until 10-15% conversion of the MMA, at which point the reaction was stopped by rapid cooling. Then the PU prepolymer, MMA prepolymer, and TMP were homogeneously mixed using a high-torque stirrer. The air entrapped during mixing was removed by applying a vacuum for 30 sec, and the mixture was poured into a mold. After gelation, the casting was cured at 60° C for 48 hours.

INSTRUMENTATION AND EXPERIMENTAL PROCEDURE. Dynamic mechanical spectra were obtained at 110 Hz using an Autovibron unit, model DDV-IIIC over a temperature range from -100° C to 200° C at a programmed heating rate of 1° C/min.; values of T_g were obtained from the temperatures corresponding to the maxima in the loss modulus (E'') peaks. The shear modulus G was also measured at room temperature using a Gehman torsional tester.

The % energy absorption, E(abs),

$$E(abs) = \frac{100 \ E(absorbed)}{[E(absorbed) + E(recovered)]} \tag{1}$$

was determined in several ways: (1) using a Zwick pendulum tester to obtain the % rebound resilience R (% energy absorbed = 100-R); (2) using a computerized Instron tester (model 1332) to obtain hysteresis loops under cyclic compression at 10 Hz; and (3) using dynamic mechanical data to calculate the ratio of energy absorbed to energy input (per quarter cycle), given by 1 / [1+(2/π tan δ)] (31).

Fatigue tests were conducted on standard-geometry compact tension specimens using a closed-loop electro-hydraulic testing machine under ambient conditions. Specimens were precracked at 100 Hz until a stable crack front was established, and the actual crack growth data were recorded at a sinusoidal frequency of 10 Hz. All testing was performed with an R value of 0.1 (R = min/max load). The crack length, a, was measured using a traveling microscope, or, in some cases, by computer (using a compliance technique).

The range of the stress-intensity factor ΔK, a measure of the driving force for crack, was calculated using the formula

$$\Delta K = Y \Delta \sigma \sqrt{a} \tag{2}$$

where: $\Delta \sigma$ = stress range = $\Delta P/tw$,
   ΔP = load range,
   t = specimen thickness,
   a = crack length, and,
   w = specimen width,
   Y = geometric correction factor, f(a/w).

The crack growth rate at a particular point was calculated to be the average crack growth rate from the previous to the succeeding point (i.e., modified secant method); that is

$$(da/dN)_n = \frac{a_{n+1} - a_{n-1}}{N_{n+1} - N_{n-1}} \tag{3}$$

where a is the crack length and N is the total number of cycles at the time of each crack-tip reading. Values of log da/dN were plotted against log ΔK. This plot reflects the Paris and Erdogan equation (32):

$$da/dN = A \, \Delta K^m \tag{4}$$

where A and m are constants.

The fracture surfaces of the SINs were examined in the ETEC Autoscan Scanning Electron Microscope at a relatively low beam current and accelerating voltage of approximately 20KV. Prior to examination, the surfaces were coated with a thin evaporated layer of gold in order to improve conductivity and prevent charging.

The fracture toughness, K_{1c}, was determined from 3-point bend specimen using ASTM standards E 399 at a strain rate of 10 mm/sec; values reported are the average for 3 specimens.

RESULTS

CHARACTERIZATION OF POLYMERS. The SIN samples from the prepolymer procedure appeared translucent to the eye, whereas the SINs from the one-shot procedure were transparent. The complex moduli are shown in Figures 1 and 2 for the one-shot and prepolymer materials, respectively. It was qualitatively evident that the two types of SIN procedures showed slight degrees of phase separation with T_g's approximately intermediate between those of the component networks, but broader. This, together with the densification observed (29), suggests that extensive interpenetration has taken place. Differences between the one-shot and the prepolymer procedure were also apparent, with slightly but consistently broader transitions with the prepolymer procedure than with the one-shot procedure, as shown in Figure 3. The shoulder in the tan δ data of prepolymer procedure suggests a greater extent of phase separation than the one-shot procedure.

Glass transition temperatures, T_g, as represented by the temperatures for the maxima in the E'' curves, are plotted against composition in Figure 4. As expected, the values varies with composition. However, the T_g's for the prepolymer procedure are always lower than those for the one-shot procedure, which fell between the predictions of Fox (33) and Pochan equations (34) (Figure 3), i.e.,

Fox equation :

$$1/T_g = w(A)/T_g(A) + w(B)/T_g(B) \tag{5}$$

Pochan equation :

$$\ln T_g = w(A) \ln T_g(A) + w(B) \ln T_g(B) \tag{6}$$

where w is the weight fraction and A and B refer to the two homopolymers involved. This may be due to the greater extent of phase separations from the prepolymer procedure than from the one-shot procedure. The modulus at 25° C also varies with composition and is compared with the predictions of several models. These include models of Takayanagi (35) (series-parallel type), Davies (36) (especially suited to dual-phase continuity), Budiansky (37) (phase inversion at the midrange

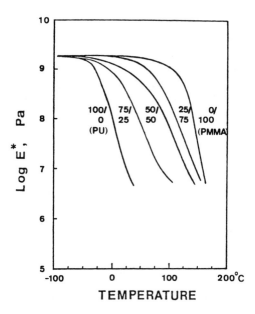

Figure 1.   Complex modulus for PU/PMMA SINs made by one-shot procedure.

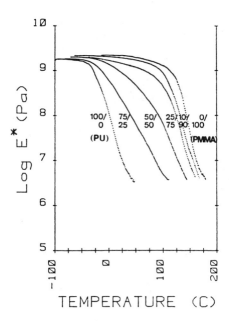

Figure 2.   Complex modulus for PU/PMMA SINs made by prepolymer procedure.

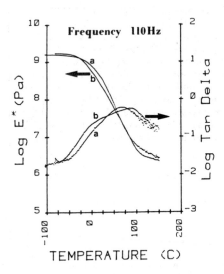

Figure 3.   Complex modulus and tan δ at 110 Hz for PU/PMMA   75/25
SINs: (a) one-shot procedure; (b) prepolymer procedure.

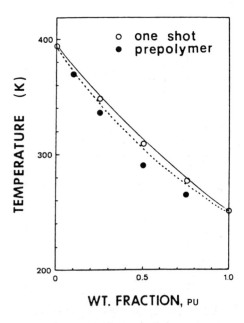

Figure 4.   Glass transition temperature as a function of composition for
PU/PMMA SINs. --- ; Fox Equation (33), —; Pochan Equation (34).

compositions), and Hourston and Zia (38) (a modified Davies equation). As shown in Figure 5, the Davies and Budiansky give the best fit to the data available for both one-shot and prepolymer procedure. This suggests that dual phase continuity exists in the mid-range compositions. Energy absorption [E(abs)] data are presented in Tables II and III, and comparison between the two procedures are made in Figure 6. It was not expected that absolute values of E(abs) obtained from such widely different tests involving quite variable states of stresses and both linear and nonlinear viscoelastic behavior to agree. In all cases, however, the energy absorption values clearly increase directly with the PU content, and the values for the SINs from prepolymer are always higher than the values from the one-shot procedure. Tables II and III summarize much of the numerical data.

Table II.   Properties of PU/PMMA SINs made by One-Shot Procedure

| | | | E(abs) % | | |
PU/PMMA	(110 Hz) T_g° C	Tan δ (25°C)	DMS	Hysteresis	K_{1c}^*
0/100	121	0.07	9.9	54	1.43
25/75	74	0.09	12.4	59	1.56
50/50	35	0.11	14.7	63	1.68
75/25	3	0.27	29.8	71	--
100/0	-24	0.93	59.4	85	--

*ASTM E399: strain rate 10 mm/sec.

Table III.   Properties of PU/PMMA SINs made by Prepolymer Procedure

| | | | E(abs) % | | |
PU/PMMA	(110 Hz) T_g° C	Tan δ (25°C)	DMS	Hysteresis	K_{1c}^*
0/100	121	0.07	9.9	54	1.43
10/90	100	0.08	11.2	58	1.53
25/75	63	0.1	13.5	64	1.65
50/50	16	0.148	18.9	70	1.73
75/25	-9	0.324	33.7	80	--
100/0	-24	0.93	59.4	85	--

*ASTM E399; strain rate 10 mm/sec.

FATIGUE CRACK PROPAGATION. As shown in Figures 7 and 8, the SINs from both the one-shot and the prepolymer procedure were significantly superior in FCP behavior to the crosslinked-PMMA control. The FCP resistance was improved with increasing PU content. Unfortunately, when the amount of PU was larger than 50

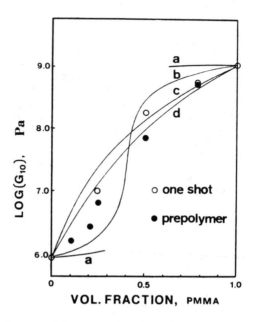

Figure 5.   Shear modulus as a function of composition for   PU/PMMA
SINs.   Models of; (a) Takayanagi(35), (b) Budiansky(37), (c) Davies(36), (d)
Hourston and Zia(38).

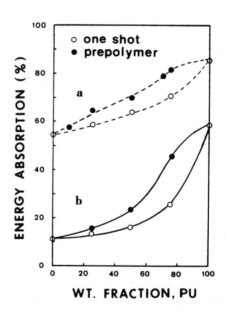

Figure 6.   Energy absorption as a function of composition.   (a) from
hysteresis, (b) average of pendulum and dynamic mechanical tests.

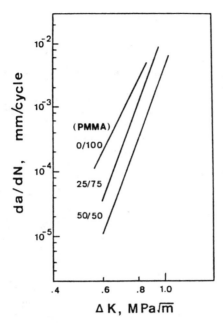

Figure 7.   Fatigue crack growth rates for PU/PMMA SINs made by one-shot procedure.

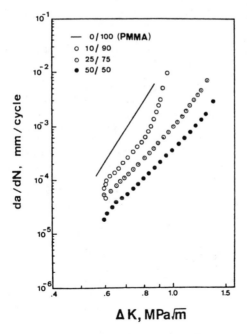

Figure 8.   Fatigue crack growth rates for PU/PMMA SINs made by prepolymer procedure.

wt-%, the FCP resistance could not be measured, since the specimen became too compliant (1,19).

With respect to the specimens that could be tested, the SINs made by the prepolymer procedure showed better FCP resistance than the SINs made by the one-shot procedure except at low ΔK values (Figure 9). Figure 10 shows that the driving force for crack extension ΔK^* (the value of ΔK corresponding to an arbitrary crack speed, in this case 10^{-3} mm/cycle) increased directly with the PU content. In order to reach the same crack growth rate, the SINs from prepolymer procedure needed much larger driving force than those from the one-shot procedure. This is consistent with the relatively higher energy absorption values exhibited by the SINs from the prepolymer procedure (see Table II and III).

FRACTURE BEHAVIOR. The fracture behavior of SINs has been examined at a displacement rate of 10 mm/sec using the 3-point bend specimen. Three basic types of fracture behavior (not shown) could be identified according to the shape of the load-displacement curve and these are associated with different types of crack growth behavior (39,40). Unmodified crosslinked PMMA showed a loading curve with no detectable decrease in gradient before the maximum load, at which point the crack propagates unstably causing a rapid load drop (brittle crack growth). In the case of 10/90, and 25/75 SINs, the load rises almost linearly until the maximum value and drops relatively slowly as the crack propagates along the specimen (transition from brittle to ductile crack growth). On the other hand, 50/50 SIN shows a different load-displacement behavior. While the load increases linearly at first, it becomes non-linear before the maximum load as the crack extends stably for a considerable distance (ductile crack growth).

FRACTOGRAPHY. Fracture surface study showed the phenomenon of stress whitening, which is caused by light scattering due to voids generated in the polymer during the fracture process. The size of the stress-whitened region increased with the PU content. In glassy thermoplastics, such microvoids arise from the initiation and growth of crazes, from shearing, or from rubber cavitation. Donald and Kramer (41) showed a transition from a crazing to a shear yielding mechanism as the length of the chains between physical entanglements decreases. Thus, in thermosetting resins, there is no evidence of crazing (40). When the crosslink density is very high or the length of chain between chain entanglement (in the present case, crosslinks and interpenetration) is comparatively short, crazing will be suppressed. Therefore, microvoid formation is expected to involve shearing or cavitation rather than crazing.

Goodier (42) showed that for a particle which possesses a considerably lower shear modulus than the matrix, the maximum stress concentration occurs at the equator of the particle. Rubbers are commonly found to undergo cavitation quite readily under the action of a triaxial tensile stress field. Thus, the microvoids are produced by cavitation around the rubbery particles during fatigue crack propagation, together with localized plastic deformation due to interaction between the stress field ahead of the crack and the rubber particles.

ONE-SHOT PROCEDURE. When the stress-whitened region was examined in detail, there was no evidence of rubber cavitation. Instead, Figure 11 showed discontinuous growth bands (DGB) whose spacings correspond to many cycles of loading before the crack jumps to a new position without an increment of crack growth from each load cycle (1,6). Indeed the crack jump was visually observed during fatigue tests. The effect of composition on the spacing of the bands, r, the yield stress, σ_y, (estimated from the Dugdale relationships, see equation 7), and the number of cycles per band was demonstrated in Figure 12. While such bands are

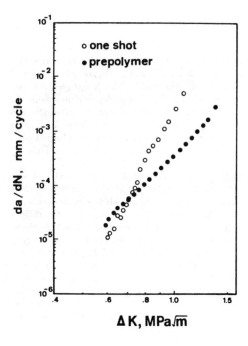

Figure 9. Comparison of fatigue crack growth rates for PU/PMMA 50/50 SINs.

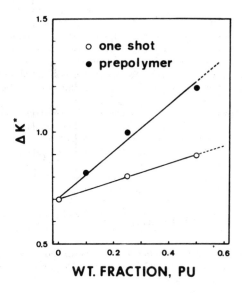

Figure 10. Effects of % energy absorption on ΔK^* (at $da/dN = 10^{-3}$ mm/cycle).

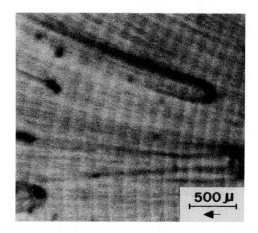

Figure 11. Scanning electron micrograph of fractured surfaces for PU/PMMA 50/50 SINs made by one-shot procedure ($\Delta K = 0.75$ MPa$\sqrt{m}$).

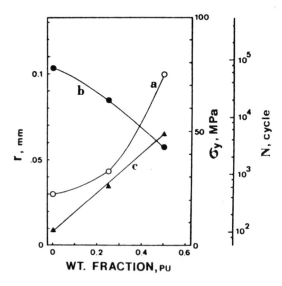

Figure 12.   Effects of composition of PU/PMMA SINs made by one-shot procedure on: (a) r, the size of DGB spacing; (b) σ_y, yield stress; and (c) log N, the number of cycles per band.

often associated with crazing, shear yielding is expected in this system because of the crosslinking and interpenetration of both networks (there is no evidence of classical void gradient). Indeed, the estimated value of yield stress for the PMMA is 77 MPa, a value more typical of shear yield, rather than crazing, stresses (1,43).

Figure 13 shows that the trend of band size in 50/50 SINs as a function of ΔK resembles that of a typical ABS (acrylonitrile-butadiene-styrene terpolymer), and that the slope of the curve conforms to the Dugdale prediction(44).

$$\gamma_y = \frac{\pi\, K_{max}^2}{8\, \sigma_y^{\,2}} \tag{7}$$

where γ_y is the damage zone dimension, K_{max} the maximum stress intensity factor, and σ_y yield strength.

PREPOLYMER PROCEDURE.   The cavitation around the PU domains is seen in Figure 14(a), but the extent is relatively small. The dominant factor for the energy dissipating deformations that occur in the vicinity of the crack tip of a propagating crack is presumably the localized plastic deformation (Figure 14).

When this localized plastic deformation is examined in detail, there are several interesting features.   Robertson and Mindroiu (45) explained complex fracture surfaces of thermosets based on the hypothesis of an instability of the propagating crack front that produces a "crack fingering" ahead of the nominal crack front. These fingers may lose coherence and move off in different direction when the stress field is inhomogeneous.   Thus, an abrupt change in modulus when the crack encounters dispersed PU particles in PU/PMMA SINs is believed to divide a crack into a pair of cracks with each on   a different plane (46,47).   Once a crack bifurcates, there is little tendency for cracks to rejoin, and therefore the pair of cracks continue on different planes for considerable distances without breaking the intervening membrane to reconnect.   The   resulting large separations between the crack planes of adjacent fingers yield large steps and welts when the intervening material is finally broken through.   The fracture surface of 10/90 SINs in Figure 14(b) shows these steps and welts as the heavy shadows and the complete whiteness, respectively.   However, 25/75 SINs shows somewhat different fracture surfaces. The fracture surface shows only irregular   patch pattern in Figure 14(a).   Finally, in the case of 50/50 SIN the fracture surfaces show clear evidence of pronounced ductility, having an almost torn appearance (data not shown).

DISCUSSION

The questions must be answered, why does the prepolymer route give consistently greater phase separation than the one-shot route, and why are the associated mechanical properties better?   Phase domain size in sequential IPNs and SINs alike is controlled by which of two phenomena happen first during the polymerization: gelation, or phase separation. If gelation happens first, the network tends to hold the two components together; the resulting domains tend to be smaller, and/or ill defined. If gelation happens after phase separation, the crosslinks tend to hold the domains apart, and the phases are larger, and/or better defined.

During the synthesis of a SIN where one of the components polymerizes by stepwise kinetics, the number average molecular weight of the stepwise component remains low until just before gelation, then things happen quickly.   The lower molecular weight of the components before gelation permits mutual solubility of the two polymers until rather late in the polymerization, then gelation and phase separation may take place at nearly the same time, but the order of events is still the critical consideration.

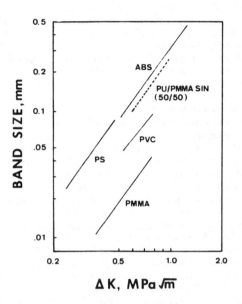

Figure 13. Comparison of r, the length of DGB bands as a function of ΔK, for PU/PMMA SINs made by one-shot procedure with values for several other polymers. (Adapted from ref. 1.)

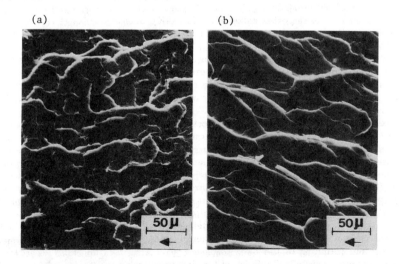

Figure 14. Scanning electron micrograph of fractured surfaces for PU/PMMA made by prepolymer procedure. (a) 25/75 SINs, ΔK = 0.70 MPa√m̄; (b) 10/90 SINs, ΔK = 0.72 MPa√m̄.

While the details of the order are not known for the present case, it may be that the longer or better defined initial chains in the chain-extended prepolymer case cause phase separation to happen slightly earlier in the polymerization, perhaps before or just at the gelation stage. As previously noted, it was observed that the viscosity was significantly increased when one equivalent weight of PPG was prepolymerized with two equivalent weights of MDI. This indicates that there was significant chain extension which resulted in the much longer initial chains. With the two identical overall recipes, the principal difference lies in the distribution of the crosslinks. In the one-shot synthesis, the crosslinks are nearly random, whereas in the prepolymer case, the crosslinks tend to be clumped up together between chain extended prepolymer segments.

In any case, the prepolymer synthesis results in somewhat better defined domains, characterized by the relative translucency of the prepolymer route material relative to the one-shot route, and as illustrated in the transition behavior, Figure 3. Better defined domains mean a purer rubber phase, which in turn may yield greater toughness to this form of the material. Important questions as to the development of intermediate T_g's with extensive interfacial boundary material (48) or on complete thermodynamic miscibility (49) remain for the future.

The increase in the FCP resistance with increasing PU content was observed in Figures 7 and 8. Manson and Hertzberg (1,11) noted that the fatigue response of rubber-toughened plastics is dominated by the characteristically higher level of viscoelastic damping relative to that of the unmodified plastics. Thus, the increased FCP resistance of the SINs is believed to arise from the ability of the PU network to dissipate energy. Indeed, Tables II and III show that the percent energy absorption determined from dynamic properties and pendulum impact tests was significantly increased with increasing PU content. Figure 9 showed that SINs by the prepolymer procedure had better fatigue resistance than SINs by the one-shot procedure. However, at low ΔK levels, the one-shot procedure was superior in the FCP resistance to the prepolymer procedure. Hertzberg (43) suggested that ductile alloys are best for high cyclic strain applications (higher ΔK) since more energy is needed to fracture them. At the other extreme (at low ΔK), ductility or toughness is not as important. Rather, resistance to deformation and damage accumulation is important in the region of low strains (43, p505). Therefore, in the present study, the fact that the percent energy absorption values for the prepolymer procedure are always larger than those for the one-shot procedure can explain why the prepolymer procedure showed better fatigue resistance than the one-shot procedure at high ΔK levels. Also, the higher moduli of SINs by one-shot procedure (Figure 5) resulted in better resistance at low ΔK levels.

Skibo et. al (19) examined the effect of rubber content on FCP behavior. While a decrease in modulus due to an increase in rubber content tends to increase crack growth rates, a decrease in yield strength, σ_y, is beneficial as a result of the associated increase in the damage zone dimension (γ_y) given by equation 7.

Also, localized heating in the damage zone would act to blunt the crack tip, thereby yielding a lower growth rate. However, at higher concentration of rubber ([PU] > 50%), the significant decrease in modulus as the PU content increases evidently overcame the beneficial effects of energy dissipation. Thus, the net effect of rubber on fatigue resistance will be determined by a dynamic energy balance between the ability of the rubber to generate energy-dissipating processes at the crack tip and softening of the bulk polymer (50). Also, for the highly elastomeric systems, specimens specifically designed for elastomers will have to be used.

Finally, the difference in the fractography between the one-shot (Figure 11) and the prepolymer procedure (Figure 14) may be due the different morphology. The nearly miscible one-shot SINs behaved like a homopolymer, since the size of the PU domain was too small (several tens of nm). On the other hand, the more separated prepolymer SINs, which are expected to have larger PU domains, showed extensive localized plastic deformations because of inhomogeneities in the stress fields.

CONCLUSIONS

The SINs prepared by a prepolymer procedure show a slightly greater phase separation than those by a one-shot procedure, while the SINs from both methods appear to have a single and broad glass-rubber transition, indicating a microheterogeneous morphology. The 25° C shear modulus behavior of the SINs are approximately described by the Davies model and the Budianski model, which suggests dual-phase continuity in the mid-range compositions.

FCP resistance for the SINs increases with PU content up to 50% and is better in the prepolymer material than in the one-shot material, since the former always has larger values of percent energy absorption. With respect to micromechanisms of failure, the generation of discontinuous growth bands associated with shear yielding is involved in the SINs from the one-shot procedure. On the other hand, the fracture surfaces of the SINs from the prepolymer procedure show extensive stress-whitening phenomenon which is associated with the cavitation around PU domains and localized shear deformation.

ACKNOWLEDGMENT

The authors wish to acknowledge support from the National Science Foundation Materials Division, Grant No. DMR-8412357, Polymers Program.

This paper is dedicated to the life and works of Dr. John A. Manson, pioneer in the field of polymer science and engineering, who passed away February, 1988, prematurely.

LITERATURE CITED

1. Hertzberg, R. W. and Manson, J. A. Fatigue in Engineering Plastics: Academic, New York, 1980.
2. Skibo, M. D.; Manson, J. A. and Hertzberg, R. W. J. Macromol. Sci. Phys. 1977, B 14(4), 525.
3. Kim, S. L.; Skibo, M. D.; Manson, J. A. and Hertzberg, R. W. Polym. Eng. Sci. 1977, 17(3), 194.
4. Janiszewski, J.; Hertzberg, R. W. and Manson, J. A. ASTM STP 743, Standard for Metric Practice; American Society for Testing and Materials: Philadelphia, Pa., 1981; P.125.
5. Kim, S. L.; Janiszewski, J.; Skibo, M. D.; Manson, J. A. and Hertzberg, R. W. Polym. Eng. Sci. 1978, 18, 1093.
6. Rimnac, C. M.; Hertzberg, R. W. and Manson, J. A. ASTM STP 733; American Society for Testing and Materials: Philadelphia, Pa., 1981; P.291.
7. Bretz, P. E.; Manson, J. A. and Hertzberg, R. W. J. Appl. Polym. Sci. 1982, 27, 1707.
8. Sheu, C. and Goolsby, R. D. Org. Coat. Appl. Polym. Sci. Proc. 1983, 48, 833.
9. Ramirez, A.; Manson, J. A. and Hertzberg, R. W. Polym. Eng. Sci. 1982, 22, 975.
10. Ramirez, A.; Gaultier, P. M.; Manson, J. A. and Hertzberg, R. W. in Fatigue in Polymers; Plastics and Rubber Institute: London, UK, 1983; Paper 3.
11. Hertzberg, R. W. and Manson, J. A. Fatigue and Fracture; Encyclop. Polym. Sci. Eng., Wiley-Interscience: New York, 1986; Vol 6, 2nd Ed..
12. Hahn, M. T. Ph.D. Dissertation, Lehigh University, Pennsylvania, 1982.

13. Hahn, M. T.; Hertzberg, R. W. and Manson, J. A. J. Mater. Sci. 1986, 21, 31.
14. Lang, R. W.; Hahn, M. T.; Hertzberg, R. W. and Manson, J. A. ASTM STP 833, Standard for Metric Practice; American Society for Testing and Materials: Philadelphia, Pa., 1984; P.266.
15. Michel, J. C.; Manson, J. A. and Hertzberg, R. W. in Deformation, Yield, and Fracture of Polymers; Plastics and Rubber Institute: London, UK, 1985; Paper 21.
16. Michel, J. C; Manson, J. A. and Hertzberg, R. W. Polym. Prepr. Am. Chem. Soc. Div. Polym. Chem. 1985, 26(2), 141.
17. Cheng, W. M.; Manson, J. A.; Hertzberg, R. W.; Miller, G. A. and Sperling, L. H. Polym. Mater. Sci. Eng. 1988, 59, 418.
18. Bucknall, C. B. Toughened Plastics; Applied Science Publisher: London, UK, 1977.
19. Skibo, M. D.; Manson, J. A.; Hertzberg, R. W.; Webler, S. M. and Collins, Jr., E. A. in Durability of Macromolecular Materials; Eby, R. K., Eds.; ACS Symp. Ser. No. 95; American Chemical Society: Washington, D.C., 1979.
20. Hertzberg, R. W.; Skibo, M. D. and Manson, J. A. ASTM STP 700, Standard for Metric Practice, American Society for Testing and Materials: Philadelphia, Pa., 1980; P.490.
21. Hahn, M. T.; Hertzberg, R. W. and Manson, J. A. J. Mater. Sci. 1986, 21, 39.
22. Manson, J. A.; Hwang, J.; Pecorini, T.; Hertzberg, R. W. and Connelly, G. M. Polym. Mater. Sci. Eng. 1987, 57, 436.
23. Manson, J. A.; Hertzberg, R. W.; Connelly, G. M. and Hwang, J. in Multicomponent Polymer Materials; Sperling, L. H. and Paul, D. R., Eds.; ACS Advances in Chemistry Series No. 211; American Chemical Society: Washington, D.C., 1986.
24. Manson, J. A. and Sperling, L. H. Polymer Blends and Composites; Plenum: New York, 1976.
25. Sperling, L. H. Interpenetrating Polymer Networks and Related Materials; Plenum: New York, 1981.
26. Sperling, L. H. in Multicomponent Polymer Materials; Sperling, L. H. and Paul, D. R., Eds.; ACS Advances in Chemistry Series No. 211; American Chemical Society: Washington, D.C., 1986.
27. Sperling, L. H. Chemtech 1988, 18, 104.
28. Qureshi, S.; Manson, J. A.; Sperling, L. H. and Murphy, C. J. in Polymer Applications of Renewable Resource Materials; Carraher, C. E. and Sperling, L. H., Eds.; Plenum: New York, 1983.
29. Hur. T.; Manson, J. A. and Hertzberg, R. W. in Crosslinked Polymers; Dickie, R. A.; Labana, S. S. and Bauer, R. S., Eds.; ACS Symp. Ser. No. 367; American Chemical Society: Washington, D.C., 1988.
30. Hur, T.; Manson, J. A.; and Hertzberg, R. W. Polym. Mater. Sci. Eng. 1988, 58, 894.
31. Ferry, J. D. Viscoelastic Properties of Polymers; Wiley: New York, 1980; 3rd Ed..
32. Paris, P. C. and Erdogan, F. J. Bas. Eng. Trans. ASME 1963, Ser. D85(4), 528.
33. Fox, T. G. Bull. Am. Phys. Soc. 1956, 1, 1.
34. Pochan, J. M.; Beatty, C. L. and Hinman, D. F. Macromolecules 1977, 11, 1156.
35. Takayanagi, M. Mem. Fac. Eng. Kyushu Univ. 1963, 23, 11.
36. Davies, W. E. J. Phys. 1971, (D)4, 318.
37. Budiansky, B. J. Mech. Phys. Solids 1965, 13, 223.
38. Hourston, D. J. and Zia, Y. J. Appl. Polym. Sci. 1983, 28, 3745.

39. Cherry, B. W. and Thomson, K. W. J. Mater. Sci. 1981, 16, 1913.
40. Kinloch, A. J.; Shaw, S. J.; Tod, D. A. and Hunston, D. L. Polymer 1983, 24, 1341.
41. Donald, A. M. and Kramer, E. J. J. Mater. Sci. 1982, 17, 1871.
42. Goodier, J. N. Trans. Am. Soc. Mech. Eng. 1933, 55, 39.
43. Hertzberg, R. W. Deformation and Fracture Mechanics of Engineering Materials; John Wiley: New York, 1983; 2nd Ed..
44. Dugdale, D. S.; J. Mech. Phys. Solids 1960, 8, 100.
45. Robertson, R. E. and Mindroiu, V. E. Polym. Eng. Sci. 1987, 27(1), 55.
46. Kinloch, A. J.; Maxwell, D. L. and Young, R. J. J. Mater. Sci. Lett. 4 1985, 1276.
47. Kinloch, A. J.; Maxwell, D. L. and Young, R. J. J. Mater. Sci. 1985, 20, 4169.
48. Lipatov, Yu. U.; Chramova, T. S.; Sergeva, L. M. and Karabanov, L. V. J. Polym. Sci. Polym. Chem. Ed. 1977, 15, 427.
49. Olabisi, O.; Robeson, L. M. and Shaw, M. T. Polymer-Polymer Miscibility; Academic: New York, 1979.
50. Lang, R. W. Ph.D. Dissertation, Lehigh University, Pennsylvania, 1984.

RECEIVED April 27, 1989

NEW MATERIALS

Chapter 13

Synthesis and Characterization of Star-Branched Ionomers Composed of Sulfonated Polystyrene Outer Blocks and Elastomeric Inner Blocks

R. F. Storey and Scott E. George

Department of Polymer Science, University of Southern Mississippi, Hattiesburg, MS 39406–0076

Star-branched ionomers composed of short sulfonated polystyrene outer blocks and hydrogenated butadiene or isoprene inner blocks, with narrow molecular weight distributions (<1.1), have been synthesized by living anionic polymerization and chlorosilane linking techniques, with subsequent hydrogenation and sulfonation reactions. The products are low molecular weight thermoplastic elastomers. Elasticity of the hydrogenated segments was obtained by varying the polymerization conditions, thereby controlling the % 1,2-addition of butadiene. Quantitative ^{13}C and ^{1}H NMR and FT-IR revealed > 99% hydrogenation using p-toluene sulfonyl hydrazide in refluxing toluene. Titration of the sulfonic acid groups revealed > 85% sulfonation of the styrene units by acetyl sulfate in methylene chloride. Characterization of the final product and the various intermediates was carried out by GPC, DSC, FT-IR, and ^{13}C NMR. A summary of the developmental work leading to these new materials is presented.

Ionomers are principally non-polar polymers which contain a small number, i.e., < 15 mole %, of attached ionic groups (1). Most commonly, the ions are derived from neutralization of carboxylic or sulfonic acids. The presence of only a small number of such groups strongly affects the properties of the polymer by creating strong intra- and intermolecular secondary bonding forces. If the polymer has a glass transition temperature (T_g) significantly below room temperature, the resulting ionomer is a thermoplastic elastomer, with strong ion aggregation serving as a thermally reversible crosslinking mechanism.

An interesting sub-class of ionomers are telechelic ionomers, of which there are several noteworthy examples. The term "telechelic" indicates that the ions are attached exclusively at the chain termini and that every chain end contains an ionic moiety. Such placement provides a network free of dangling chain ends, and minimizes melt viscosity, since the telechelic polymer molecular weight is of the same order as the elastically effective molecular weight between crosslinks. This is in direct contrast to "random"

0097–6156/89/0395–0330$06.75/0
© 1989 American Chemical Society

ionomers in which the polymer molecular weight is relatively high, and the ions are attached at random along the polymer backbone. A very important feature of telechelic ionomers is that the concentration of ions in the polymer is inversely proportional to the number average molecular weight (M_n). Two examples of telechelic ionomers are the carboxy-terminated polybutadienes, which have been actively investigated by Broze, Jerome, and Teyssie, (2-3) and the sulfonated polyisobutylene (PIB) ionomers which were developed by Kennedy and Storey (4-5) and later characterized extensively by Wilkes et al. (6-8)

It is appropriate to briefly discuss telechelic PIB ionomers, the structures of which are shown in Figure 1, since their study brought into focus the effect of branching on the properties of telechelic ionomers. Due to the relationship between molecular weight and ion content in telechelic ionomers, samples containing chain segments sufficiently long to display elasticity have relatively low ion contents, i.e., < 3 mole %. Thus it is thought that higher organizational structures (clusters) are absent, and that ion pair multiplets, especially ion pair dimers, are the most important structures. This hypothesis is supported by the differences observed in the properties between linear and three-arm star PIB's. Shown in Figure 2 are stress vs. strain curves for four three-arm star samples of varying M_n and one linear sample, all neutralized with KOH. The linear sample, which lacks a covalent branch point, is very weak in relationship to the star-branched samples. Apparently, ion aggregation in the form of ion pair dimers produces only chain extension in the linear sample, and a network is not formed. In contrast, a strong network exists in the star-branched samples. Finally, it is noteworthy that the lowest molecular weight sample (M_n = 6,900), which possesses the highest ion content, is relatively stiff and inelastic. It is difficult to know whether this is due to low molecular weight or high ion content (or both), since neither parameter can be adjusted without changing the other.

Given the potential usefulness and very interesting properties of elastomeric, star-branched, telechelic ionomers, it is appropriate to explore further the possibilities for such systems. This paper discusses the synthesis and characterization of poly(ethylene-co-1-butene)-based star-branched ionomers carrying variable-length outer blocks of sulfonated styrene units. As shown schematically in Figure 3, they are in effect star-block copolymers comprised of elastic inner blocks and polyelectrolyte outer blocks. The principal advantage of this block ionomer system is that ion concentration can be varied independently of molecular weight while limiting placement of the ion-containing units to the ends of the polymer molecules.

Experimental

Sulfonation of Model Compounds.   Suppliers of materials were ICN Biomedicals and Wiley Organics for *cis*-3-hexene and *cis*-3-methyl-3-hexene, respectively, and Aldrich Chemical Co. for *trans*-3-hexene, decalin, and decane.

The model compounds and decalin were weighed into a small glass vial containing a magnetic stir-bar. A small, arbitrary amount of n-decane was added to serve as internal standard for GC analysis. Acetyl sulfate was generated *in situ* by adding equimolar amounts of first acetic anhydride and then concentrated (> 96%) sulfuric acid. Upon formation, acetyl sulfate separated into a lower layer. As the reaction proceeded the appearance of the lower layer changed from a pale yellow color to a dark yellow-brown. A representative procedure was as follows: To the glass vial were charged 0.055 g (0.56 mmole) *cis*-3-methyl-3-hexene, 0.467 g (5.6 mmole) *cis*-3-hexene,

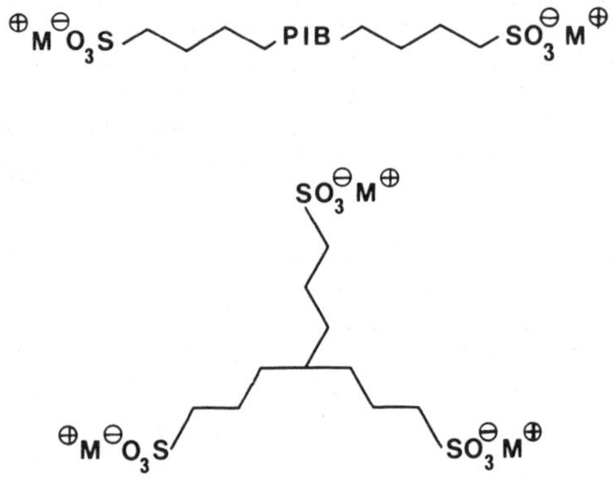

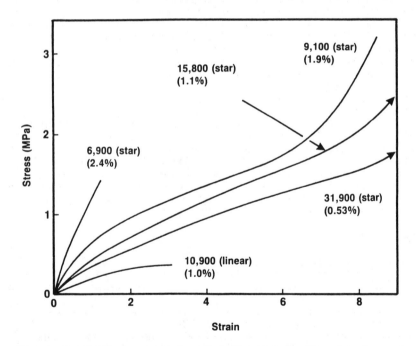

Figure 1. Linear and three-arm star telechelic polyisobutylene-based metal sulfonate ionomers.

Figure 2. Stress vs. strain for telechelic polyisobutylene-based metal sulfonate ionomers. Numbers at each curve indicate sample molecular weight and mole % ion content (in parenthesis).

5.0 g decalin, and 0.5 g n-decane. Acetyl sulfate was generated *in situ* by the dropwise addition of 0.57 g (5.6 mmole) acetic anhydride and 0.55 g sulfuric acid. The stirred mixture was allowed to react for 2 h, during which time samples were taken from the upper hydrocarbon layer every 15 min and analyzed for loss of olefins by GC.

<u>Polymer Synthesis</u>. All materials were obtained from Aldrich Chemical Co., and rigorous purification was performed using standard high vacuum techniques. (9) The general polymerization procedure was as follows: A 1 L all glass reactor was constructed, equipped with a magnetic stir bar, and checked for pinholes with a tesla coil. The apparatus was then flamed with a hand torch several times to remove surface moisture. Approximately 500 ml of benzene was distilled into the vessel from an oligostyryl lithium solution, and 6.4 ml (0.056 mole) of styrene was released into the reactor from an ampule fitted with a breakseal. In some cases, triethyl amine was added at this juncture to serve as polar cosolvent. An ampule containing 0.0112 mole of s-butyl lithium as a 1.60 N solution in benzene was then added to form the red-orange living oligostyryl solution. After the reaction had proceeded for 36 h at 24°C an ampule of 1,3-butadiene was slowly added over a period of several hours, during which time a cold bath was used to maintain the polymerization temperature below 30°C (the temperature was maintained at 10°C for reactions employing a polar cosolvent). During the next 48 h color progressed from red-orange to a pale yellow. The linking agent, methyl trichlorosilane, (3.373 mmole) was then added from an ampule as a 20% solution in benzene. An additional 48 h was then allowed for the linking reaction to proceed before the reactor was opened and the contents precipitated into methanol. Redissolution into hexanes and precipitation into methanol was then performed two times more, and the polymer was stored until use at 0°C in a hexanes solution under a blanket of argon. Using gel permeation chromatography (GPC), the molecular weight of the polymer was found to be three times that of the arm when compared to polystyrene standards.

<u>Hydrogenation</u>. A diimide technique (10) was used to selectively hydrogenate the butadiene block of the copolymer under mild conditions. The transitory diimide species was generated from p-toluene sulfonyl hydrazide (TSH) which was obtained from Aldrich Chemical Co. and used without further purification. Thus, into a 2 L round bottom flask fitted with a magnetic stir bar, reflux condenser, and heating mantle was charged 50 g polymer (0.860 mole butadiene repeat units). Sufficient toluene was then added to make a 5-7% (w/w) solution, and 480 g (2.58 mole) TSH was then added to obtain a 3:1 molar ratio of TSH to butadiene repeat units. When the originally two-phase reaction mixture was brought to reflux at 110°C, its appearance gradually changed from slightly turbid yellow-green to a clear yellow-green over the span of 20 h. The reaction mixture was then cooled to 60°C and poured into 2 L of methanol. A viscous white precipitate was then collected and dissolved into 350 ml of hexanes and re-precipitated into methanol. The reprecipitated material was dissolved in hexanes, washed twice with water, and dried for 24 h over $CaCl_2$. The solution was filtered and stored until use at 0°C under a blanket of argon.

<u>Fractionation</u>. Mono-functional arm material was removed by the following procedure: The hydrogenated polymer, a clear, viscous liquid, was dissolved in a sufficient amount of toluene to make a 3-6% solution. The mixture was heated to 45°C, and titrated with methanol to the point of turbidity. Increasing the temperature to 50°C re-established solution clarity, and the

mixture was then transferred to a warm 4 L separatory funnel, which was placed in a 20 L water bath. The initial temperature of the bath was 58°C, and it was allowed to fall to ambient over the course of 24 h, during which time the solution separated into two layers. After precipitation of the bottom layer into methanol, GPC was employed to determine the effectiveness of the fractionation. The procedure was then repeated if necessary, to decrease the amount of mono-functional arm material to less than 5%. To illustrate, the lower GPC trace in Figure 4 shows a polystyrene/polybutadiene three-arm star block copolymer which was linked using insufficient methyl trichlorosilane. The sample contains about 75% wt star polymer and 25% wt arm polymer, with no detectable di-arm species. The upper trace is of the same sample after hydrogenation and selective solvent fractionation; the arm impurity has been lowered to about 4-6% wt. The shoulder which appears on the high molecular weight side of the main peak represent polymer of just about twice the molecular weight of the star. As discussed in a later section, this is believed to be caused by a small amount of chain coupling which occurs during hydrogenation.

The efficiency of the fractionation is highly dependant upon the amount of methanol that is added to the toluene solution. We have found that employing a tubidity point of 50°C instead of 45°C results in poorer resolution of the star/arm mixture. A turbidity point of 40°C provides good resolution but substantially lowers the yield of recoverable material.

Sulfonation. Homogeneous sulfonations were conducted in methylene chloride solutions (10 g/100 ml) at room temperature. A representative procedure was as follows: Into 100 ml stirred methylene chloride was dissolved 10.15 g of polymer, with a molecular weight based on stoichiometry $(M_s) = 18,600$ g/mole (6.5 mmole styrene units). Acetyl sulfate was prepared in a separate vessel by mixing 2.14 g (0.0210 mole) of acetic anhydride, 1.95 g (0.0192 mole) of 96% H_2SO_4, and 150 ml of methylene chloride. After stirring this mixture for 30 min, the polymer solution was slowly added at 0°C. This procedure yielded a molar ratio of acetyl sulfate to styrene units of 3:1. After complete addition the ice water bath was removed and the solution was allowed to stir for 8 h, during which time the color changed from yellow to a dark yellow-brown. The reaction mixture was then slowly added to 1 L stirred distilled water at 80°C. The precipitate was collected by vacuum filtration, washed with fresh distilled water, and pressed to remove excess water. The wet crumb was then transferred to a soxhlet apparatus and extracted with distilled water for 36 h to remove excess acetic and sulfuric acid. After extraction the polymer was again vacuum filtered and pressed to remove excess water.

Titration. The wet polymer crumb (~ 1 g) was added to a tared 250 ml erlenmeyer flask and dissolved in 75 ml of THF. The sample was then titrated with ethanolic KOH (0.05 N) to a phenolphthalein endpoint. Removal of the solvents was then accomplished by passing a stream of nitrogen over the solution, followed by drying in-vacuo at 70°C until constant weight was reached. The original dry weight of the polymer sample was then calculated and the percent sulfonation obtained.

Instrumentation. Molecular weights were determined with a Waters GPC system equipped with a 6000A pump, 100, 500, 10^4, and 10^5 Å ultrastyragel column set, and a model 410 refractive index detector. ^{13}C NMR spectroscopy was performed on a Bruker MSL 200 MHz spectrometer at room temperature and 50°C. Glass transition temperatures (T_g's) were

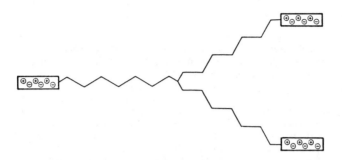

Figure 3. Star-block copolymer comprised of elastic inner blocks and polyelectrolyte outer blocks.

Fractionated sample

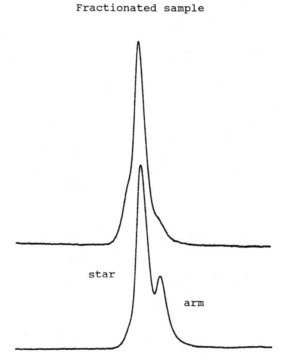

star

arm

Figure 4. Gel permeation chromatograms of a polystyrene/ polybutadiene block copolymer (SBD-4): Before hydrogenation and fractionation (lower) and after (upper).

determined using a DuPont DSC 910 interfaced to a 9900 data analysis system.

Results and Discussion

The synthetic strategy adopted in this work is one of regiospecific sulfonation of preformed polymer molecules using acetyl sulfate. Acetyl sulfate is a versatile sulfonating reagent which has been successfully employed for the sulfonation of polystyrene, (11) poly(ethylene-co-propylene-co-diene) (EPDM), (12-13) and polyisobutylene (4-5). The general technique, of which two main variations are discussed, is the formation of living block copolymers by anionic polymerization techniques, i.e., sequential monomer addition, as shown in Figure 5. The living block copolymer arms are then linked together using a trifunctional or higher chlorosilane linking agent. The molecules are designed so that the first-polymerized monomer, i.e., that which becomes the outer block of the star polymer (block A in Figure 5), is capable of undergoing sulfonation, while the inner block segment (block B) is inert toward sulfonation. The versatility of this general method lies in the ability to control the average number of block A monomer units which are placed at the end of each arm. With this synthetic flexibility, the ion content of the ionomer can be varied independently of the overall star-polymer molecular weight.

Table I is a master list of the polymers synthesized. It includes sample designations used in this paper, and the experimental conditions under which each polymer was produced. Data in Table I will be referred to during the course of the discussion.

TABLE I

Polybutadiene Microstructure[a] as a Function
of Polymerization Conditions

Sample	Temp. (°C)	Solvent		$[I]^b$ x 10^3	$[M]^c$	Microstructure (mole%)	
		Non Polar	Polar			1,2	1,4
IBD-2	25	Benzene	-	0.92	0.086	10	90
SBD-1	24	Benzene	-	11.2	1.15	8	92
SBD-2	10	Benzene	TEA[d]	16.5	1.70	43	57
SBD-3	10	Benzene	TEA	16.5	2.29	42	58
SBD-4	10	Benzene	TEA	14.1	2.52	-	-

[a] by quantitative ^{13}C NMR.
[b] living chain end concentration (mole/L).
[c] butadiene concentration (mole/L).
[d] triethyl amine. [TEA]/[I] = 44.

Polyisoprene/Polybutadiene (IPB) Star-Block Copolymers. The observed selective sulfonating capabilities of acetyl sulfate was the basis for the first attempted variation of the synthetic strategy. Regardless of the exact nature of the sulfonating reagent, it is well documented that α-olefins containing a 2-methyl branch undergo fast, high-yield sulfonations to produce 2-alkenesulfonic acids as the near-exclusive product.(4,14,15) Assuming that the mild acetyl sulfate would make more pronounced the reactivity differences between olefins, it seemed reasonable that the unsaturated repeat unit found in *cis/trans*-1,4-polybutadiene should resist sulfonation by acetyl sulfate under appropriate conditions, while the unsaturated repeat unit derived from isoprene should be readily sulfonated due to the presence of the methyl branch. This presented a prospective route toward the desired ionomer with block A composed of isoprene units and block B composed of 1,4-butadiene units.

The sulfonation of low molecular weight model olefins was undertaken to determine the feasibility of this approach. Competitive sulfonations using acetyl sulfate were carried out on the model compounds below, representing the repeat structures of *cis*-1,4-polyisoprene (PIP), *cis*-1,4-polybutadiene (*c*-PBD), and *trans*-1,4-polybutadiene (*t*-PBD), respectively. It was necessary to model both the *cis* and *trans* isomeric forms of 1,4-polybutadiene, since they have a nearly equal probability of occurrence when the anionic polymerization (Li counterion) is conducted in a nonpolar hydrocarbon medium.

$$CH_3CH_2 \diagdown \diagup CH_3$$
$$C$$
$$\|$$
$$C$$
$$CH_3CH_2 \diagup \diagdown H$$

PIP

$$CH_3CH_2 \diagdown \diagup H$$
$$C$$
$$\|$$
$$C$$
$$CH_3CH_2 \diagup \diagdown H$$

c-**PBD**

$$CH_3CH_2 \diagdown \diagup H$$
$$C$$
$$\|$$
$$C$$
$$H \diagup \diagdown CH_2CH_3$$

t-**PBD**

Model sulfonations were conducted in decalin, a high boiling aliphatic solvent, to facilitate subsequent GC analysis. In separate experiments, each of the two polybutadiene models was reacted competitively against the polyisoprene model. The molar ratio of the model compounds was constant at ten mole equivalents of *c*-PBD or *t*-PBD per one mole equivalent of PIP. This was a typical ratio envisioned for the star-block copolymer. Sulfonation of the olefins was monitored as a function of time for several temperatures and various molar ratios of sulfonation reagent to PIP. Representative % sulfonation vs. time curves are shown in Figure 6. It may be seen that sulfonation is selective to a degree; sulfonation of the PIP was > 90% complete after 63 min, while the ten fold greater excess of *c*-PBD had been sulfonated only ~10%. Table II lists the results obtained under various conditions. In no case could selectivity be obtained which was judged satisfactory.

Although the model studies indicated that selectivity of sulfonation was considerably less than desired, it was decided to proceed with the synthesis of the polyisoprene/polybutadiene star-block copolymer. It was felt that selectivity was good enough so that an appreciable fraction of the isoprene units could be sulfonated while limiting sulfonation of the butadiene units to some negligible level. For example, the data in Table II indicates that at 21°C, with a 1:1 mole ratio of sulfonating reagent, one should obtain about 50% sulfonation of the isoprene units with virtually no sulfonation of the butadiene units. This was deemed acceptable since the number of isoprene units at the chain ends could be simply doubled, and

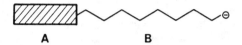

Figure 5.   General synthetic strategy utilized.

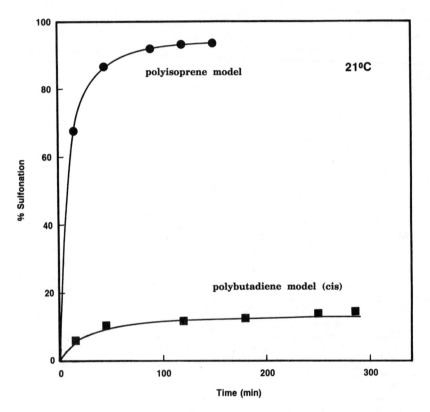

Figure 6.   Competitive sulfonation of polyisoprene model compound
vs. *cis*-1,4-polybutadiene model compound using acetyl sulfate.   PIP:
c-PBD: acetyl sulfate = 1:10:2 (mole).   Temperature = 21°C.   Solvent
= decalin.

TABLE II

Competitive Sulfonation of *cis*-1,4-Polyisoprene Model (PIP)
vs. *cis* or *trans*-1,4-Polybutadiene (*c*-PBD or *t*-PBD)
Model Compounds in Decalin Solution

	Model Compounds	Mole Ratio[a]	Time[b] (min)	% Sulfonation PBD Model	Maximum % Sulfonation PIP Model
	PIP	5	< 5	24	> 99
21°C	vs.	2	123	14	> 99
	c-PBD	1	205	9	95
	PIP	10	33	-	> 99
21°C	vs.	2	192	4	> 99
	t-PBD	1	-	< 1	56
	PIP	5	142	48	> 99
0°C	vs.	2	-	6	75
	c-PBD	1	-	< 1	33
	PIP	10	133	13	> 99
0°C	vs.	5	-	8	61
	t-PBD	2	-	8	45
	1	-	3	21

[a] moles acetyl sulfate/moles isoprene model
[b] time to reach >90% sulfonation of PIP model

incomplete sulfonation would still yield a tight grouping of the desired number of ion groups at the ends of the chains.

The synthesis of IBD star-block copolymers is outlined in Figure 7. Benzene was used as the solvent to maximize 1,4-enchainment of the butadiene, and the values of n and m were 5 and 95, respectively. After chlorosilane linking, but prior to sulfonation, the product copolymer was characterized extensively. Number average molecular weight of the copolymer was determined by vapor-phase osmometry (VPO) to be 14,800 g/mole; the M_s was 16,200 g/mole. GPC of the arm and star products, shown in Figure 8, indicated the linking reaction was virtually complete except for a small amount of unreacted arm. This small amount of monofunctional contaminant was considered negligible. Microstructural analysis of the polybutadiene portion of the molecules was obtained by ^{13}C NMR using the method of van der Velden et al.; (16) the results were 10% 1,2 and 90% 1,4 enchainment (sample IBD-2 in Table I). This is in excellent agreement with previously reported results (17) for polymers prepared under nearly identical conditions.

Sulfonation of the IBD copolymer was attempted in both hexanes and methylene chloride under various conditions of temperature and acetyl sulfate concentration. Using accepted polymer isolation procedures,(12) which involve precipitation of the sulfonation reaction mixture into rapidly-stirred, hot water, a white emulsion was obtained rather than the elastic, water-swollen crumbs which are characteristic of sulfonated polymers of this type. In hexanes solvent, the sulfonation was also ac companied by a significant amount of charring. This is believed to occur because the sulfonation reaction is being localized to a few polymer chains. This is most likely a phase transfer phenomenon in which the partially sulfonated polymer chains migrate from the non-polar, hexanes layer to the polar, acetyl sulfate layer where they undergo rapid, nonselective sulfonation resulting in char. Homogeneous sulfonations in methylene chloride eliminated charring, but an emulsified product was always obtained which indicated insufficient selectivity of sulfonation.

Polystyrene/Polybutadiene (SBD) Star-Block Copolymers. After the above described experiments it was decided that the selective sulfonation strategy would only work if the B block repeat units could be made absolutely nonreactive toward the sulfonating reagent. The obvious approach was to utilize a fully saturated elastomeric inner block, since this was the key to the successful regiospecific sulfonation of PIB.(4,5) Since the monomers involved must be anionically polymerizable, it was decided to base the saturated soft segments on either hydrogenated polyisoprene or hydrogenated polybutadiene. In either case, the ion-containing outer blocks would be based on sulfonated styrene units.

The synthesis scheme employed is depicted in Figures 9 and 10. s-BuLi was used to oligomerize short segments of living polystyrene; these may be made to any desired length by appropriate choice of stoichiometry. Initial experiments have involved segments averaging four styrene units. The styrene oligomers were then used to initiate butadiene or isoprene, as desired. Initial experiments have utilized stoichiometries resulting in addition of 60-110 diene units to each chain. This produces an arm molecular weight in the range 3600 < M_n < 7500. The resulting living arms were coupled with trichlorosilane to yield the star-block copolymer. Figure 11 shows GPC traces for a SBD block copolymer (SBD-1) before and after the linking reaction. In general, slightly less than the theoretically required chlorosilane was used. This ensured that the only products were three-arm star and a slight excess of unlinked arm. The latter can either be ignored if

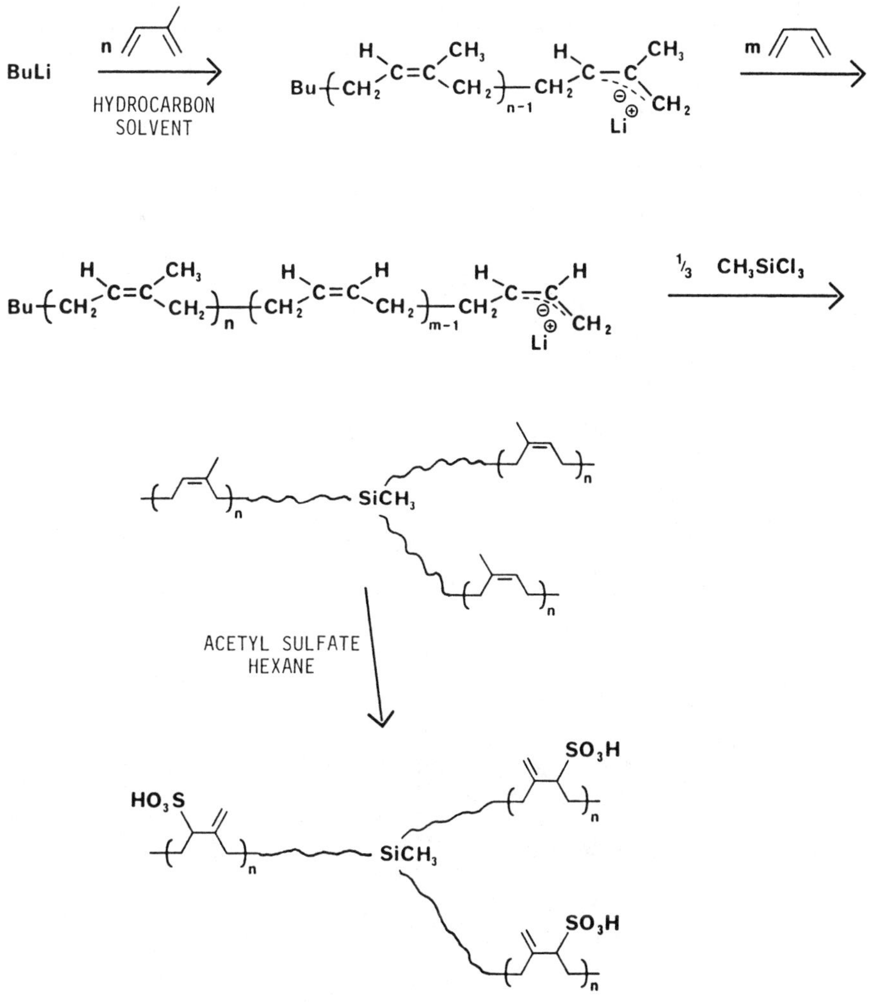

Figure 7. Synthesis scheme for three-arm star polyisoprene/polybutadiene (IBD) block copolymers.

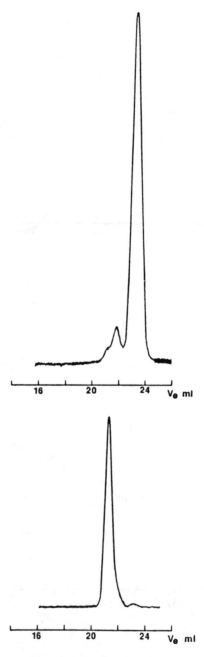

Figure 8. Gel permeation chromatograms of polyisoprene/
polybutadiene block copolymers (IBD-2): arm (top) and star.

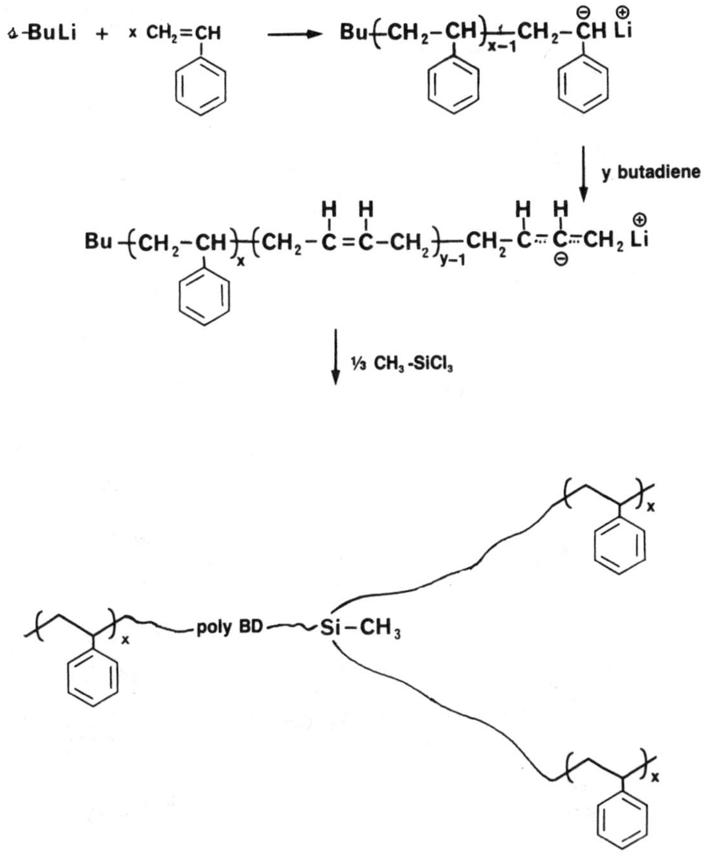

Figure 9. Synthesis scheme for polystyrene/polybutadiene (SBD) star-block copolymers.

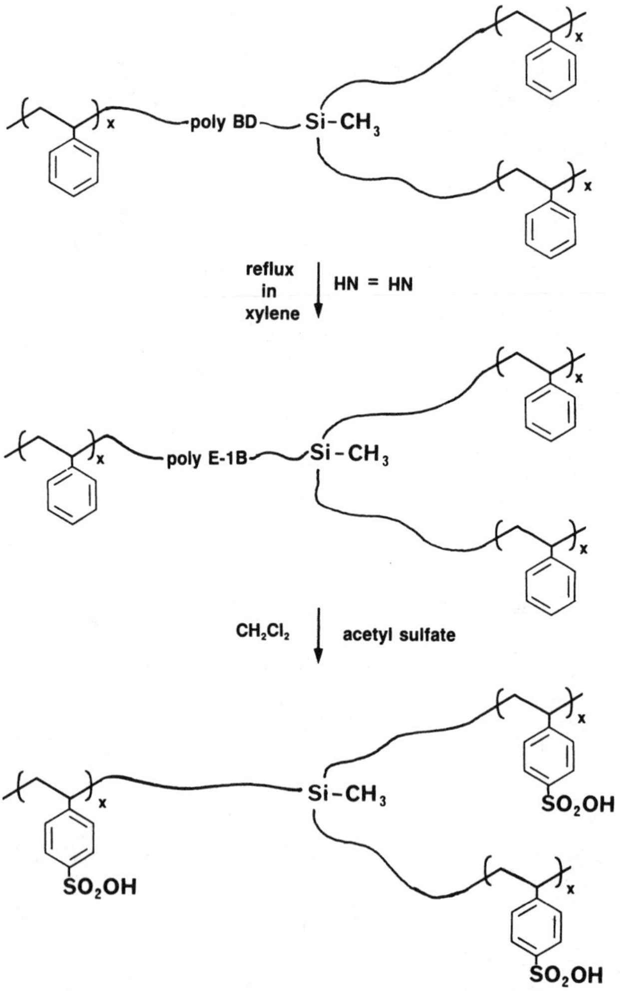

Figure 10. Hydrogenation and sulfonation reactions of star-block copolymers.

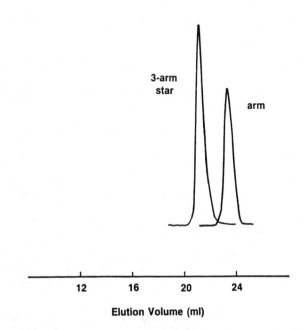

Figure 11. Gel permeation chromatograms of a polystyrene/
polybutadiene block copolymer (SBD-1): arm and three-arm star.
The star sample was fractionated once.

its concentration is negligible or removed by solvent fractionation (18) or preparative GPC.(19) The three-arm star sample pictured in Figure 11 was fractionated one time. Experience has shown that the linking of polybutadienyl chains is generally quite successful, while polyisoprene-based living polymers often suffer incomplete reaction with the chlorosilane. This is apparently due to steric difficulties associated with the methyl branch present in the isoprene unit. If the isoprene-based ionomer is desired, the living arms should be end-capped with a few units of butadiene prior to linking.

To achieve absolutely regiospecific sulfonation, the diene inner blocks have been hydrogenated to yield a fully saturated poly(ethylene-alt-propylene) elastomer, in the case of isoprene, and poly(ethylene-co-1-butene) in the case of butadiene. Due to ease of the linking reaction, most of the latest work has dealt with polybutadiene-based samples, and the remaining discussion will deal with them, although the reactions and techniques are also applicable to the polyisoprene-based samples.

Diimide, produced from p-toluene sulfonyl hydrazide, is a convenient laboratory hydrogenating reagent.(10) It produces clear, white hydrogenated products, and the course of the reaction may be followed visually by the gradual loss of turbidity of the characteristic yellow-green reaction mixture. The extent of hydrogenation by diimide appears to be > 99% as evidenced by FT-IR, ^{1}H NMR, and ^{13}C NMR. The FT-IR spectra shown in Figure 12 suggest that a high degree of hydrogenation was achieved, evidenced by the large reduction in peak size of the C=C stretch around 1640 cm^{-1} (peak A) and the out-of-plane olefinic C-H bend absorbances in the 900-1000 cm^{-1} range (peak group B). More substantial evidence is obtained from the ^{13}C NMR spectra in Figure 13, which show complete loss, upon hydrogenation, of the side-vinyl olefinic carbon resonances at 114 and 143 ppm, and of the cis/trans-1,4 units which appear as a complex series of peaks in the 127-132 ppm range. The hydrogenated sample spectrum (bottom) clearly shows the presence of the aromatic carbons of the short polystyrene outer blocks. Examination of the ^{1}H NMR spectra in Figure 14 yields conclusive evidence that essentially quantitative hydrogenation has occurred, since the olefinic resonances between 5 and 6 ppm are completely removed by the hydrogenation reaction, and there is no overlap of the aromatic and olefinic resonances.

The ^{13}C NMR spectra are also quite useful with regard to characterization of the polybutadiene microstructure. Microstructure of the polymer is very critical, especially the extent of 1,2-enchainment, and must be controlled by selection of the proper polymerization conditions. Obviously, high 1,4-addition of butadiene will yield a crystalline polymer upon hydrogenation. Van der Velden (16) has reported specific conditions whereby ^{13}C NMR can be performed on polybutadiene to yield quantifiable peak areas which are unaffected by differences in nuclear Overhauser effects and spin-lattice relaxation times. Table I lists polymerization conditions and resulting butadiene microstructure determined by this method for several samples. The sample spectrum depicted in Figure 15 (sample SBD-2 in Table I) has a side vinyl content of 43 mole %, and it yielded a totally amorphous elastomer upon hydrogenation. By contrast, sample SBD-1 in Table I, with a side vinyl content of 8%, yielded a semicrystalline polymer as evidenced by its powdery morphology and the presence of a melting endotherm which is apparent in the DSC thermogram shown in Figure 15.

Sulfonation of the outer styrene units of the hydrogenated polymers has been conducted using methylene chloride solutions at room temperature. In order to sulfonate every phenyl ring in the sample, a large excess of acetyl sulfate has been used. The reactions appear homogeneous with the

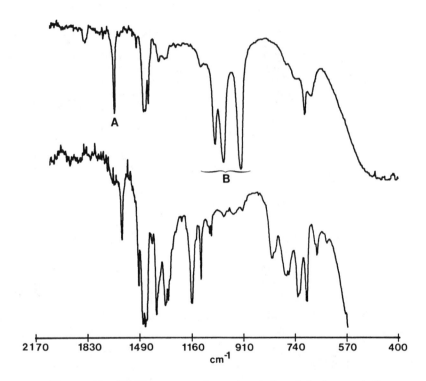

Figure 12. FT-IR spectra of unsaturated and hydrogenated polystyrene/polybutadiene star-block copolymer (SBD-2).

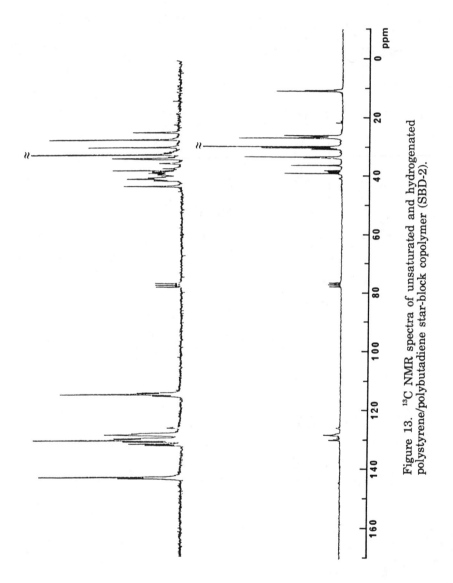

Figure 13. ^{13}C NMR spectra of unsaturated and hydrogenated polystyrene/polybutadiene star-block copolymer (SBD-2).

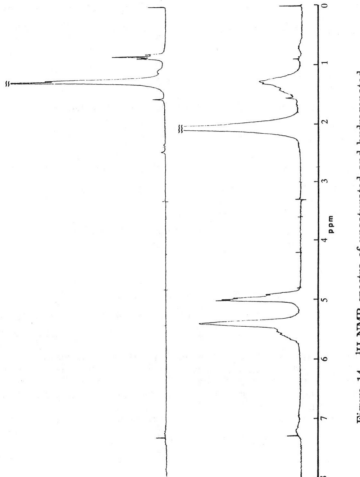

Figure 14. ^{1}H NMR spectra of unsaturated and hydrogenated polystyrene/polybutadiene star-block copolymer (SBD-2).

gradual development of a dark brown color. The minimum reaction time has not been established, but 8 h gives good results. Precipitation of the reaction mixture into hot water yields light brown, spongy crumbs which are contaminated by residual acetic acid, and perhaps sulfuric acid. The most efficient way to remove these acidic contaminants is by continuous warm water extraction in a Soxhlet extraction apparatus. For example, Table III lists meq acid/g polymer determined by titration after various extraction times. Clearly the polymer is initially contaminated, but after extracting for 24-48 h, a constant acid content is obtained. This particular sample, SBD-2, has an M_s, after linking and hydrogenation, of 18,600 g/mole and an average of four styrene units per arm. Based upon these values, the final titration value reached in Table III indicates that > 85% of the styrene units were sulfonated.

TABLE III

Acid Content of Sulfonated Star-Block Copolymer[a]
During Continuous Warm-Water Extraction

Extraction Time (h)	meq H$^{\oplus}$/g	% Styrene Units Sulfonated (Theoret.)
0	0.726	113
24	0.585	91
48	0.596	93
72	0.559	87

[a]Sample SBD-2. Three-arm star, M_s = 18,600.
Average styrene units per arm = 4.

The ionomer which was isolated from the neutralization of sample SBD-2 was a brown-colored elastic network of moderate strength. Ionomer samples SBD-1 and SBD-2, neutralized to the stoichiometric end point using KOH, were compression molded at 140°C and examined for tensile properties. The results, as shown in Figure 16, illustrate the profound influence of crystallinity on the elastomeric inner block. The semi-crystalline material (SBD-1) behaves much like a rigid plastic, while the amorphous sample (SBD-2) is an elastomer of moderate strength.

Conclusions

Star-branched ionomers composed of short sulfonated polystyrene outer blocks and hydrogenated diene inner blocks have been synthesized and characterized. The techniques employed yield polymers of well-defined micro-architecture and narrow molecular weight distribution. Polybutadiene microstructure can be controlled by proper choice of polymerization conditions, and is in fact critical in determining the elastomeric properties of the hydrogenated ionomer. Polyisoprene-based ionomers can be made by this

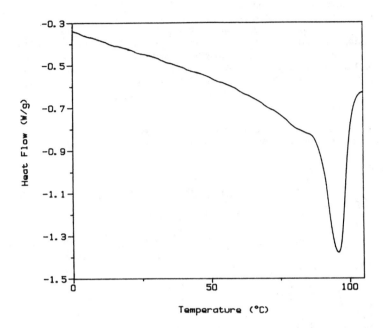

Figure 15.   DSC thermogram of star-branched poly(ethylene-co-1-butene) ionomer, K salt (SBD-1).   1-Butene content = 8 mole %.

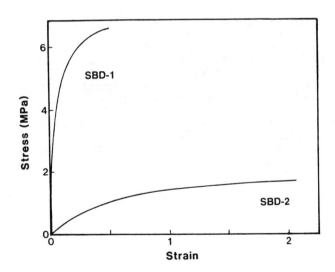

Figure 16.   Tensile properties of semi-crystalline (SBD-1) and totally amorphous (SBD-2) three-arm star ionomers (K salt).

technique, but for best results the living chains should be end-capped with butadiene units prior to linking.

The work presented outlines the sucessful development of a new model ionomer system. Future efforts will be directed toward exploitation of the technique to produce a variety of well-defined ionomers in which the micro-architectural features of the molecules are systematically varied and their effects upon physical properties investigated.

Acknowledgments

Financial support for this research by Morton Thiokol, Elkton Division, Elkton, VA, is gratefully acknowledged.

Literature Cited

1. Eisenberg, A.; King, M. Ion-Containing Polymers: Physical Properties and Structure, Academic Press, New York, 1977.
2. Broze, G.; Jerome, R.; Teyssie, Ph.; Marco, C. Polym. Bull. 1981, 4, 241.
3. Broze, G.; Jerome, R.; Teyssie, Ph. Macromolecules 1981, 14, 224.
4. Kennedy, J.P.; Storey, R.F. ACS Org. Coat. Appl. Polym. Sci. Proc. 1982, 46, 182.
5. Storey, R.F. Ph.D. Dissertation, U. of Akron, 1983.
6. Mohajer, Y.; Tyagi, D.; Wilkes, G.L.; Storey, R.F.; Kennedy, J.P. Polym. Bull. 1982, 8, 47.
7. Bagrodia, S.; Mohajer, Y.; Wilkes, G.L.; Kennedy, J.P.; Storey, R.F. Polym. Bull. 1983, 9, 174.
8. Bagrodia, S.; Pisipati, R.; Wilkes, G.L.; Storey, R.F.; Kennedy, J.P. J. Appl. Polym. Sci. 1984, 29(10), 3065.
9. Morton, M.; Fetters, L. J. Rubb. Chem. Tech. 1975, 48, 359.
10. Mango, L.; Lenz, R. Makromol. Chem. 1973, 163, 13.
11. Makowski, H.S.; Lundberg, R.D.; Singhal, G.S. U.S. Patent 3,870,841 to Exxon Research and Engineering Company, 1975.
12. Makowski, H.S.; Lundberg, R.D.; Westerman, L.; Bock, J. Polym. Prepr., ACS Div. Polym. Chem., 1978, 19(2), 292.
13. Makowski, H.S.; Lundberg, R.D.; Bock, J. U.S. Patent 4,184,988 to Exxon Research and Engineering Co., 1980.
14. Robbins, M.D.; Broaddus, C.D. J. Org. Chem. 1974, 39(16), 2459.
15. Boyer, J.L.; Gilot, B.; Lanselier, J.P. Phosphorus and Sulfur 1984, 20, 259.
16. van der Velden, G.; Didden, C.; Veermans, T.; Beulen, J. Macromolecules 1987, 20, 1252.
17. Young, R.N.; Quirk, R.P.; Fetters, L.J. Adv. Polym. Sci. 1984, 56, Springer-Verlag, New York.
18. Hadjichristidis, N.; Roovers, J.E.L. J. Polym. Sci., Polym. Phys. Ed. 1974, 12, 2521.
19. Gadkari, A.C.; Zsuga, M.; Kennedy, J.P. Polym. Bull. 1987, 18, 317.

RECEIVED February 2, 1989

Chapter 14

Grafted-Block Copolymer Networks Formed by Transition Metal Coordination of Styrene- and Butadiene-Based Polymers

A. Sen[1,4], R. A. Weiss[1,2], and A. Garton[1,3]

[1]Polymer Science Program, University of Connecticut,
Storrs, CT 06269-3136
[2]Department of Chemical Engineering, University of Connecticut,
Storrs, CT 06269-3136
[3]Department of Chemistry, University of Connecticut,
Storrs, CT 06269-3136

Blends of functionalized polystyrene and polybutadiene were prepared using transition metal coordination as a means of improving the interaction between the two polymers. The polystyrene contained 4.2 mole percent of 4-vinyl pyridine comonomer and the polybutadiene chains were terminated at both ends with copper carboxylate groups. Fourier transform infrared spectroscopy, electron spin resonance spectroscopy and small angle x-ray scattering evidence are presented for the formation of molecular interactions between the Cu-carboxylate and the vinyl pyridine groups. Although the blends were phase separated, improvements in miscibility were realized when the complex was formed. A molecular architecture similar to that of a physically crosslinked grafted-block copolymer is proposed. Thermal mechanical and dynamic mechanical analyses demonstrated a significant improvement in the mechanical properties of the blends compared with a blend in which only an acid-base type interaction was possible. The formation of the transition metal complex increased the rubbery modulus between the two glass transitions and gave rise to a new plateau region in the modulus above the glass transition of the polystyrene-rich phase.

Polymer blends have received considerable industrial and academic attention in recent years. Ideally, two or more polymers may be blended to form a wide variety of morphologies that offer potentially desirable combinations of properties. However, it is often not possible to achieve useful compositions because of the unfavorable thermodynamics of mixing of polymers. The entropy of

[4]Current address: Texaco, Inc., Beacon Research Laboratories, P.O. Box 509, Beacon, NY 12508

0097-6156/89/0395-0353$06.00/0
© 1989 American Chemical Society

mixing of polymers is very small and, therefore, the free energy of mixing is dominated by the enthalpy of mixing, which is generally positive in the absence of specific intermolecular interactions (1). Mixing on a molecular level may occur if exothermic interactions such as hydrogen-bond formation, proton transfer, charge transfer, or dipole-dipole interactions take place between specific functional groups on the two polymers (2).

A comprehensive review of the literature describing the use of specific functional groups to enhance miscibility of polymer blends is beyond the scope of this paper. It is instructive, however, to mention several studies that used amine functional groups to enhance specific interactions. Clas and Eisenberg (3) studied blends of poly(vinyl chloride) and poly(ethyl acrylate-co-4-vinylpyridine), and observed that miscibility was enhanced by hydrogen bonding, dipole-dipole interactions, or a combination of the two. Similarly, Eisenberg and coworkers (4, 5) achieved miscibility in several polymer blends by incorporating an acid-base interaction between polymers containing vinyl pyridine and sulfonic acid groups. Horrian et al. (6) used ionic interactions between end-functionalized (telechelic) polystyrene and polybutadiene containing acid and tertiary amine functionalities, respectively. Ion-pair formation between the polymer endgroups resulted in a "multiblock copolymer".

Agnew (7) reported a number of transition metal complexes with poly(vinylpyridine), and these generally exhibited high thermal stability. Peiffer et al. (8) used transition metal complexation between a zinc neutralized sulfonated EPDM and poly(styrene-co-vinyl pyridine) to improve the mechanical properties of the blend. In the literature mentioned so far, the authors' intentions were to improve miscibility of the blends. With the exception of the study by Horrian et al. (6) the polymers used were randomly functionalized. In the work described herein the goal was to prepare "grafted-block" copolymer networks through the interaction between a randomly functionalized polymer and a telechelic polymer. Although complete miscibility of the two polymers was not an objective of this research, it was anticipated that improved adhesion between the phases would result. Both acid-base interactions and transition metal coordination were evaluated for promoting the graft, though the latter was considerably more successful as judged by the properties of the blends. In particular, this chapter describes blends of carboxyl-terminated polybutadiene (CTB) and its Cu(II) salt (CTB-Cu) with poly(styrene-co-4-vinyl pyridine).

EXPERIMENTAL SECTION

Carboxyl-terminated polybutadiene (CTB) was obtained from Scientific Products Inc., and had a reported number average molar mass of 4,600 and a weight average molar mass of 9,000. Carboxyl group concentration was determined by titrating a toluene-tetrahydrofuran solution of the CTB to a phenolphthalein endpoint with alcoholic sodium hydroxide. The CTB contained 0.51 meq COOH per gram of polymer, which corresponds to a number average functionality of 2.39 based on a molar mass of 4600. The copper salt (CTB-Cu) was formed by refluxing a 20% CTB solution in toluene for 24 hours with a stoichiometric amount of Cu(II)-acetate (i.e, one-half mole of Cu(II) acetate per mole of carboxylic acid endgroups) dissolved in a 50/50 mixture of water and methanol. The solvents were distilled off under reduced pressure, and the polymer was dried under vacuum at 75-80°C for 24 hours. The as-received CTB was a clear viscous liquid, the CTB-Cu was a highly viscous dark brownish-green liquid, and the blends were rigid, dark brown solids.

The poly(styrene-co-4-vinylpyridine) (PSVP) was prepared by a free radical emulsion copolymerization process, following the method described by Lundberg et al. (9). The styrene was washed with 10% sodium hydroxide to remove the inhibitor and then with distilled water. The 4-vinyl pyridine was vacuum distilled at 5mm Hg and 30°C. Potassium persulfate was used as initiator, sodium lauryl sulfate as surfactant, and dodecylthiol as chain transfer agent. The reaction was carried out for 24 hours and terminated with hydroquinone. The copolymer was precipitated in a large excess of acetone, washed with methanol, air-dried at 60°C for two days, and vacuum dried at 50°C for 3 days. The PSVP had a number average molar mass of 142,000 and a weight average molar mass of 291,000 as determined by GPC using polystyrene standards for calibration. The vinyl pyridine content of the copolymer was 3.4 mole percent based on nitrogen analysis.

Blends were prepared by slowly adding a solution of the CTB-Cu in a mixed solvent of toluene and THF (20/80 v/v) to a solution of PSVP in THF. This mixture was then refluxed for 24h and the solvent distilled off under reduced pressure. The isolated blends were dried under vacuum at 80°C for 24 hours. Three different blends were prepared (the ratio in parentheses following each blend represents the nominal molar ratio of vinyl pyridine groups to CTB endgroups): (1) PSVP/CTB (1:1), (2) PSVP/CTB-Cu (1:1), and (3) PSVP/CTB-Cu (2:1). The compositions of these blends are summarized in Table I. Note that blends (1) and (2) contained nominally stoichimetrically equivalent amounts of pyridinyl and carboxyl groups, while in blend (3) the equivalence ratio between the pyridinyl concentration and the Cu(II) ion concentration was nominally 2:1.

TABLE I
BLEND COMPOSITION

Sample	wt% PSVP	CTB	CTB cation	equiv. VP equiv. COOH	equiv. VP equiv. Cu
1	56.5	43.5	H	0.8	
2	56.5	43.5	Cu		0.8
3	72.4	27.6	Cu		1.7

A Perkin Elmer differential scanning calorimeter, DSC-2, was used to obtain the glass transition temperatures of the samples in two different temperature ranges: -100°C to +10°C and -20°C to

140°C. In the low temperature range, the measurements were performed under a helium atmosphere using liquid nitrogen as a coolant, while a nitrogen atmosphere and a mechanical cooler were used in the upper temperature range. Glass transition temperatures were defined as the midpoint of the change in specific heat. Thermal mechanical analyses (TMA) were made with a Perkin Elmer TMA-7. The measurements were made under a helium atmosphere from -100°C to 300°C at a heating rate of 10°C/min. The specimens were approximately 0.9 mm thick compression molded films, and a penetration probe with a 0.5 mm radius flat tip and a force of 500 mN were used. A Perkin Elmer thermogravimetric analyzer, TGA-7, was used to measure the thermal oxidative stability of the polymers. The specimens were heated in air at a rate of 10°C/min.

Dynamic mechanical properties of the blends were measured at 1 Hz with a Rheometrics System 4 mechanical spectrometer using rectangular or parallel plate geometries in torsion or tension. For tension, the linear motor and rectangular specimens were used and a temperature range of -100°C to 50°C was covered. The auto-tensioning capability of the System 4 linear motor was used in order to compensate for changes in the sample length as temperature was varied. Torsional experiments were conducted between 50°C and 250°C using 1 mm thick discs between oscillating parallel plates. The test specimens were compression molded at 150-200°C and annealed overnight under vacuum at 130°C between spring-loaded parallel plates in order to eliminate residual molded-in stresses.

SAXS measurements were made at the National Center for Small Angle Scattering Research (NCSASR) at Oak Ridge National Laboratory using the 10m SAXS camera. The instrument used a rotating anode CuK$_\alpha$ (λ = 0.1542 nm) x-ray source, crystal monochromatization of the incident beam, pinhole collimation, and a two dimensional, position sensitive, proportional counter. Sample to detector distance was 1.12 m. Disc-shaped samples were cut from the molded films and dried in a vacuum oven at 65°C for four weeks prior to the SAXS analyses.

FTIR spectra were obtained with four wavenumber resolution using either a Nicolet 60-SX or a Mattson Cygnus spectrometer. Specimens were cast as thin films on sodium chloride discs, and the solvent was removed under vacuum at 80°C. The spectrometers were purged with dry, carbon dioxide-free, air. To explore the effect of exposure to moisture, selected specimens were left in the laboratory environment at about 25°C and 40% humidity for several days before being re-examined by FTIR spectroscopy.

ESR measurements were made at x-band frequency with a Varian E-3 electron spin resonance (ESR) spectrometer using 4mm quartz tubes. The magnetic field was varied from 0 to 5,000 G. The spectra were taken at room temperature and stored in a Nicolet LAS 12/70 Signal Averager.

RESULTS AND DISCUSSION

Both the PSVP/CTB and PSVP/CTB-Cu blends exhibited good thermal oxidative stability as shown by the TGA curves in Fig. 1. There was a small amount of degradation that occurred between 350 and 450°C, but the major weight loss occurred about 450°C. Pyrolysis GC/MS was used to identify the degradation products, which for both blends were found to be styrene, hydrocarbons and 1,3-diphenylpropane. The latter product was most likely due to the presence of stabilizer in the CTB.

DSC thermograms of the component polymers and of the blends are shown in Fig. 2 and the glass transition temperatures (Tg) are summarized in Table II. Conversion of the CTB to the copper salt broadened the transition region, but did not significantly affect the Tg, which agreed with previous reports on CTB salts (10, 11). Two Tg's corresponding to polybutadiene-rich and polystyrene-rich phases were observed for each blend, which indicated that all were phase-separated. A small increase in the Tg of the rubbery phase was observed in blend #1 (PSVP/CTB), though this may have been due to broadening of the transition. This may indicate some miscibility of the PSVP in the CTB phase or alternatively, it may have resulted from a distribution of phase sizes. Substituting the CTB-Cu for CTB in the blend raised the Tg of the rubbery phase by about 15°C, which indicated that the PSVP was more miscible with the salt than the acid. This adds weight to the argument for the formation of a transition metal complex between the Cu(II) ion and the pyridinyl nitrogen since this was the only difference between blends #1 and #2. This complex was expected to be stronger than the acid-base interaction between the carboxylic acid group of CTB and the pyridinyl group. The addition of CTB to PSVP broadened the glass transition of the styrene-rich phase and decreased Tg. This was probably due to limited miscibility of the low molar mass CTB in PSVP. The conversion of the CTB to the Cu-salt resulted in further broadening of the transition and an apparent increase in Tg. Because the styrene-phase transition region extended over about 40°C for the blends, the designation of a single value for Tg is not particularly meaningful. The qualitative changes in the transition region, however, may be explained by improved miscibility of the components and/or physical crosslinking of the PSVP by the difunctional CTB-Cu.

TABLE II
GLASS TRANSITION TEMPERATURES OF THE BLENDS

| Sample | Wt % | | | Tg (C) | |
	PSVP	CTB	CTB-CU	DSC	DMA*
CTB		100.0		-85	
CTB-Cu			100.0	-86	
PSVP	100.0			107	
1	56.5	43.5		-81/82	-76/109
2	56.5		43.5	-66/90	-69/116
3	72.4		27.6	**/93	-68/118

* taken as the maximum in tanδ
** not measured

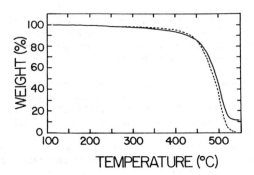

Figure 1.   TGA thermograms in air of (----) PSVP/CTB (1:1) and
(———) PSVP/CTB-Cu (1:1).

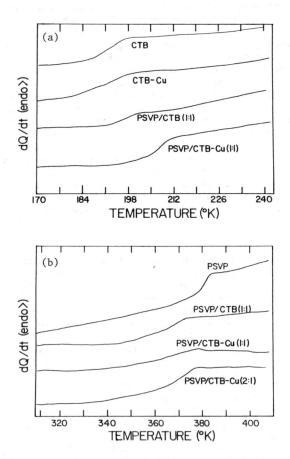

Figure 2.   DSC thermograms of component polymers and their
blends.   (a) 170 - 240 K and (b) 310 - 400 K.

A comparison of the softening behavior of blends #1 and #2 is shown by the TMA thermograms in Fig. 3. Blend #1, PSVP/CTB (1:1), began to soften at about 0°C and the probe had essentially penetrated through the sample by about 120°C. In contrast, under an identical load, the PSVP/CTB-Cu (1:1), blend #2, did not begin to deform until about 100°C and exhibited a more gradual deformation. In this case, the penetration of the specimen was not complete until about 200°C. Although these results were strictly qualitative, they did demonstrate that the blend containing the CTB-Cu remained stiffer at elevated temperatures than the blend with CTB. The higher resistance to deformation showed that the physical interactions between the CTB-Cu and the PSVP were stronger than those in the blend containing the free-acid derivative of CTB. This, again, strongly suggested the formation of a intermolecular complex between the polymers.

The strongest evidence for association of the CTB-Cu with the PSVP is the DMA results shown in Fig. 4. Like the DSC data, the DMA curves exhibited two Tg's, indicating a two-phase morphology in the blends. A comparison of the Tg's measured by DSC and DMA (taken as the maximum in tanδ) is given in Table I. In general, the lower Tg's were in agreement. However, the Tg's of the polystyrene phase measured by DMA were significantly higher than those measured by DSC. This may be attributed in part to experimental rate effects, but the difference may also be a consequence of the broadening of the higher temperature transition region due to the interactions of the two polymers. This makes the assignment of a single value for Tg by either technique considerably more ambiguous. In fact, if the onset of the drop in the dynamic modulus at the transition were taken as Tg, much better agreement was found achieved between the values obtained by DSC and DMA.

Blend #1, PSVP/CTB, maintained a dynamic modulus $>10^7$ dynes/cm^2 to about 100°C, Fig. 4. This was most likely a consequence of an acid-base interaction between the CTB and the PSVP, such that the PSVP glassy phase acted to reinforce or crosslink the rubbery phase. The material flowed above the Tg of the PVSP phase. Conversion of the CTB to the Cu-salt increased the modulus between the two Tg's and gave rise to a new plateau region above the glassy phase Tg. The increase in the modulus above the rubbery phase Tg for the blend containing the salt was a consequence of the greater strength of the transition metal coordination compared with the interaction between the relatively weak carboxylic acid and the basic pyridinyl group. Thus, the blends containing the CTB-salt had a higher effective crosslink density. In both systems the crosslinks were thermally labile as evidenced by the decline in modulus with temperature and the fact that the samples could be compression molded.

The greater temperature stability of the transition metal complex versus the ionic complex was also demonstrated by the modulus above the higher Tg. In the case of the blends containing the Cu-salt, a relatively high modulus was maintained to 200°C and no distinct viscous flow regime was observed. The highest values were achieved in the blend containing a 1:1 ratio of Cu(II) ions and vinyl pyridine groups. The maintenance of modulus above the PSVP Tg was due to the persistence of the transition metal complex

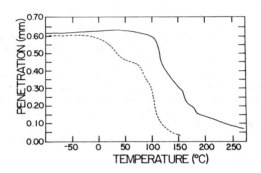

Figure 3.   TMA thermograms of (----) PSVP/CTB (1:1) and
(——) PSVP/CTB-Cu (1:1).

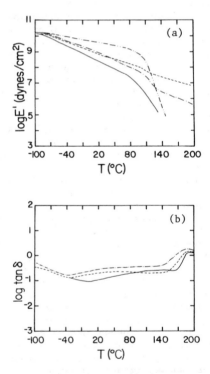

Figure 4.   (a) Dynamic storage moduli and (b) tan δ versus
temperature for (—··—) PSVP, (——) PSVP/CTB (1:1),
(----) PSVP/CTB-Cu (1:1) and (—·—) PSVP/CTB-Cu (2:1).

responsible for the physical crosslinks. The labile aspect of these physical crosslinks was demonstrated by the fact that when relatively high stresses were imposed, such as during compression molding of the samples, the materials flowed, even though viscous flow was not observed during the DMA experiment. The development of a plateau region above Tg, which is not directly attributable to simple chain entanglements, has been observed in other associating polymer systems and is due to a broadening of the relaxation time distribution with more emphasis on longer time processes (12). In the case of the blends considered here, this also was manifested by the high degree of molded-in stress in the compression molded films. These residual stresses were eliminated by annealing the molded films. Otherwise, severe shrinkage of the films occurred at elevated temperatures.

Fig. 5 shows the IR spectral changes associated with the reaction of CTB to form the Cu-salt. The acid had a characteristic carbonyl absorption at 1713 cm^{-1}, Fig. 5a, that was replaced by the broad carboxylate absorption at about 1560 cm^{-1}, Fig. 5b. However, on exposure of the carboxylate salt to the laboratory atmosphere for three days, there appeared to be partial reversion to the carboxylic acid, Fig. 5c. Fig. 6c shows that the IR spectrum of the 1:1 PS-VP/CTB-Cu blend was not simply a superposition of the IR spectrum of its component parts, PSVP, Fig. 6a, and CTB-Cu, Fig. 6b. The carboxylate absorption of CTB-Cu at 1560 cm^{-1} was moved to >1600 cm^{-1}. The frequencies of the carboxylate absorptions are highly sensitive to structural changes in the local environment. Spectral shifts in the blends may therefore be taken as evidence that there was sufficient association between the phases that the local environment of the carboxyl group was changed on a molecular level. This observation was confirmed by subtraction of the PSVP spectrum from the spectrum of the the blend, Fig. 7. The residual spectrum was superficially similar to CTB-Cu, except that the carboxyl absorption was shifted significantly from its position in the pure material.

The concern over the hydrolytic stability of the CTB-Cu salt that was raised by Fig. 5 was confirmed by exposure of the PSVP/CTB-Cu blend to the laboratory environment for several days. Fig. 8 shows that with time there was an apparent reversion to carboxylic acid functionality (>1700 cm^{-1}) and a reduction in carboxylate functionality (ca. 1600 cm^{-1}). Surprisingly, though, a limited study revealed no noticeable deterioration of the mechanical properties of the PSVP-CTB-Cu blends after exposure to the atmosphere.

Electron spin resonance spectroscopy (ESR) has been used by several research groups to characterize the local structure of CTB-Cu (10, 11, 13). For the neat ionomer, both isolated and dimeric copper complexes, as shown in Fig. 9, have been reported. Fig. 10 compares the ESR spectra for CTB-Cu and Blend #1, PSVP/CTB-Cu (1:1). The strong signal near 3160 G was due to isolated copper ions with a square planar structure as in Fig. 9a. The measured g-Lande factor and hyperfine interaction parameters were $g_{||}$ = 2.320, $g_{\perp}$ = 2.059, $A_{||}$ = 145 G, and $A_{\perp}$ = 30 + 5 G, which agreed with those reported for isolated Cu(II) ions in model compounds (14).

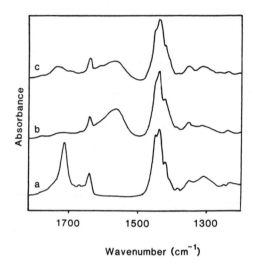

Figure 5. FTIR spectra for (a) CTB, (b) CTB-Cu, and (c)   CTB-Cu
exposed to the laboratory atmosphere for three days.

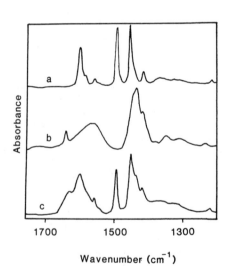

Figure 6. FTIR spectra for (a) PSVP, (b) CTB-Cu, and (c)
PSVP/CTB-Cu (1:1).

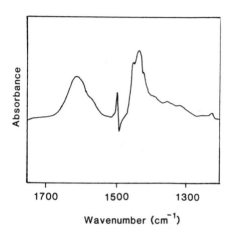

Figure 7. FTIR difference spectrum of PSVP/CTB-Cu (1:1) minus PSVP.

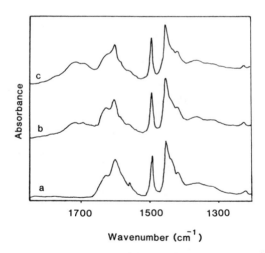

Figure 8. FTIR spectra of (a) PSVP/CTB (1:1), freshly made, (b) after one day exposure to the laboratory atmosphere, and (c) after three days exposure.

a Monomeric isolated complexes

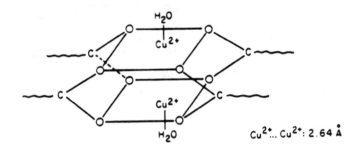

b Dimers (monohydrated copper acetate type)

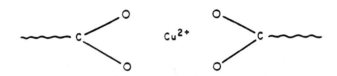

$Cu^{2+}...Cu^{2+}: 2.64 \text{ Å}$

c Dimers (anhydrous copper formate type)

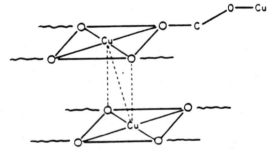

$Cu^{2+}...Cu^{2+}: 3.44 \text{ Å}$

Figure 9. Structures of Cu^{2+} carboxylate salts, (a) monomeric isolated complexes, (b) dimeric complex (monohydrated copper acetate type), and (c) dimeric complex (anhydrous copper formate type). (Reproduced with permission from ref. 10. Copyright 1986 Butterworth)

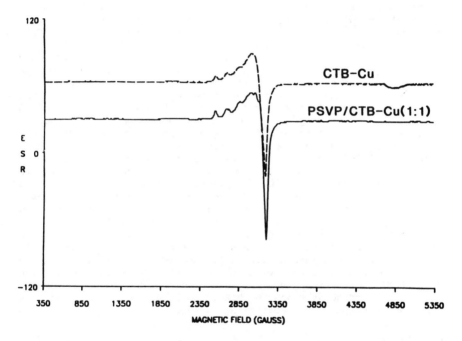

Figure 10. ESR spectra of CTB-Cu and PSVP/CTB-Cu (1:1).

There was some broadening of the signal that was probably due to dipolar interactions. The CTB-Cu spectrum also exhibited a signal at higher field that was due to a Cu-Cu dimer complex similar to the structure found in copper acetate, Fig. 9c. No evidence of a copper formate-type dimeric structure, Fig. 9b, was found. The major difference in the ESR spectrum of the blend was the disappearance of the dimeric copper complex. This may be due to the formation of the transition metal complex. Since the copper dimer complex requires the association of four carboxylate endgroups, association of the PSVP is likely to sterically hinder its formation.

Most ionomers are known to form microphase-separated aggregates of the ionic species ($\underline{15}$). The major evidence for these structures is a maximum in the small angle x-ray scattering (SAXS) profiles, which corresponds to a characteristic size of 1-3 nm. Studies of the effect of diluents on the structure of ionomers have shown that compounds that interact with the ionic species perturb the SAXS pattern, while compounds that preferentially interact with the non-ionic polymer matrix have a marginal effect on the scattering curves ($\underline{16}$). The SAXS curves of CTB-Cu and blend #1 are shown in Fig. 11. The abscissa is the scattering wavevector, $k = 4\pi\sin\theta/\lambda$, where λ is the wavelength of the x-ray radiation and 2θ is the scattering angle. The scattering curve of the neat CTB-Cu exhibited a broad, but distinct peak at about $k = 1.0$ nm^{-1} that corresponded to a distance in real space of about 6.3 nm. No clear peak was observed in the SAXS of the blend, though the increase in scattering intensity at low angles may have been due to a shift of the peak to lower wavevector. This result was consistent with the effect of polar plasticizers such as water, methanol or glycerol on the peak in the scattering curves of other ionomers ($\underline{17}$), and is taken here as further evidence for the interaction of the Cu-carboxylate and vinyl pyridine groups. One interpretation of the shift of the peak to lower wavevector or larger characteristic distance is that the vinyl pyridine-containing segments of the chain swelled the ionic microdomains. This interaction would require some mixing of the two phases, which would also explain the Tg data. Based on the previous SAXS studies, in the absence of this interaction, one would expect the addition of polystyrene to CTB-Cu to have no influence on the qualitative shape of the scattering curve.

CONCLUSIONS

Transition metal coordination of Cu(II) carboxylate groups and pyridine groups was employed as a means of coupling a telechelic butadiene-base polymer with a randomly functionalized styrenic polymer. Dynamic mechanical analysis (DMA) and differential scanning calorimetry (DSC) indicated partial miscibility of the two polymers and Fourier transform infrared (FTIR) spectroscopy demonstrated that interactions occurred on a molecular level. When compared with blends of PSVP and the free acid derivative of CTB, the compositions based on the transition metal complex had improved dimensional stability at elevated temperatures, though there remains some question as to the stability of the copper salt to hydrolysis. Electron spin resonance (ESR) spectroscopy showed that only the

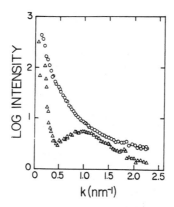

Figure 11. SAXS intensity versus wavevector for (Δ) CTB-Cu and (o) PSVP/Cu (1:1).

monomeric (isolated) copper complexes were present in the blend whereas dimer formation occurred in the neat CTB-Cu. This supports the presence of the Cu-N complex. Small angle x-ray scattering (SAXS) indicated that the PSVP disrupted the microphase separated ionic domains found in the neat CTB-Cu.

Based on these results, we propose a molecular architecture for the blend analogous to that of a grafted-block copolymer network as shown schematically in Fig. 12. The CTB-rich phase is drawn as spherical strictly for convenience. The important feature of this model is that the Cu-carboxylate and the vinyl pyridine groups interact at the phase boundary or, more likely, in a broadened interphase region. The physical network is formed as a result of vinyl pyridine groups from more than one PSVP molecule interacting with a single CTB-rich domain. It should be emphasized, however, that in the absence of microscopic evidence, this model is only speculative, though it is consistent with the data presented here. Such a structure would be expected to yield a morphology and properties similar to that of an SB block copolymer. Future work will focus on microscopic examination of the morphology of these blends as well as their mechanical properties.

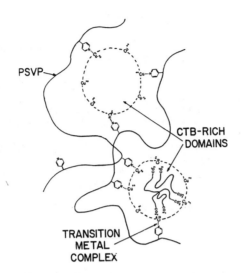

Figure 12. Schematic of transition metal complex-based grafted copolymer network.

ACKNOWLEDGMENTS

We gratefully acknowledge the donors of the Petroleum Research Fund, administered by the American Chemical Society, the Polymer Program of the National Science Foundation (Grant DMR-8407098), and Exxon Research and Engineering Co. for support of this research. We also wish to thank Dr. J. S. Lin of the National Center for Small Angle Scattering Research at Oak Ridge National Laboratory for his help with the SAXS experiments, Dr. H. A. Frank for his help with the ESR experiments and Dr. D. G. Peiffer of Exxon for his helpful comments and suggestions.

LITERATURE CITED

1. Polymer Blends; Paul, D. R.; Newman, S., Eds.; Academic: New York, 1977.
2. Olabisi, O; Robeson, L. M.; Shaw, M. T. Polymer-Polymer Miscibility; Academic: Mew York, 1979.
3. Clas, S. D.; Eisenberg, A. J. Polym. Sci.: Polym. Phys. Ed. 1984, 22, 1529.
4. Eisenberg, A.; Smith, R.; Zhou, Z. L. Polym. Eng. Sci. 1982, 22, 1117.
5. Smith, P; Eisenberg, A. J. Polym. Sci.: Polym. Phys. Ed. 1983, 21, 595.
6. Horrian, J.; Jerome, R.; Teyssie, Ph. J. Polym. Sci.: Polym. Let. Ed. 1986, 24, 69.
7. Agnew, N. H. J. Polym. Sci.: Polym. Chem. Ed. 1976, 14, 2819.
8. Peiffer, D. G.; Duvdevani, I.; Agarwal, P. K.; Lundberg, R. D. J. Polym. Sci.: Polym. Let. Ed. 1986, 24, 581.
9. Lundberg, R. D.; Peiffer, D. G.; Phillips, R. R. U.S. Patent 4 480 063, 1984.

10. Galland, D.; Melakhovsky, M.; Medrignac, M.; Pineri, M. Polymer 1986, 27, 883.
11. Broze, G.; Jerome, R.; Teyssie, Ph. J. Polym. Sci.: Polym. Phys. Ed. 1983, 21, 2205.
12. Kim, D. Ph. D. Thesis, University of Connecticut, Storrs, 1987.
13. Sen, A.; Weiss, R. A.; Oh, J.; Frank, H. A. Polym. Preprints 1987, 28(2), 220.
14. Maki, A. H.; McGarvey, B. R. J. Chem. Phys. 1958, 29, 31.
15. Eisenberg, A.; King, M. Ion-Containing Polymers: Academic: New York, 1977.
16. Fitzgerald, J. J.; Kim, D.; Weiss, R. A. J. Polym. Sci.; Polym Let. Ed. 1986, 24, 263.
17. Fitzgerald, J. J.; Weiss, R. A. J. Macromol. Sci.: Rev. Macromol. Chem. Phys. 1988, C28(1), 99.

RECEIVED April 27, 1989

Chapter 15

Structure and Properties of Short-Side-Chain Perfluorosulfonate Ionomers

Martin R. Tant[1], Kevin P. Darst, Katherine D. Lee, and Charles W. Martin

Texas Applied Science and Technology Laboratories, Dow Chemical USA, Freeport, TX 77541

An equivalent weight series of short-side-chain per-fluorosulfonic polymers has been studied in the sulfonyl fluoride, sulfonic acid, and sodium sulfonate forms by dynamic mechanical spectroscopy, differential scanning calorimetry, and wide angle X-ray scattering. Results indicate that the functional form of the polymer as well as the side chain length and equivalent weight strongly affect the morphology of the materials and therefore their dynamic mechanical and thermal properties. Evidence is also presented for the presence of two different types of crystal structures in these materials. Side chain length thus provides an additional variable by which the structure and properties of perfluorinated ionomers may be controlled.

Perfluorinated ionomers have reached a high level of industrial importance due to their outstanding performance as membranes in applications such as chlor-alkali cells and fuel cells (1). The Nafion material synthesized by duPont more than twenty years ago (2) has been widely used in such applications. The unusual transport properties of the acid and salt forms of these materials, such as ion permselectivity in solutions of high ionic strength, are a direct result of both the molecular and morphological structure (3). Strong coulombic associations lead to the formation of ionic regions typically referred to as clusters. The fact that the per-fluorinated backbone is crystallizable results in the formation of crystalline regions as well. Thus these materials really consist of at least three different phases: an amorphous phase, a crystalline phase, and an ionic phase. Recent results to be reported here

[1]Current address: Research Laboratories, Eastman Chemicals Division, Eastman Kodak Company, Kingsport, TN 37662

suggest the presence of two different types of crystalline regions, the relative proportions of which are dependent upon such variables as side chain length, equivalent weight, and sequence distribution. This further complicates the task of deciphering the morphology of these materials. It also underscores the fact that, for materials of such complexity, an understanding of the structure-property behavior must certainly be attained if one is to be able to design, synthesize, and fabricate materials with optimum properties.

A great deal of effort by both industrial and academic scientists has been directed at gaining an understanding of the morphological structure and the mechanism of ion transport in perfluorinated ionomers. The molecular response to an imposed force field, be it mechanical or electromagnetic, is highly dependent upon not only the molecular structure, but the morphological structure as well. Study of the dynamic mechanical properties yields useful information about the effects of morphology upon molecular response mechanisms. When correlated with other information, such as a knowledge of molecular structure, thermal behavior, and percent crystallinity, an understanding of the overall structure-property behavior begins to evolve. Eisenberg and coworkers have studied molecular motions in Nafion using dynamic mechanical spectroscopy (4-7), dielectric spectroscopy (5), and solid-state NMR (8). This work has been reviewed by Kyu (9). MacKnight and coworkers (10,11) have made similar studies of the perfluorocarboxylate systems. More recent work by Mauritz, Fu, and Yun (12-14) has focused on gaining an understanding of ion transport in solvent-swollen Nafion using dielectric spectroscopy. Morphological structure has been probed using small and wide-angle X-ray scattering (15-18), electronic absorption spectroscopy (19,20), extended X-ray absorption fine structure (21,22), as well as other techniques.

Nearly all of this work has focused on Nafion, a long-side-chain polymer as shown in Figure la. Difficult synthetic problems have limited the availability of perfluorinated ionomers having differing side chain lengths. In 1982 a new synthetic route was reported (23) which led to the short-side-chain polymer shown in Figure lb. The polymers in Figure 1 are shown in their thermoplastic precursor (SO_2F) form. The membranes are fabricated in this form and are then hydrolyzed to the ionomer form. Little data has appeared in the literature concerning the properties of the short-side-chain material, but it is now clear that side chain length may have a significant effect upon ion transport properties. Significant increases in selectivity of ion transport in chlor-alkali (24) and maximum power density in fuel cell applications have been reported (25) using membranes derived from short side chain polymer. In this paper we report the initial results of a study directed at gaining an understanding of the structure-property behavior of the short-side-chain materials. The effect of both equivalent weight and functional form (i.e. precursor, sulfonic acid, or sodium sulfonate) on the dynamic mechanical properties are discussed. The effect of side chain length upon the dynamic mechanical behavior is also probed by comparing the properties of the short-side-chain polymers with those of the long-side-chain analog having the Nafion structure. Crystallinity data obtained from variable temperature wide angle X-ray scattering and differential

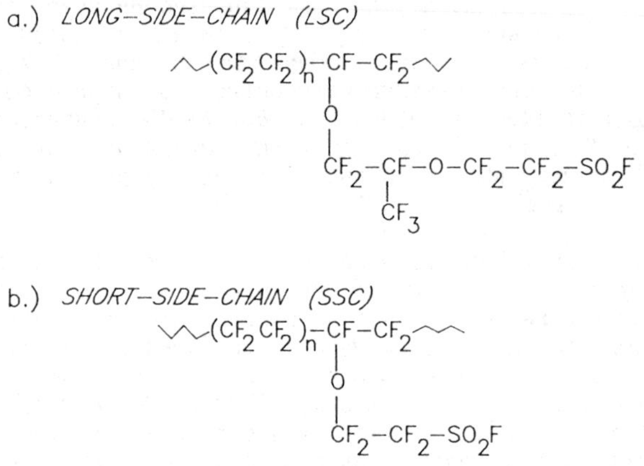

a.) *LONG–SIDE–CHAIN (LSC)*

$\sim(CF_2CF_2)_n-CF-CF_2\sim$
$|$
O
$|$
$CF_2-CF-O-CF_2-CF_2-SO_2F$
$|$
CF_3

b.) *SHORT–SIDE–CHAIN (SSC)*

$\sim(CF_2CF_2)_n-CF-CF_2\sim$
$|$
O
$|$
$CF_2-CF_2-SO_2F$

Figure 1. Chemical structures of the (a) long-side-chain and (b) short-side-chain perfluorosulfonyl polymers.

scanning calorimetry experiments are used to support the interpretation of the dynamic mechanical data.

Experimental

Synthesis. The short-side-chain vinyl ether comonomer, $CF_2=CFOCF_2CF_2SO_2F$, was prepared in the laboratory by literature procedures ($\underline{23}$). Emulsion polymerization was carried out using standard techniques in a 500 ml laboratory reactor similar to those previously described in the literature ($\underline{26-27}$). The equivalent weight was controlled by adjusting the pressure of tetrafluoroethylene (TFE) to vary the comonomer:TFE feed ratio in the reactor. The reactor was charged under a nitrogen atmosphere with an aqueous solution of initiator, buffers, and surfactant, followed by comonomer. The synthesis was then carried out at 60°C with a continuous feed of TFE to maintain the desired pressure in the stirred reactor. After the required quantity of TFE was added, the reactor was vented and the copolymer was isolated by acid coagulation, washed, and dried. The comonomer:TFE ratio was adjusted to a level that would produce 80-90 grams of polymer at about 50% conversion of comonomer. The TFE pressure was controlled to produce polymers which analyzed at 1200, 1000, 800 and 600 (+-26) equivalent weight, respectively. Equivalent weights were determined by functional group hydrolysis and titration and confirmed by percent sulfur analysis.

Sample Preparation. Samples were prepared from the thermoplastic polymer powders at 282°C and 5 tons pressure. The molded thermoplastic samples were converted to the acid form by hydrolysis in 17% KOH/DMSO/water at 90°C for 4 days followed by acidification in 50% nitric acid at 90°C for 4 hours. The sulfonic acid form was converted to the sodium sulfonate form by neutralization in 10% NaOH at 90°C for 4 hours. Both sulfonic acid and sodium sulfonate forms of the polymer were rinsed with water at 80°C for 20 hours and then vacuum dried at 120°C for 16 hours.

Differential Scanning Calorimetry. Differential scanning calorimetry (DSC) experiments were performed using a Perkin-Elmer DSC-7 interfaced with a 7700 series microprocessor/controller. Scans were made on samples weighing 30 mg in the temperature range from room temperature to 340°C at a heating rate of 20°C/min.

Dynamic Mechanical Properties. Dynamic mechanical properties were measured with a Rheometrics Dynamic Spectrometer Model 7700 in the torsional rectangular mode. Temperature sweeps were run from -160 to 340°C using a strain of 10% and frequencies of 1, 10, and 100 radians per second (rad/s). Data was obtained at 5 degree temperature increments with at least a one minute thermal equilibration time at each temperature. Only the 10 rad/s data are reported in this paper.

Wide Angle X-ray Scattering. Percent crystallinity determinations were made using wide angle X-ray scattering (WAXS). Samples were compression-molded and then slow-cooled in the mold to promote maximum crystallization. Each sample was analyzed using a pinhole camera having a 700 mm collimator and a 70 mm camera length. Two 0.8 mm square pinholes were used. The camera was mounted vertically atop a Rigaku RU200 rotating anode generator operated at 45 kV and 150 mA. A Braun OEM-50 linear position sensitive detector was mounted at a 25 degree angle to the sample. The detector was operated with a resolution of 0.036 degree 2 θ per channel. The data acquisition time was 20 minutes. Because of the lack of pure standards, the method of using differential intensity measure-ments to calculate crystallinity indices was used (30). Relative weight percent crystallinities were obtained using curve fitting results generated from the 1200 EW short-side-chain precursor sample.

Variable temperature wide angle X-ray scattering data were also obtained on several of the samples. A Mettler FP82 hot stage with FP80 controller was placed inside the camera. WAXS patterns were collected at 10°C intervals from 30 to 290°C using a heating rate of 2°C/min. Detector resolution was decreased to 0.168 degree 2 θ and the acquisition time was 4.75 minutes. All other conditions were the same as described earlier.

Results And Discussion

The materials studied in this work are summarized in Table I, which also includes the side chain content in terms of weight percent and mole percent comonomer. In the following section, the effect of side chain length on the dynamic mechanical behavior of the per-fluoro-sulfonyl fluoride precursor is addressed by comparing the properties of the long-side-chain (LSC) precursor with those of the short-side-chain (SSC) precursor at equal weight percent and mole percent comonomer content. The effect of equivalent weight on the dynamic mechanical behavior of the SSC perfluorosulfonyl fluoride precursor in the range between 600 and 1200 EW is then explored. Next, the effect of conversion to the sulfonic acid and sodium sulfonate forms on the dynamic mechanical behavior is considered. Finally, the effect of side chain length on the properties of the sulfonic acid and sodium sulfonate forms of the material is examined.

Table I. Summary of Materials Evaluated

SSC EW	Comonomer Weight %	Content Mole %	T_g (G"peak) degrees C	% Crystallinity by WAXS
600	47	24	13	0
800	35	16	26	7
1000	28	12	41	17
1200	23	10	50	24
LSC EW				
1140	39	13	6	8

Effect of Side Chain Length for Precursor. Figure 2 shows both the storage modulus, G', and loss modulus, G", as a function of temperature for the LSC 1140 EW and SSC 800 EW polymers in the thermo-

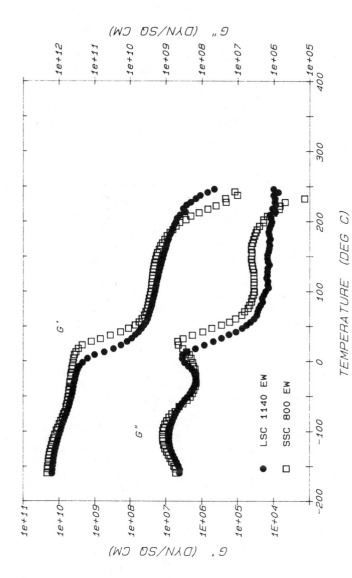

Figure 2. Comparison of the dynamic mechanical behavior of the long-side-chain and short-side-chain precursors of similar weight percent comonomer content.

plastic precursor (sulfonyl fluoride) form at a frequency of 10 rad/s. These two materials have about the same comonomer content in terms of weight percent (39% for the LSC and 35% for the SSC). Therefore the SSC material has more side chains. These equivalent weights are also those typically used in applications requiring high ionic conductivity. The response of the materials below their glass transition is quite similar. The shorter, less mobile side chains of the SSC material result in an increase in the glass transition temperature of about 20°C. Polymer molecules with shorter side chains are able to pack more closely, decreasing the free volume available for main chain motion--the origin of the glass transition ($\underline{29}$). Vincent ($\underline{30}$) found similar results for a series of n-alkyl ethers, where T_g was found to decrease in the order polyvinyl methyl ether (-10°C) > polyvinyl ethyl ether (-17°C) > polyvinyl n-propyl ether (-27°C) > polyvinyl n-butyl ether (-32°C) due to the increasing length of the side chains. It is also likely that, in addition to the effects of improved packing efficiency, the lower mobility of the short side chains also contributes to the observed trend in T_g.

Above T_g the behavior of the LSC and SSC materials is also very similar. Both sustain a strong rubbery plateau to well above 150°C. WAXS experiments have revealed relative crystallinity levels of about 8% for the LSC polymer and 7% for the SSC polymer, suggesting that crystallinity is likely the major contributor to the rubbery plateau. (The crystallinity data are also summarized in Table I.) Since randomly placed pendant groups act to reduce the extent of main chain crystallization, both the equivalent weight and side chain length would be expected to affect the crystallinity level. Increasing the number of side chains and increasing side chain length should both act to reduce crystallinity. From the fact that both materials have about the same level of crystallinity, it is clear that the increased side chain length of the LSC 1140 EW precursor essentially compensates for the greater number of side chains of the SSC 800 EW material in reducing the extent of crystallization. Also, the fact that the degree of crystallinity is similar for both materials suggests that the observed difference in glass transition temperature is not due to crystallinity differences.

In Figure 3 the storage modulus, G', and loss modulus, G", of the LSC 1140 EW and SSC 1000 EW precursors are shown as a function of temperature at a frequency of 10 rad/s. In this case both materials contain about the same number of side chains, or about 12 to 13% comonomer content on a molar basis. If the materials were amorphous so that crystallinity effects could be ignored, the SSC material would be expected to have a higher T_g due to its shorter chains. Of course, the SSC material should also be more crystalline, since its shorter side chains should pack more efficiently than the longer side chains of the LSC material. WAXS experiments confirm this expectation, with the SSC material being about 17% crystalline compared to the 8% crystalline LSC material. Again, crystallinity is known to raise T_g and broaden the transition region, so this factor would be expected to further elevate the T_g

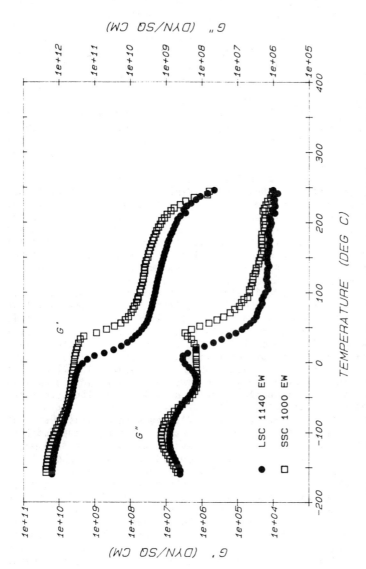

Figure 3. Comparison of the dynamic mechanical behavior of the long-side-chain and short-side-chain precursors of similar mole percent comonomer content.

of the SSC precursor compared to that of the LSC precursor. In fact, the T_g of the SSC material is about 35°C above that of the LSC material, thus agreeing with the expected result.

Effect of Equivalent Weight for Precursor. Figure 4 shows the effect of equivalent weight on the storage and loss modulus curves of the precursor form of the SSC polymer. Below the glass transition temperature the storage and loss moduli of the three highest equivalent weight materials are all quite similar. However, the moduli of the 600 EW material are substantially lower, probably due to the fact that the material is amorphous. Clearly, there is a very strong increase in the T_g as the equivalent weight is increased. This is likely to be due to two effects. The increasing crystallinity tends to reduce main chain mobility, thus resulting in an increase in T_g and a broadening of the transition region. The effect of decreasing the number of side chains is to increase the packing efficiency in the amorphous regions as well with the result again that main chain motion is restricted and T_g is increased. As equivalent weight is increased from 600 to 1200, the T_g increases from 13 to 50°C (Table I).

There has been some question in the past concerning whether the strong rubbery plateau observed for Nafion procursor is due to association of the sulfonyl fluoride groups or to crystallinity (5). Figure 4 shows clearly that, as the equivalent weight increases from 600 to 1200, the rubbery plateau modulus increases substantially; and the temperature at which the material begins to flow increases as well. If the plateau was due to association of nyl fluoride groups, the trend would be reversed since more associating groups (lower equivalent weights) would result in a higher effective crosslink density. Figure 5 shows WAXS patterns for each of the four precursors. It is quite clear that the crystalline diffraction peak becomes much stronger with increasing equivalent weight. Analysis of the diffraction patterns reveals that relative crystallinity varies from 0 to 24% as equivalent weight is increased from 600 to 1200. This data is summarized in Table I and plotted in Figure 6. For the three crystalline polymers, the % crystallinity is a linear function of mole % vinyl ether comonomer and the regression extrapolates to zero crystallinity at 18.9 mole % or 709 EW. The fact that the relative crystallinity level as determined by WAXS increases so substantially supports the conclusion that the rubbery plateau is due to the effect of crystallinity instead of association of sulfonyl fluoride groups. Based on these results it seems quite likely that the rubbery plateau in Nafion precursor is also due to crystallinity.

Figure 7 shows the results of the variable temperature WAXS experiments on the SSC precursors of 1000 and 1200 EW and the 1140 EW LSC precursor. Similar data has been presented by Gierke et al. (15) and Fujimura et al. (16) for Nafion precursor. In Figure 7 the crystalline melt is observed to begin at 130°C for both SSC polymers and terminate at 250 and 270°C for the 1000 EW and 1200 EW SSC polymers, respectively. Certainly this melting range agrees with the dynamic mechanical data since the rubbery plateau extends to these same upper temperatures. The melting range for the LSC polymer is observed in Figure 7 to occur between 110 and 230°C.

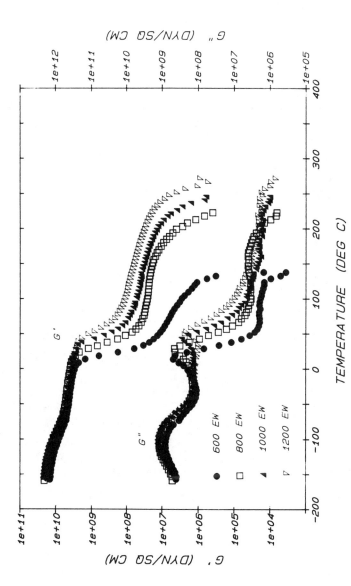

Figure 4. Effect of equivalent weight on the dynamic mechanical behavior of the short-side-chain perfluorosulfonyl precursor.

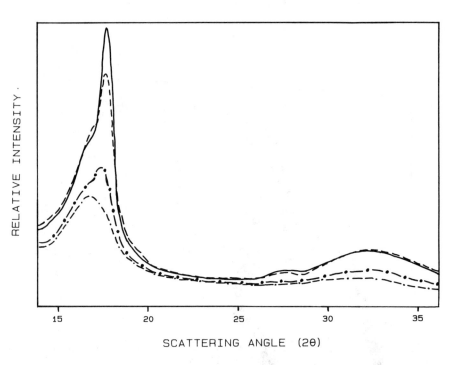

Figure 5.   Effect of equivalent weight on the wide angle X-ray scattering curves of the short-side-chain precursor.

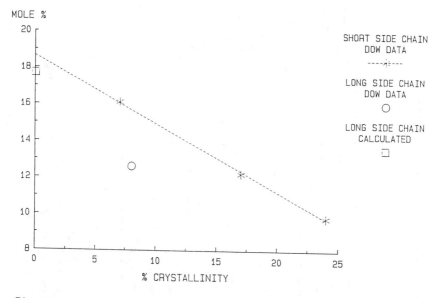

Figure 6. Effect of side chain length and mole percent comonomer on the percent crystallinity by WAXS.

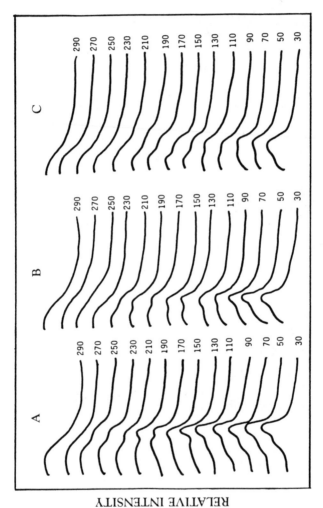

SCATTERING ANGLE (2Θ)

RELATIVE INTENSITY

Figure 7. Variable temperature wide angle X-ray scattering curves for (A) 1200 EW short-side-chain precursor, (B) 1000 EW short-side-chain precursor, and (C) 1140 EW long-side-chain precursor.

Differential scanning calorimetry curves for the 600 to 1200 EW SSC polymers are shown in Figure 8. Two peaks appear in the scans of the 1000 and 1200 EW materials: a broad endotherm in the temperature range from 120 to 260°C and a sharper one appearing near 323°C. The high-temperature peak occurs at about the same temperature where melting is observed for pure PTFE. The lower temperature endotherm occurs over the temperature range where the crystalline diffraction peak gradually reduces in magnitude. The fact that a crystalline WAXS diffraction peak is not observed above 270°C for these polymer is probably due to the low percent crystallinity at this temperature, coupled with small crystallite size and the low sensitivity of the WAXS conditions. The presence of the high temperature DSC peak in the region of the PTFE crystalline melt is probably not due to the presence of gross PTFE blocks in the copolymers, since a large excess of vinyl ether monomer was present at all times during the polymerization. Rather, it is likely the result of the random distribution of side chains giving rise to longer segments of C_2F_4 as the equivalent weight is increased. For qualitative comparison, a very old sample of 1500 EW Nafion precursor (circa 1972) was analyzed by DSC. The same type of dual endotherm described above was observed.

If the polymers are random and homogeneous, then one must conclude that there are two distinct crystal structures present in the 1000 and 1200 EW SSC polymers. If the polymer had a single, PTFE-like crystal structure with all of the side chain material in the amorphous phase, one would expect a steady rise in the crystalline melt point to that of the homopolymer crystal. On the contrary, the existence of a second crystal structure incorporating side chain material is consistent with a single endotherm at lower equivalent weights and the appearance of a second, higher temperature endotherm at higher equivalent weights. The broad skewed shape of the lower endotherm, coupled with the fact that the shape and position of the endotherm, is similar for SSC and LSC polymers with the same mole percent ionomer is consistent with a structure that can incorporate variable amounts of side chain. Such a structure was proposed by Starkweather (18) based upon WAXS data for Nafion.

Table II summarizes the relative heats of fusion for the two types of crystallinity for each polymer. The 600 EW polymer showed

Table II. Summary of DSC Data

	1st Transition		2nd Transition		Total	% Crystallinity
	Peak	dHf	Peak	dHf		
SSC EW	Deg C	(J/g)	Deg C	(J/g)	J/gram	by WAXS
600	---	---	---	---	---	0
800	143	4.90	---	---	4.90	7
1000	201	12.88	320	0.45	13.33	17
1200	218	17.34	323	2.81	20.15	24
LSC EW						
1140	152	8.12	---	---	8.12	8

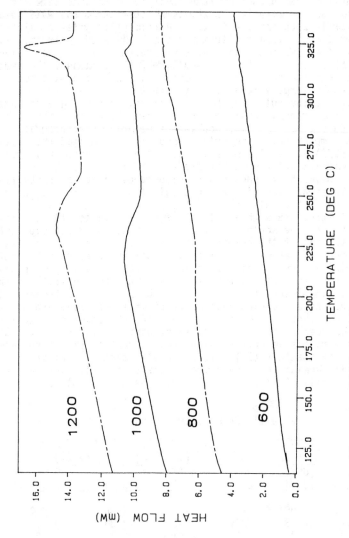

Figure 8. Differential scanning calorimetry curves for the short-side-chain precursors.

no discernible melting peak, while a broad crystalline transition
was observed in the region of 120 to 260°C for the 800, 1000, and
1200 EW materials. The sharper high-temperature crystalline melt
at 320-323°C was observed only in the 1000 and 1200 EW polymers.
The DSC data agrees well with the WAXS and dynamic mechanical
results. Figure 9 shows the effect of SSC comonomer content on the
observed heat of fusion for the broad crystalline endotherm.
Extrapolation of the SSC regression line gives the same zero crystalline point, 18.9 mole %, as was obtained from the WAXS data.
The corresponding literature data for Nafion (18) is also plotted
in Figure 9 along with our data for the 1140 EW LSC polymer. The
current LSC data falls on the Nafion regression line which gives
zero crystallinity at 16.1 mole % (965 EW) LSC comonomer. The heats
of fusion for the two polymers converge as the mole % comonomer
decreases, as one would expect.
     The lower temperature crystalline transition has a greater heat
of fusion in every case than the higher transition. If the molar
heats of fusion for the two crystalline endotherms were equal, the
low melting crystal would be 86% and 96% respectively of the total
crystallinity of the 1200 and 1000 EW SSC polymer. Evidently,
since the rubbery modulus in the dynamic mechanical data is controlled by crystallinity, it is the low temperature crystalline
component which has the greatest influence.

Effect of Conversion to Acid and Salt Forms. Figure 10a-d shows
the effect of conversion to sulfonic acid and sodium sulfonate
forms for all four SSC materials. There is a strong increase in
the apparent T_g following conversion of the precursor to the sulfonic acid form. This increase occurs due to the strong association of the sulfonic acid groups which limits the mobility of the
main chain. Upon neutralization to form the sodium sulfonate, a
further increase in the apparent T_g is observed. This is due to
the stronger association of the sodium sulfonate groups. Similar
behavior has been observed for Nafion materials by Kyu et al. (6).
We also observe such behavior for the LSC material, as illustrated
in Figure 11. Table III shows relative amounts of crystallinity,
determined by WAXS, for the precursor, acid, and salt forms of all

Table III. Percent Crystallinity for Precursor,
Acid, and Salt Forms

	Percent Crystallinity		
SSC EW	Precursor	Sulfonic Acid	Sodium Sulfonate
600	0	0	0
800	7	0	0
1000	17	9	7
1200	24	13	11
LSC EW			
1140	8	0	0

the materials. The large decrease in percent crystallinity upon
conversion from precursor to sulfonic acid occurs due to the competing nature of the ionic association phenomenon. The further

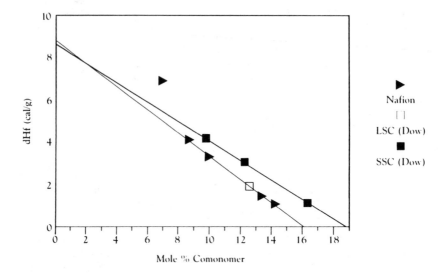

Figure 9. Heat of fusion as a function of side chain length and mole percent comonomer for the lower melting crystallite.

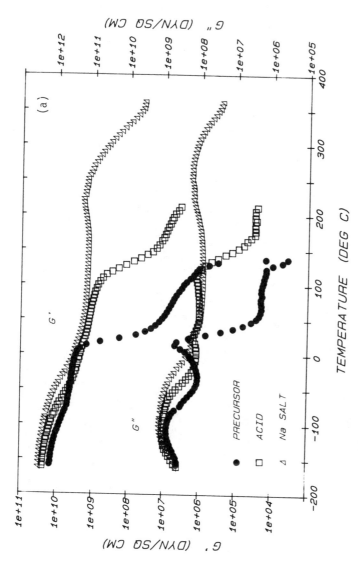

Figure 10a. Effect of conversion from precursor to sulfonic acid and sodium sulfonate on the dynamic mechanical behavior of the short-side-chain polymer with equivalent weight of 600. *Continued on next page.*

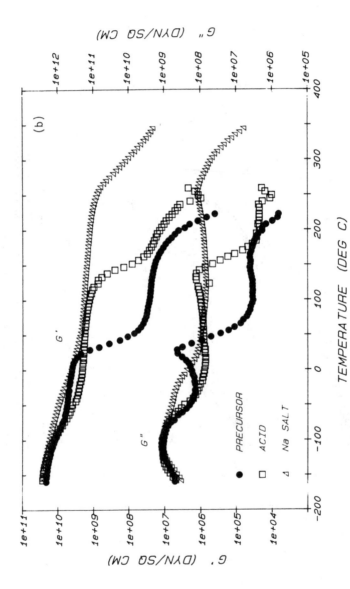

Figure 10b. Effect of conversion from precursor to sulfonic acid and sodium sulfonate on the dynamic mechanical behavior of the short-side-chain polymer with equivalent weight of 800. *Continued on next page.*

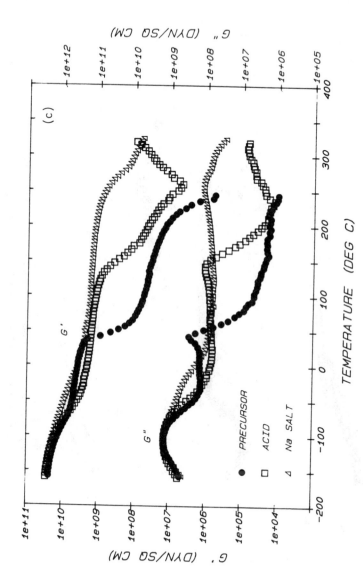

Figure 10c. Effect of conversion from precursor to sulfonic acid and sodium sulfonate on the dynamic mechanical behavior of the short-side-chain polymer with equivalent weight of 1000. *Continued on next page.*

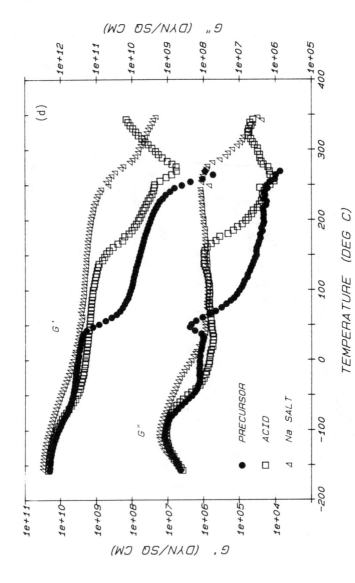

Figure 10d. Effect of conversion from precursor to sulfonic acid and sodium sulfonate on the dynamic mechanical behavior of the short-side-chain polymer with equivalent weight of 1200.

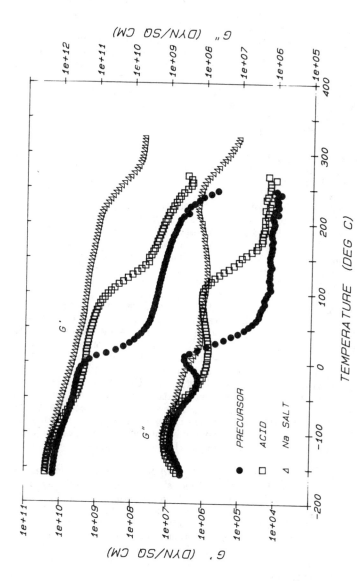

Figure 11. Effect of conversion from precursor to sulfonic acid and sodium sulfonate on the dynamic mechanical behavior of the 1140 EW long-side-chain polymer.

decrease observed upon neutralization is additional evidence that
sodium sulfonate groups associate more strongly than sulfonic acid
groups. For both the SSC and LSC sulfonic acids, a weak plateau
above the apparent glass transition temperature is evident. For
all except the 600 EW SSC material, the sulfonic acid storage modulus curve conforms well to the precursor curve, suggesting that the
weak plateau may be due to residual crystallinity. A very weak
plateau is also observed for the 600 EW material. Since there is
no evidence of crystallinity in the 600 EW polymer before hydrolysis and hydrolysis results in even lower crystallinity, it is
likely that weak hydrogen bonding is contributing to the plateau.
The large increase in storage modulus which is observed for several
of the acid forms at very high temperatures results from a crosslinking reaction which occurs at these temperatures. This reaction is currently under study.

Effect of Equivalent Weight for Acid and Salt Forms. The dynamic
mechanical properties of the SSC sulfonic acid materials are compared in Figure 12 in order to investigate the effect of equivalent
weight. Clearly, as equivalent weight increases, the storage modulus curves are shifted higher in both modulus and temperature. Even
though the sulfonic acid groups associate quite strongly, it is
still apparent that crystallinity has more of an effect upon the
flow temperature than ionic association. However, the trend with
equivalent weight reverses when the sulfonic acids are converted to
the sodium sulfonate form as shown in Figure 13a. The high temperature region of this graph is expanded in Figure 13b so that the
effects of equivalent weight may be seen more clearly. It can be
seen that the trend is reversed in this case from that observed for
the precursor and sulfonic acid forms. That is, the higher equivalent weight materials begin to flow at lower temperatures. The
reason for this is that the higher equivalent weight materials have
a lower ionic content and therefore a lower ionic crosslink density. It is also noted that, in the storage modulus curves for the
two highest equivalent weight materials, i.e. the 1000 and 1200 EW
materials, a slight high-temperature plateau is observed which is
similar to those observed for the acid forms. Of course, those for
the acid forms were attributed to crystallinity effects. A similar
cause seems likely here, since these were the only two salt forms
of the material which displayed a crystalline WAXS peak.

Effect of Side Chain Length for Acid and Salt Forms. Figures 14
and 15 illustrate the effect of side chain length on dynamic
mechanical behavior of the sulfonic acid and sodium sulfonate forms
by comparing the behavior of the LSC 1140 EW material with that of
the SSC 800 and 1000 EW materials. Again the LSC 1140 EW material
has about the same comonomer content as the SSC 800 EW material on
a weight basis and about the same as the SSC 1000 EW material on a
mole basis. For both forms, the apparent glass transition temperature of the LSC material is substantially lower than that of both
of the SSC materials. The most likely reason for this lower transition is the fact that the longer side chains of the LSC material
increase the effective molecular weight between ionic crosslinks
thus rendering the main polymer chain more mobile. This difference

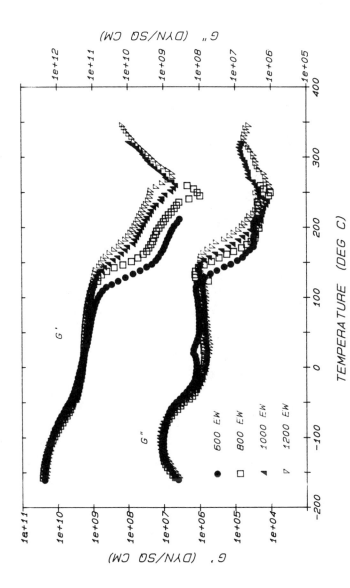

Figure 12. Effect of equivalent weight on the dynamic mechanical behavior of the short-side-chain sulfonic acid polymer.

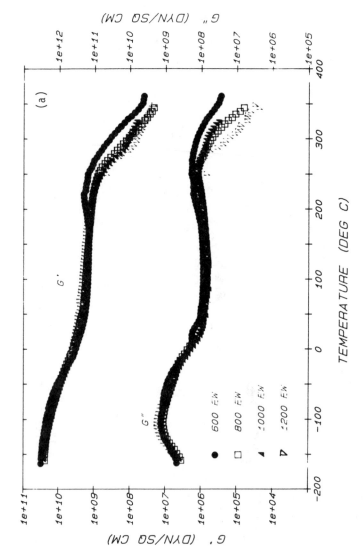

Figure 13a. Effect of equivalent weight on the dynamic mechanical behavior of the short-side chain sodium sulfonate polymer.

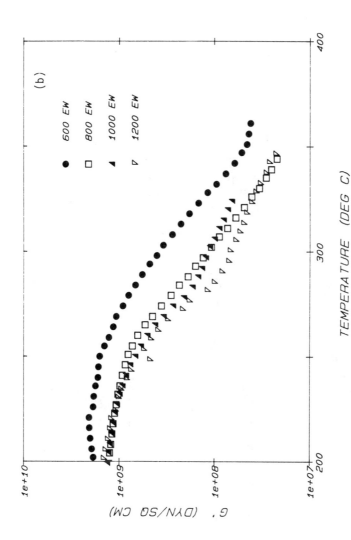

Figure 13b. Expansion of the high temperature region showing the effect of equivalent weight on the storage modulus.

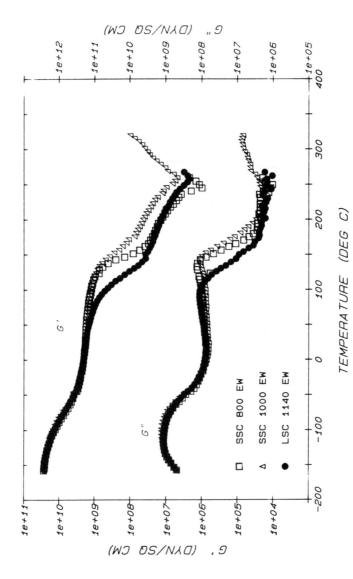

Figure 14. Effect of side chain length on the dynamic mechanical behavior of the sulfonic acid polymer.

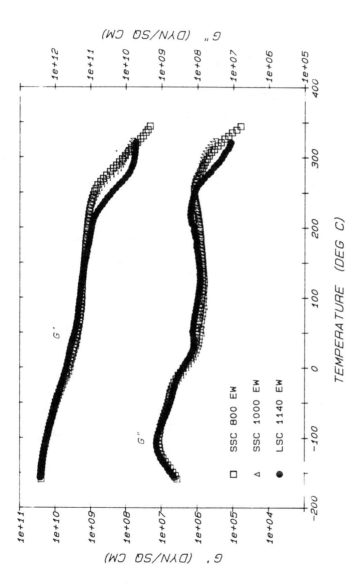

Figure 15. Effect of side chain length on the dynamic mechanical behavior of the sodium sulfonate polymer.

has been previously noted for the acid form (31), but not for the
sodium sulfonate form.

Conclusions

This study has demonstrated that both side chain length and equivalent weight strongly affect the structure and properties of these
materials in the sulfonyl fluoride, sulfonic acid, and sodium sulfonate forms. In the precursor, or sulfonyl fluoride form, the
dynamic mechanical behavior is controlled primarily by the extent
of crystallization, which is directly controlled by both equivalent
weight and side chain length. This is evidenced by the fact that
the rubbery plateau modulus, the melt flow temperature, and the
percent crystallinity all increase with increasing equivalent
weight. Since side chains act to reduce the packing efficiency of
polymer molecules, either increasing the number of side chains or
increasing their length decreases the extent of crystallization.
When the precursor is converted to the sulfonic acid form, there is
a dramatic change in the dynamic mechanical behavior due to the
strong intermolecular association of the acid groups. This association results in a very large increase in the apparent glass
transition temperature due to the fact that the ionic association
limits the extent of main chain motion. Conversion to the sulfonic
acid form also substantially reduces the degree of crystallinity,
since ionic association and crystallization are competing processes.   Conversion of the sulfonic acid form to the sodium sulfonate form results in even stronger association, a decrease in crystallinity, and a further large increase in the apparent glass transition temperature.
    Differential scanning calorimetry experiments suggest the pres-
ence of two different types of crystalline structures in materials
of sufficiently high equivalent weight.   At lower temperatures,
there is a very broad melting peak which apparently results from
crystalline regions possessing a high content of side chains. These
side chains tend to limit crystallization and reduce crystal perfection. At 320-325°C there is a sharp crystalline melting peak
which is likely due to more perfect crystalline regions possessing
a more PTFE-like crystalline structure. The high temperature peak
is more likely to appear for materials whose chemical structure
approaches more closely the structure of PTFE, i.e. materials with
higher equivalent weight and shorter side chains.

Acknowledgments

The authors would like to express their gratitude to Mr. T. K. Tran
for synthesizing the materials and to Mr. M. J. Castille, Jr. for
conducting   the   dynamic   mechanical   experiments.

Literature Cited

1. Perfluorinated Ionomer Membranes, A. Eisenberg and H. L.
    Yeager, Eds., ACS Symposium Series 180, American Chemical
    Society, Washington, D.C., 1982.

2. D. J. Connolly and W. F. Gresham, U.S. Patent 3,282,875 (1966).
3. K. A. Mauritz and A. J. Hopfinger, in Modern Aspects of Electrochemistry, J. O'M. Bockris, B. E. Conway, and R. E. White, Eds., Plenum, New York, 1982, Chapter 6.
4. S. C. Yeo and A. Eisenberg, J. Appl. Polym. Sci. 21, 875 (1977).
5. I. M. Hodge and A. Eisenberg, Macromolecules 11, 289 (1978).
6. T. Kyu, M. Hashiyama, and A. Eisenberg, Can. J. Chem. 61, 680 (1983).
7. T. Kyu and A. Eisenberg, in Perfluorinated Ionomer Membranes, A. Eisenberg and H. L. Yeager, Eds., ACS Symposium Series 180, American Chemical Society, Washington, D.C., 1982, Chapter 6.
8. N. G. Boule, V. J. McBrierty, and A. Eisenberg, Macromolecules 16, 80 (1983).
9. T. Kyu, in Materials Science of Synthetic Membranes, D. R. Lloyd, Ed., ACS Symposium Series 269, American Chemical Society, Washington, D.C., 1985, Chapter 18.
10. Y. Nakano and W. J. MacKnight, Macromolecules 17, 1585 (1984).
11. G. Smyth, V. J. McBrierty, and W. J. MacKnight, Macromolecules 20, 1019 (1987).
12. K. A. Mauritz and R. M. Fu, Macromolecules, in press.
13. K. A. Mauritz and H. Yun, Macromolecules, in press.
14. K. A. Mauritz and H. Yun, Macromolecules, submitted.
15. T. D. Gierke, G. E. Munn, and F. C. Wilson, in Perfluorinated Ionomer Membranes, A. Eisenberg and H. L. Yeager, Eds., ACS Symposium Series 180, Washington, D.C., 1982, Chapter 10.
16. M. Fujimura, T. Hashimoto, and H. Kawai, in Perfluorinated Ionomer Membranes, A. Eisenberg and H. L. Yeager, Eds., ACS Symposium Series 180, Washington, D.C., 1982, Chapter 11.
17. E. J. Roche, M. Pineri, R. Duplessix, and A. M. Levelut, J. Polym. Sci. Polym. Phys. Ed. 19, 1 (1981).
18. H. Starkweather, Jr., Macromolecules 15, 320 (1982).
19. M. Falk, in Perfluorinated Ionomer Membranes, A. Eisenberg and H. L. Yeager, Eds., ACS Symposium Series 180, Washington, D.C., 1982, Chapter 8.
20. J. M. Kelly, in Structure and Properties of Ionomers, M. Pineri and A. Eisenberg, Eds., Reidel, Dordrecht, Holland, 1987, p. 127.
21. H. K. Pan, G. S. Knapp, and S. L. Cooper, Coll. Polym. Sci. 262, 734 (1984).
22. H. K. Pan, A. Meagher, M. Pineri, G. S. Knapp, and S. L. Cooper, J. Chem. Phys. 82, 1529 (1985).
23. B. R. Ezzell, W. P. Carl, and W. A. Mod, U.S. Patent 4,358,412, Nov. 9, 1982.
24. B. R. Ezzell; W. P. Carl; and W. A. Mod; in Industrial Membrane Processes; R. E. White and P. N. Pintauro, Eds., AIChE Symposium Series 248, American Institute of Chemical Engineers, New York, NY, 1986; p 45.
25. D. Watkins; K. Dircks; and D. Epp; Canadian Solid Polymer Fuel Cell Development, presented at the 1988 Fuel Cell Syminar, Oct. 23-26, 1988, Long Beach, CA, Sponsored by the National Fuel Cell Coordinating Group.

26. B. R. Ezzell, W. P. Carl, and W. A. Mod, U.S. Patent
     4,330,654, May 18, 1982.
27. B. R. Ezzell, W. P. Carl, and W. A. Mod, U.S. Patent
     4,515,989, May 7, 1985.
28. L. E. Alexander, X-ray Diffraction Methods in Polymer Science,
     Wiley, New York, 1969, Reprinted by Krieger Publishing Co.,
     Huntington, NY, 1985, pp. 176-180.
29. I. M. Ward, Mechanical Properties of Solid Polymers, Wiley,
     New York, 1983, pp. 168-170.
30. P. I. Vincent, in The Physics of Plastics, P. D. Ritchie, Ed.,
     Iliffe Books, London, 1965.
31. G. A. Eisman, Proceedings of the 169th Electrochemical Society
     Meeting, Vol. 83-13, 156 (1986).

RECEIVED February 17, 1989

Chapter 16

Perfluorinated-Ionomer-Membrane-Based Microcomposites

Silicon Oxide Filled Membranes

K. A. Mauritz, R. F. Storey, and C. K. Jones[1]

Department of Polymer Science, University of Southern Mississippi, Hattiesburg, MS 39406–0076

The objective of this work was to affect the in situ growth of silicon oxide microclusters or interpenetrating networks in Nafion membranes via the sol-gel reaction for tetraethoxysilane (TEOS). The underlying hypotheses are: 1) The resultant morphology of the inorganic phase is ordered by the polymer's phase-separated morphology. 2) TEOS/ alcohol/H_2O solutions will preferentially incorporate within pre-existing polar clusters. 3) TEOS hydrolysis in clusters is catalyzed by the fixed SO_3H groups. Reproducible procedures for the in situ growth of "silica" in hydrated membranes contacting TEOS/alcohol solutions were established. Solids uptake vs. immersion time curves are a function of alcohol solvent type. Tensile properties vs. solids uptake indicate an initial strengthening, followed by a ductile-brittle transition suggesting a silicon oxide phase percolation threshold. FT-IR and ^{29}Si solid state NMR analyses of microstructural evolution portray an inorganic network that is not as highly crosslinked as that of sol-gel-derived free silica and in fact becomes less coordinated and connected with increasing solids uptake.

The remarkable physicochemical stability and outstanding ion transport properties of perfluorinated ionomer membranes have stimulated their use mainly in electrochemical cells targeting the efficient production of industrial chemicals and energy (1). It is becoming clearer, although not totally understood, as to how the molecular details of the ionic-hydration structure within the microphase-separated morphology can be manipulated to affect material design and process optimization (2). Nafion (E.I. Dupont de Nemours & Co.), in the SO_3H form, has also been interesting as a strong acid catalyst for a number of reactions (3). In the work reported herein, we have taken advantage of both aspects of the cluster microphase morphology and strong acid nature of perfluorosulfonic acid films.

[1]Current address: Exxon Research and Engineering Company, Clinton Township, Annandale, NJ 08801

0097–6156/89/0395–0401$06.00/0
© 1989 American Chemical Society

In a distinctly different materials realm, the sol-gel process for metal alkoxides $(M(OR)_x, \chi = 3,4)$ has received considerable attention as a novel low-temperature route toward producing glass monoliths, coatings, fibers, gel powders and ceramic precursor particles (4-9). Tetraethoxysilane (or TEOS for tetraethylorthosilicate), $Si(OC_2H_5)_4$, the most common alkoxide, is soluble in lower alcohols. When water is added to these solutions, hydrolysis followed by polyfunctional condensation polymerization results in alcohol and water-swollen gels. Low concentrations of added water (i.e. $\leq 1H_2O/3SiOR$) produces mainly short linear polymers and small cyclic structures (10). The degree of network connectivity strongly depends on acid or base addition although molecular structural evolution beyond the trisilicate stage is presently understood in but general terms (11). An acidic component in solution catalyzes the hydrolysis reaction (7, 12, 13). Schaefer et al., on the basis of intermediate angle x-ray scattering results, reported that an $H_2O/TEOS$ mole ratio of 4 (in propanol) at pH 1 will generate essentially linear or at most randomly-branched chains with little crosslinking (14). The degree of branching is less than that as would be expected from random bond percolation.

Englehardt et al. studied polymerization in aqueous acidic solutions using ^{29}Si NMR spectroscopy and reported a first intermediate of cyclotrisilicic acid with subsequent evolution of linear polymers and higher-order rings (15, 16).

Balfe and Martinez, in ^{29}Si NMR studies of acid-catalyzed aqueous tetramethoxysilane (TMOS)-methanol solutions observed a cyclic trimer and other cyclic intermediates early in the reaction, followed by linear chains (10), in harmony with the results of Englehardt.

Of course, further condensation occurs during the collapse of the gel network upon subsequent drying and annealing. Klemperer et al. have shown, using ^{29}Si NMR spectroscopy, that when TMOS in methanol is reacted with water such that $H_2O:TMOS = 4:1$ (uncatalyzed), a considerable relative population of uncondensed SiO_4 units remain even at 400°C (17).

Infrared spectroscopy has been used to monitor the kinetics of the sol-gel reaction by Prassas and Hench (18). In particular, the time-dependent behavior of bands they observed at 1060 and 950 cm^{-1}, assigned to the asymmetric Si-O-Si stretch and Si-OH vibrations, respectively, provides on interpretive baseline for the infrared spectroscopic studies of the more complex sol-gel reaction that takes place within the polymer matrix as reported in our work. It was reported that, as the reaction proceeds, the 1060 cm^{-1} peak shifts to lower wave numbers, with new bands appearing in its neighborhood, and that the appearance of cyclic structures in the later stages is evidenced by a band at around 1080 cm^{-1}.

Yoldas and Partlow also used IR spectroscopy to investigate the structure of gels derived from TEOS with various initial H_2O/alkoxide ratios which were subsequently dried and pyrolyzed at 500°C (19). Their studies had shown a splitting of the asymmetric Si-O-Si stretch that was interpreted as a reflection of at least two major Si-O-Si geometries. The overall relative increase in absorbance of this composite band indicates an increased relative population of bridging oxygens. The shift of this band to higher/lower wavenumbers, being a measure of bond distortion, indicates a strengthening/weakening of the gel structure. Gel strengthening, monitored in this way, can be induced by heating at elevated temperatures (20).

On a larger scale as seen by electron microscopy, the sol/gel reaction followed by drying-heating, produces structural units of a spherical morphology with diameters in the 30 to 300 Å range, depending on alkoxide type and conditions of hydrolysis (21).

Our goal was the creation of unique microcomposite membranes by affecting the low temperature in situ growth of either isolated silicon oxide (ca. 30-50Å-in-diameter) microclusters or continuous silicon oxide interpenetrating networks in Nafion sulfonic acid films via the hydrolysis and polyfunctional condensation of TEOS sorbed from alcohol solutions. Our underlying working hypotheses are as follows: (1) The resultant morphology of the inorganic phase will be ordered by the existing Nafion phase-separated morphology so that the in-growths will necessarily have a high surface/volume ratio. (2) The mutually-soluble TEOS/alcohol/water components will diffuse to and preferentially incorporate within the polar clusters. This assumption is motivated not only by the expectation of solution-cluster thermodynamic compatibility but also by anticipating that the known crystalline packing of the TFE backbone (22) will render these hydrophobic regions impermeable to these molecules. While the ionic clusters will be expected to have the long side chains interdigitated so as to realize electrostatic energetic stability (2), their enforced packing, however, must be considerably less efficient than that of the stereoregular backbone sections. Hence, we reasonably expect a generous concentration of free volume in these regions within which the sol/gel reaction can take place, albeit in confined fashion. (3) TEOS hydrolysis will then proceed in hydrated cluster "microreactors" as catalyzed by the polymer-affixed SO_3H groups in much the same way as in acidic bulk TEOS/alcohol solutions to which water is added as depicted in Figure 1.

It should be acknowledged that Risen utilized the concept of the ionic domains in ionomers (Nafion sulfonates, sulfonated linear polystyrene) as microreactors within which transition metal particles can be grown and utilized as catalysts (23-25). Transition metal (e.g. Rh, Ru, Pt, Ag) cations were sorbed by these ionomers from aqueous solutions and preferentially aggregated within the pre-existing clusters of fixed anions. Then, the ionomers were dehydrated, heated and reduced to the metallic state with H_2. Risen discussed the idea of utilizing ionomeric heterophasic morphology to tailor the size and size distributions of the incorporated metal particles. The affected particle sizes in Nafion were observed, by electron microscopy, to be in the range of 25-40 Å, which indeed is of the established order of cluster sizes in the pre-modified ionomer.

In a similar vein, Sakai, Takenaka and Torikai produced Nafion/metal microcomposites by reducing palladium, rhodium, platinum and silver ion-exchanged membranes with H_2 at high temperatures (100-300°C) (26). The affected metal clusters, distributed homogeneously, were about 50 Å in size. As in Risen's work, a polymer matrix template effect appears to be operative. These systems were considered and evaluated within the context of gas separations technology.

Pineri, Jesior and Coey have reported methods of precipitating iron oxide (α-Fe_2O_3, α-FeOOH) particles in Nafion sulfonate membranes (27). Transmission electron microscopy of thin sections indicated spherical α-Fe_2O_3 particles around 100 Å in diameter that form larger aggregates up to 1000 Å in size, and α-FeOOH acicular or blade-shaped particles, up to 1000 Å long and around 100 Å wide, that are well-separated. While the nucleation and growth habit of these particles must be influenced by the heterophasic morphology of the host polymer, to be sure, a direct template effect wherein the in-growths are of sizes and spacings that are similar to those of the ionic domains within unmodified membranes, is not present in these cases.

There are clear technological incentives for generating novel microcomposite membranes based on perfluorinated ionomer matrices. To begin with, the remarkable chemical inertness and high temperature stability of the base polymer would allow applications in a range of environments that is broader than that for most polymeric systems. Technical opportunities in the areas of catalysis, gas and ionic separations, unique dielectric materials and ceramics are envisioned using these unique hybrid materials.

Experimental-Membrane Preparation

Procedure A. All the membranes utilized in this work were 5 mil-thick 1200 equivalent weight Nafion (E. I. DuPont Co.) which were originally in the SO_3K^+ form. First, all the films were initialized to the dry sulfonic acid form in the following way. In order to ensure complete sidechain ionization and hydrolyze any possible SO_2F groups remaining from the precursor form, the films were equilibrated in stirred 7M aqueous NaOH at 60°C for 48 hours. The films were then removed and placed in stirred deionized water at 60°C for 24 hours to leach out the excess base. Then, the samples were equilibrated in stirred 7M aqueous HCl at 60°C for 48 hours to ion exchange the sulfonate groups to the H^+ form. Finally, the membranes were placed in stirred deionized water at 60°C for 24 hours to leach out excess acid and then vacuum dried for 2 hours at 130°C. As it is well-understood that these materials exhibit swelling hysteresis and that their physical properties are very sensitive to ionic form and water content, it is critical to reduce all samples to this standard initial state in the interest of reproducibility and meaningful comparison of results.

All initialized membranes were pre-swollen (equilibrated) in stirred 2:1 (vol/vol) alcohol/water solutions for 5 hours at 25°C. The alcohols utilized were methanol, ethanol and 1- and 2-propanol. The pre-sorbed water serves to initiate the in situ sol-gel reaction upon the subsequent uptake of TEOS which is introduced in amounts such that H_2O:TEOS = 4:1 (mole/mole). While this ratio theoretically provides for complete alkoxide hydrolysis, there is naturally additional water generated during the ensuing network condensation to support hydrolysis. Solutions consisting of TEOS/alcohol = 1.5 (vol/vol) were added to the flasks that already contained the membranes in equilibrium with the alcohol/water solutions. Reaction flasks were always stoppered after liquids were added.

The membranes were then removed from these multicomponent solutions after time intervals spaced two minutes apart. Of course, the sol-gel reaction continues within the membranes beyond this point, but the diffusion-controlled exchange of reactants and solvent across the membrane/solution interface ceases. Hence, we refer to the time the membrane resides in solution after TEOS has been added, as "immersion time" rather than as reaction time. Upon removal from solutions, the films were surface-blotted dry, then dried under vacuum for 24 hours and subsequently heated at 130°C in the same vacuum for 2 hours to remove trapped volatiles as well as promote further silicon oxide network condensation. The entire procedure up to the last step was carried out at 25°C.

Finally, the membranes were weighed to determine the net dried uptake from the initial dry acid form.

Procedure B. This method is identical to procedure A in all respects except for the fact that the TEOS/methanol solution, rather than being added to the flask contents all at once, was added at 4 ml/min (from 37.5 ml TEOS, 25 ml methanol) for 16 minutes. This slow addition procedure was only applied for methanol at this single immersion time.

Experimental-Membrane Characterization

Mechanical Tensile Studies. The simple stress vs. strain characteristics of two-phase materials and their variation with relative phase composition can usually be given straightforward interpretation in terms of critical underlying structural factors. Among these factors can be listed topological connectivity of, and interfacial interaction between the phases.

An Instron Model 1130 was employed to run samples at 50% relative humidity at 25°C with a crosshead speed of 5 cm/min.

Infrared Spectroscopy. Owing to the high infrared absorbance of these 5 mil-thick films, samples had to be studied in reflectance. To be sure, one should always be concerned over whether spectra taken in the ATR mode are also representative of the bulk structure. In particular, within the context of this study, the question as to whether the silicon oxide phase will be distributed uniformly across the membrane thickness needs to be answered by future morphological investigation. Let us state at this time that we have in fact performed limited FT-IR studies of 1 mil-thick microcomposite films in transmission and have not observed significant qualitative deviations from ATR spectra of thicker films prepared under identical conditions. Also, in an earlier report of the ATR spectra of Nafion membranes, Lowry and Mauritz argued that a considerable number of morphological unit structures of around 50 Å periodicity will be sampled along a line, from the surface, of length approximately equal to the IR beam penetration depth (28).

Spectra were obtained using a Nicolet 5-DX FT-IR spectrophotometer with 4 cm^{-1} resolution and a KRS-5 ATR plate. A total of 1000 interferograms were taken for each sample. Membranes having very high silicon oxide uptakes were too brittle to be applied against the plate with pressure to affect good sample contact.

^{29}Si Solid State NMR Spectroscopy. ^{29}Si solid state NMR spectra for the methanol series of uptakes affected by both procedures A and B were obtained with a Bruker MSL-200 spectrometer. Membrane samples were immersed in liquid nitrogen and subsequently ground with mortar and pestle. Particle sizes sufficient for NMR analysis resulted within three such cooling-grindings. Particulation was more easily accomplished with the high uptake membranes as they were already brittle.

All ^{29}Si NMR spectra were recorded in a 4.7T field at 39.756 MHz. The standard proton-enhanced MAS probe was used to acquire high-resolution spectra. Samples were packed in fused alumina rotors equipped with Kel-f caps. The sample spinning rate was approximately 3 KHz which effectively eliminated spinning sidebands from the region of interest. A cross-polarization pulse sequence was used to enhance sensitivity of the ^{29}Si nucleus. Although high-power decoupling was used, the linewidths of the silicon resonances remained large (5-10 ppm) due to the polymeric nature of the glasses. All measurements were carried out at 300°K with hexamethyl cyclotrisiloxane as a secondary chemical shift reference. The chemical shift of the molecule was set to -10.5 ppm from liquid tetramethyl silane. The contact pulse was set at 2 ms with the Hartmann-Hahn condition fulfilled. The ^{29}Si 90° pulse was set at 5 microseconds. A 40-second delay was inserted between successive rf pulses to allow for spin-lattice relaxation of the ^{29}Si nuclei. Each experiment acquired 1000 to 2000 pulses.

Results and Discussion

Silicon Oxide Solids Uptake. In Figure 2 are plotted percent dried weight uptake of the membranes vs. immersion time for the four solvents. The

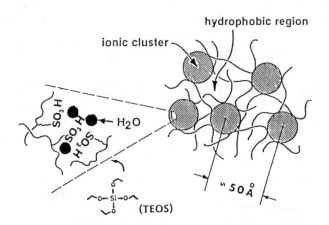

Figure 1. Nafion hydrated polar domains as microreactors in which TEOS hydrolysis is catalyzed by pendant SO_3H groups.

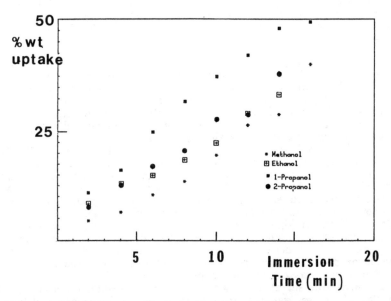

Figure 2. Net silicon oxide percent dried weight uptake vs. time of immersion of Nafion membranes in indicated TEOS/alcohol solutions.

following general observations are made. (1) The percent uptake increases very uniformly and reproducibly to considerable levels without tapering off within the experimental times. (2) The curves for the four alcohols tested are not in the same vertical order as the known percent membrane swellability of untreated membranes in these pure solvents (29). It is quite clear that the net solids uptake cannot be rationalized in simple terms of, say, a uniform progression of Nafion cluster morphological expansion over a spectrum of polymer-solvent interactions. To be sure, a sorting-out of interactions between the host polymer matrix and an incorporated multicomponent solution, whose composition and microstructure is continuously evolving in complex fashion during the time of immersion in reactants, would be formidable. Nonetheless, the curves of Figure 2 are obviously well-ordered, solvent-sensitive and useful in a predictive sense. (3) The affected $SiO_{2[1-1/4x]}$ $(OH)_x$ in-growths cannot be leached out of the membranes by the pure solvents in which the reactants are soluble. This implies strong inorganic/polymer structural binding in general terms.

Mechanical Tensile Studies. The stress vs. strain curves for membranes having various silicon oxide solid uptakes for the four alcohol solvents are displayed in Figures 3-6. Also shown, for reference, on each of these figures is the tensile behavior of untreated Nafion in the dry acid form.

The general mechanical trends observed in these figures are as follows. Ultimate tensile strength and Young's modulus increases and elongation-at-break decreases with increasing solids content for all alcohol solvents. There appears to be a distinct ductile-to-brittle transformation that occurs as the region of plastic deformation disappears with increasing solids uptake. These results, coupled with the knowledge of Nafion microphase separation, have stimulated the following tentative morphological interpretation. Consider silicon oxide clusters as growing in relative isolation at low uptakes resulting in a gradually increasing conventional filler-strengthening effect. Toward the later growth stages, however, as the ionic domains become filled, the clusters necessarily become progressively interconnected forming an invasive network. One might imagine a percolation threshold occurring at an inorganic solids content, say at volume fraction ϕ_c, at which single (-Si-O)$_n$ chains begin to span macroscopic distances throughout the membrane. The inorganic interpenetrating network then becomes a direct load-bearing phase giving enhanced strength arising from Si-O bonds, however, also being brittle owing to its glassy nature. This concept is roughly illustrated in Figure 7.

In a practical vein it was observed that the structural integrity of the microcomposite membranes (from methanol solvent) is very poor after about 45 min. treatment in that the films are susceptible to brittle fracture within the context of simple handling. In fact, samples having an immersion time of 60 min. (88 wt%) can be easily crumbled which is an interesting fact when compared with the conventional experience of Nafion behaving as an extremely cohesive material.

It is also significant that, for the methanol series using procedure A, the highest membrane loading that is allowed before the loss of mechanical integrity (within a stress vs. strain experimental context) is 16.4% (10 min. immersion time), whereas strengthened membranes can in fact be produced by the slow TEOS addition method of procedure B at 21.7% (16 min.), as seen in Figure 3. It is reasonable to assume that this contrast between films prepared according to procedures A and B are reflective of significant differences between the degrees of branching and connectivity of the long range structures of the silicon oxide in-growths.

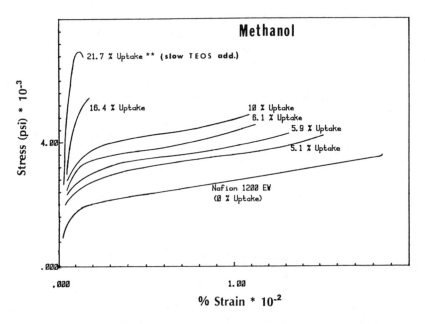

Figure 3. Tensile stress vs. strain for Nafion microcomposite membranes having indicated silicon oxide solids uptakes from TEOS/methanol solutions.

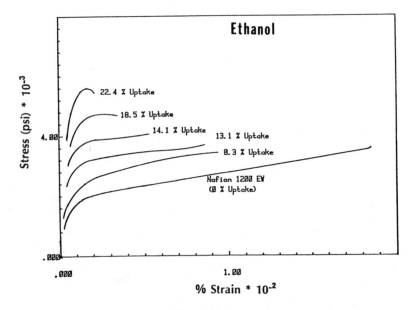

Figure 4. Same as Figure 3, but for ethanol solutions.

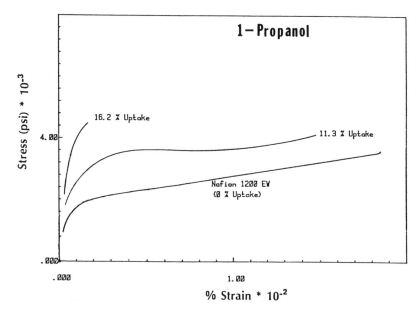

Figure 5. Same as Figure 3, but for 1-propanol solutions.

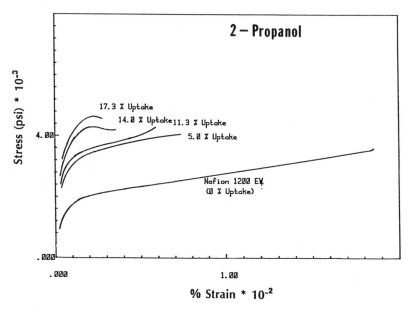

Figure 6. Same as Figure 3, but for 2-propanol solutions.

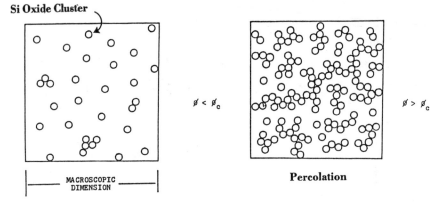

Figure 7. Silicon oxide cluster aggregate distribution below and above percolation volume fraction threshold ϕ_c.

Infrared Spectroscopy. An inspection of the infrared spectra of dry or hydrated pure Nafion in the sulfonic acid or various cationic salt forms reveals a multiplicity of bands (28, 30, 31) some of which are inconveniently located in close proximity to the aforementioned peaks characteristic of silicon oxide structures. The Nafion contribution to the composite spectra was subtracted in each case using the 2860 cm^{-1} band (combination: 1140 + 1720 cm^{-1}, both CF$_2$CF$_2$) as an internal thickness standard. While this band appears weak and may not be an ideal internal standard (Membranes were not available to test absorbance vs. thickness linearity.), it is backbone-related, lies in a region of peak noninterference and the resultant subtractions do appear effective.

Subtraction spectra for films having been subjected to 2, 4, 6 and 8 minute immersion times using methanol solvent are superimposed in Figure 8. All spectra have the same absorbance scale but those for greater than 2 minutes immersion time have been vertically shifted by an arbitrary amount.

The Si-O-Si asymmetric and symmetric stretching vibrations (v_a(SiOSi) and v_s(SiOSi) respectively, as well as the Si-OH stretching vibration, are seen and appear to be rather broad, indicating a wide range of local molecular environments about these groups. The considerable magnitude of the absorbance of the Si-OH vibration relative to that of v_a(SiOSi) would seem to reflect a poorly-connected network having a considerable relative population of uncondensed OH groups. The existence of v_s(SiOSi) at around 800 cm^{-1} is interpreted to be indicative of a significant degree of distortion from ideal tetrahedral coordinative symmetry about the silicon atoms. v_a(SiOSi) is the greatest (1043 cm^{-1}) and the v_a(SiOSi)/v(Si-OH) peak intensity ratio is the highest for the lowest uptake. These facts might reflect a network that is actually tightest and most bridged in the early stages of the in situ sol-gel process. It is seen in Figure 8 that a distinct shoulder develops on the high wavenumber side of the asymmetric Si-O-Si stretching deformation at high uptakes. The appearance of this feature, also observed by Yoldas and Partlow (19) and discussed by Prassas and Hench (18) within the context of bulk sol-gel-derived silica may be due to the formation of cyclic vs. predominantly linear structures.

An absorbance also appears at somewhat greater than 600 cm^{-1} that becomes more pronounced with increasing solids uptake. At this time, however, we are uncertain as to its group vibrational assignment.

The same general observations were made for spectra taken of microcomposites produced using the other alcohol solvents.

Lastly, for comparison, a spectrum for a sol-gel-derived free silica sample prepared in our laboratory is seen in Figure 9. The same TEOS/water/alcohol composition, as well as the same drying conditions as those used in the Nafion membrane treatments existed in the preparation of this sample. The general observations are as follows. (a) Aside from detail, the bands for free silica are the same as those appearing in the subtraction spectra representative of membrane in situ-grown silica, save for the existence of a prominent bond at around 1105 cm^{-1} indicative of vitreous silica, seen in Figure 9. (b) The v_a(SiOSi)/v(SiOH) intensity ratio is higher for free silica than for membrane in situ-grown silica. The latter fact may arise from the restraints on the mobility of reacting TEOS molecules imposed by the confining polymer matrix.

^{29}Si Solid State NMR Spectroscopy. ^{29}Si NMR spectra for microcomposite samples having the indicated solids percent weight uptakes affected according to procedure A (methanol solvent) are shown in Figure 10.

As is reasonable, the spectra have a high noise/signal characteristic and present a broad distribution of chemical shifts indicating that a variety

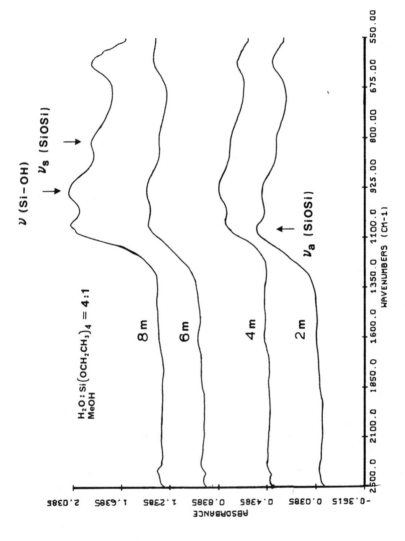

Figure 8. FT-IR (ATR) difference spectra of Nafion/silicon oxide microcomposites produced by immersion of hydrated membranes in TEOS/methanol solutions for the indicated times.

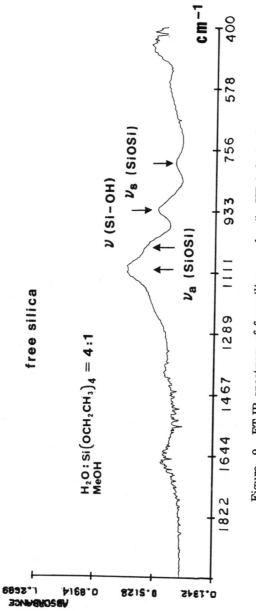

Figure 9. FT-IR spectrum of free silica powder (in KBr) derived from the acid-catalyzed sol-gel reaction for TEOS + water in methanol for the same concentration and drying conditions used in the Nafion <u>in situ</u> sol-gel reactions.

of Si nuclei are incorporated throughout the invasive oxide microstructure. As expected, ^{29}Si chemical shifts in SiO_4 systems are primarily determined by the state of condensation about the tetrahedra (32-33). A number of investigators have assigned chemical shifts to Si atoms having from two to four bridging, i.e. Si-O-Si, oxygens (33, 35). These assignments are as follows:

^{29}Si Chemical Shift (ppm) SiO_4 Coordination
-110 $Si(OSi)_4$
-99.8 $HOSi(OSi)_3$
-90.6 $(HO)_2Si(OSi)_2$

A progression of structure from predominantly linear chains to branched chains to a network of tetrahedral coordination is thus manifested by increasing diamagnetic shielding which moves the chemical shifts to higher fields. In this way, solid state ^{29}Si NMR spectroscopy is the most direct molecular-specific probe of SiO_4 network connectivity.

In Figure 10 however it is seen that the overall relative silicon oxide network connectivity in fact diminishes with increasing solids uptake as evidenced by a uniform displacement of the distribution of chemical shifts to lower fields. This result is quite consistent with the interpretation of our infrared spectroscopic results.

In Figure 11 is the ^{29}Si NMR spectrum for a microcomposite sample produced by procedure B having a 22% solids uptake. In comparing this spectrum with those in Figure 10 we observe that the high load sample of procedure B is actually more similar to the low load sample of procedure A. Our tentative conclusion is that a more highly developed silicon oxide network is generated when a given amount of TEOS is added incrementally rather than all at once. This view is also in harmony with our microstructural interpretation of the relatively greater mechanical strength exhibited by microcomposites produced by procedure B, as discussed earlier.

Yoldas has determined by IR spectroscopy that the degree of silicon oxide network connectivity increases with increasing H_2O/TEOS mole ratio for acid-catalyzed polymerizations in bulk solutions (21). These results, if applicable to the membrane in situ acid-catalyzed polymerizations described herein, serve to reinforce our conclusion that a more highly-coordinated silicon oxide structure exists within microcomposites produced with short immersion times according to procedure A, or according to the slow stepwise TEOS addition in procedure B. In either case, more initial hydrolysis water molecules per alkoxide molecule are available to promote this situation.

General Conclusions

Low temperature procedures for generating unique microcomposite membranes via the in situ hydrolysis and polycondensation of tetraethoxysilane in Nafion sulfonic acid films have been developed. For procedure A, the percent dried silicon oxide weight uptake of membranes, initially containing hydrolysis water and alcohol and subsequently immersed in TEOS/alcohol solutions, increases quite uniformly and reproducibly with immersion time. The uptake vs. immersion time curves are distinctly solvent-sensitive. Furthermore, the final dried in-growths cannot be leached out by soaking the membranes in the pure alcohol carrier solvents, which implies a strong silicon oxide/Nafion structural binding within this general context.

The general progression of an increasing Young's modulus as well as increasing tensile strength and decreasing elongation-at-break with increasing silicon oxide content was seen using all four alcohol solvents. The

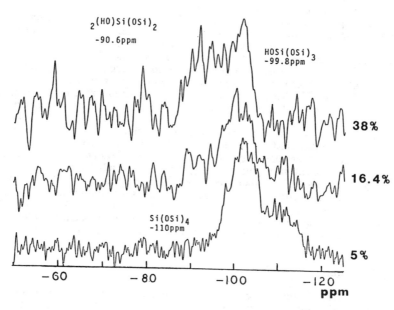

Figure 10. ^{29}Si solid state NMR spectra of Nafion/silicon oxide microcomposites having indicated percent dried weight uptakes from TEOS/methanol solutions using procedure A. The chemical shifts assigned to 2-, 3- and 4- coordinated SiO$_4$ tetrahedra are displayed for comparison.

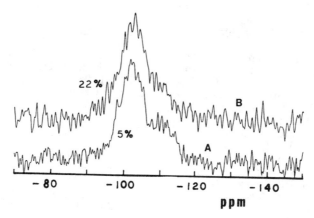

Figure 11. Same as Figure 10, but for a 22% solids uptake by procedure B. Included for comparison is the similar spectrum for 5% uptake by procedure A.

distinct ductile-to-brittle transformation observed with increasing solids uptake is interpreted as a consequence of ($-Si-O)_n$ percolation over the Nafion microphase-separated morphology in the following way. Silicon oxide clusters are initiated within effective microreactors consisting of the discrete, nanometer-in-scale, polar regions wherein growth proceeds by polycondensation in single isolation in the early stages. Eventually, adjacent silicon oxide clusters become knitted into an inorganic interpenetrating network that then becomes the predominant load bearing phase.

FT-IR and ^{29}Si NMR spectroscopic investigations reveal an invasive silicon oxide network structure that becomes less interconnected or coordinated with increasing solids uptake according to procedure A. However, incorporated networks of greater connectivity can be affected using the slow, incremental, TEOS addition method of procedure B.

In short, a new class of microcomposite membranes, based upon the rational morphological tailoring of an in situ grown silicon oxide phase by the sol-gel process for alkoxides has been created. Technological opportunities in the vital areas of gas or liquid separations, heterogeneous catalysis, electronics materials and ceramics are envisioned. Aside from the issue of specific function, these materials, owing to their perfluorinated and inorganic composition, can be expected to exhibit good environmental stability.

Acknowledgments

The Authors greatly appreciate the assistance given by Douglas G. Powell in obtaining and interpreting the ^{29}Si solid state NMR spectra. Acknowledgment is made to the donors of the Petroleum Research Fund, administered by the American Chemical Society, and the Plastics Institute of America for support of this research.

Literature Cited
1. (a) Hora, C.J.; Maloney, D.E. 152nd National Meeting, Electrochem. Soc., Atlanta, Oct. 1977. (b) "Perfluorinated Ionomer Membranes," ACS Symp. Ser. 180, Eisenberg, A; Yeager, H.L., Eds.; Am. Chem. Soc.: Washington, D.C., Chs. 14-17.
2. Mauritz, K.A. J. Macromol. Sci. Rev. Macromol. Chem. Phys. 1988, C28(1), 65.
3. See, for examples, (a) Olah, G.A.; Prakash, G.K.S.; Sommer, J. Science 1979, 206, 13. (b) Olah, G.A.; Meidar, D.; Mulhotray, R.; Olah, J.A.; Narang, S.C. J. Catal. 1980, 61, 96.
4. Ultrastructure Processing of Ceramics, Glasses, and Composites. Hench, L.I.; Ulrich, D.R., Eds., John Wiley & Sons, New York, 1984, Chs. 5, 6, 9.
5. Yoldas, B.E. J. Mat. Sci. 1977, 12, 1203.
6. Ibid. 1979, 14, 1843.
7. Yamane, M.; Aso, S.; Sakaino, T. J. Mat. Sci. 1978, 13, 865.
8. Mukherjee, S.P. J. Non-Cryst. Solids 1980, 42, 477.
9. Dislich, H. Angewandte Chemie 1971, 10(6), 363.
10. Balfe, C.A.; Martinez, S.L. Mat. Res. Symp. Proc. 1986, 73, 27.
11. Klemperer, W.G.; Ramamurthi, S.D. ACS Polym. Preprints 1987, 28(1), 432.
12. Aelion, R.; Loebel, A.; Eirich, F. J. Am. Chem. Soc. 1950, 72, 5705.
13. Yoldas, B.E. J. Mat. Sci. 1975, 10, 1856.
14. Schaefer, D.W.; Keefer, K.D.; Brinker, C.J. IUPAC 28th Macromol. Symp. Proc. (Amherst) 1982, 490.

15. Englehardt, V.G.; Altenburg, W.; Hoebbel, D.; Wieker, W.   Z. Anorg. Allg. Chem. 1977, 428, 43.
16. Hoebbel, D.; Garzo, G.; Englehardt, G.; Till, A.   Z. Anorg. Allg. Chem. 1979, 450, 5.
17. Klemperer, W.G.; Mainz, V.V.; Millar, D.M.   Mat. Res. Soc. Symp. Proc. 1986, 73, 15.
18. Prassas, M.; Hench, L.L. in Ref. 4, pp. 102-8.
19. Yoldas, B.E.; Partlow, D.P. "Formation of Monolithic Oxide Materials by Chemical Polymerization," U.S. Army Res. Rep. 16824.1-MS, 1982.
20. Sakka, S.; Kanichi, K.   J. Non-Cryst. Solids 1980, 42, 403.
21. Yoldas, B.E.   J. Non-Cryst. Solids 1982, 51, 105.
22. Starkweather, H.W., Jr.   Macromolecules 1982, 15, 320.
23. Risen, W.M., Jr.   In NATO ASI Series C, Vol. 198: "Structure and Properties of Ionomers"; Pineri, M.; Eisenberg, A., Eds.; Reidel: Amsterdam, 1987, p. 87.
24. Mattera, V.D., Jr.; Squattrito, P.J.; Risen, W.M., Jr.   Inorg. Chem. 1984, 23, 3597.
25. Shim, I.W.; Mattera, V.D., Jr.; Risen, W.M., Jr.   J. Catal. 1985, 94, 531.
26. Sakai, T.; Takenaka, H.; Torikai, E.   J. Membrane Sci. 1987, 31, 227.
27. Pineri, M.; Jesior, J.C.; Coey, J.M.D.   J. Membrane Sci. 1985, 24, 325.
28. Lowry, S.R.; Mauritz, K.A.   J. Am. Chem. Soc. 1980, 102, 4665.
29. Yeo, R.S. In Ref. 1(b), p. 66.
30. Falk, M. In Ref. 1(b), p. 139.
31. Heitner-Wirguin, C.   Polymer. 1979, 20, 371.
32. Lippmaa, E.; Magi, M.; Samoson, A.; Englehardt, G.; Grimmer, A.   J. Am. Chem. Soc. 1980, 102, 4889.
33. Klinonski, J.; Thomas, J.M.; Fyfe, C.A.; Hartmann, J.S.   J. Phys. Chem. 1980, 85, 2590.
34. Oldfield, E.; Kirkpatrick, R.J.   Science 1985, 227, 4694.
35. Ramdas, S.; Klinowski, J.   Nature (London) 1984, 308, 521.

RECEIVED January 9, 1989

MORPHOLOGY

Chapter 17

X-Ray Analysis of Ionomers

Richard A. Register[1], Y. Samuel Ding[1,4], Marianne Foucart[2], R. Jérôme[2],
Stevan R. Hubbard[3,5], Keith O. Hodgson[3], and Stuart L. Cooper[1]

[1]Department of Chemical Engineering, University of Wisconsin,
Madison, WI 53706
[2]Laboratory of Macromolecular Chemistry and Organic Catalysis,
University of Liège, Sart Tilman B6, B–4000 Liège, Belgium
[3]Department of Chemistry and Department of Applied Physics, Stanford
University, Stanford, CA 94305

X-ray analysis techniques, such as small-angle x-ray
scattering (SAXS), have been profitably applied in
probing ionomer morphology for many years. We discuss
below the application of two newer techniques, extended x-ray absorption fine structure (EXAFS) spectroscopy and anomalous SAXS (ASAXS), in two recent
studies of ionomer morphology and structure-property
relationships. The combination of SAXS and EXAFS was
used to explain the modulus enhancement, due to interlocking loops of polymer chain, and high-strain
behavior, due to aggregate cohesion, of five
carboxy-telechelic polyisoprenes neutralized with
divalent cations. ASAXS was used to show that the upturn near zero angle commonly observed in the SAXS
patterns of ionomers is due to scattering from cations, possibly reflecting an inhomogeneous distribution of dissolved ionic groups.

The incorporation of a small amount of bound ionic functionality,
typically less than ten mole percent, has a profound effect on the
properties of a polymer. These materials, termed "ionomers", can
exhibit marked increases in such important material properties as
modulus, adhesive strength, tear and abrasion resistance, melt
viscosity, and impact strength[1-3]. It is now generally accepted
that these effects result from aggregation of the ions into
microdomains[4,5], which act as physical crosslinks in the material.
The ionic groups may be spaced randomly along the polymer chain, located at regular intervals along the chain (such as in the polyurethane ionomers[6]), or located only at the chain ends (the

[4]Current address: Baxter Healthcare Corporation, Round Lake, IL 60073
[5]Current address: Department of Biochemistry and Molecular Biophysics, College of Physicians
and Surgeons of Columbia University, New York, NY 10032

"telechelic" ionomers[7,8]). While the placement of ionic groups is expected to modify the aggregation process, all of these materials exhibit microphase separation under certain conditions.

Small-angle x-ray scattering (SAXS) has been used for twenty years to characterize ionomer morphology, since the pioneering observation by Wilson et al.[4] that ethylene/methacrylic acid ionomers, upon neutralization, developed an intense peak at $2\theta \simeq 4°$. This peak was immediately taken as evidence for ionic aggregation, and has since been found to be a nearly universal feature of ionomers. Also, many researchers have noted a strong upturn in scattered intensity as zero angle is approached[6,8,9]. The origin of this feature has remained unknown until recently and will be discussed below. The prominence of SAXS in the analysis of ionomers arises from the strong electron density contrast between the ionic aggregates and the polymer matrix, giving a strong signal, as well as the difficulty in obtaining useful information by electron microscopy[10]. The use of morphological models, through which analytical expressions may be obtained for the SAXS intensity, allow investigators to estimate parameters such as the aggregate size and interaggregate spacing.

With the recent availability of high-intensity, energy-tunable x-rays from synchrotron sources, the number of practical x-ray analysis techniques has increased dramatically. In the area of scattering, the technique of anomalous small-angle x-ray scattering (ASAXS) has emerged recently as a powerful technique for isolating the scattering due to a particular element in a material. By tuning the x-ray energy near an absorption edge of that element, its scattering power can be strongly altered, while leaving that of the other elements unchanged. ASAXS is particularly useful for ionomers, since the elements of greatest interest are the cations, many of whose absorption edges lie in an energy range available at current synchrotron sources.

Another valuable tool is extended x-ray absorption fine structure (EXAFS) spectroscopy. This absorption technique makes use of the modulation in absorption coefficient above an element absorption edge, due to backscattering of the ejected photoelectron by neighboring atoms. By suitable data treatment, it is possible to determine the elemental type, number, and distance of the atoms coordinated to the absorbing cation. Thus, EXAFS probes the atomic-scale structure in the material, while SAXS and ASAXS probe the microdomain-scale structure. By using both scattering and absorption techniques, a full picture of ionomer morphology may be developed.

To demonstrate the utility of the x-ray analysis techniques in the study of ionomers, two recent research projects will be discussed. The first[11] employs SAXS and EXAFS to explain unexpected features in the mechanical behavior of a series of carboxy-telechelic polyisoprene ionomers neutralized with divalent cations. The second project[12] uses ASAXS to identify the origin of the zero-angle upturn in scattered intensity commonly observed for ionomers.

Structure-Property Relationships in Telechelic Polyisoprenes

Synthesis and Characterization. The synthesis of carboxy-telechelic polyisoprenes was carried out as described previously[7], in tetrahydrofuran at -78°C using sodium naphthalide as initiator. Approximately 2-3 units of α-methylstyrene per living end were added after completion of the isoprene polymerization to reduce the reactivity of the chain ends, which were then terminated by the addition of gaseous CO_2. The functionality of these polymers is 1.90, as determined by potentiometric titration, and the molecular weight was 8000 as determined by gel permeation chromatography. The materials were neutralized with 95% of the stoichiometric metal methoxide (Ca, Sr) or acetate (Ni, Zn, Cd) in toluene solution, and the methanol or acetic acid byproduct was removed by azeotropic distillation. Finally, approximately 1% by weight of the antioxidant Irganox 1010 was added to prevent chemical crosslinking of the polyisoprene units. The materials were found by ^{1}H FTNMR to contain the possible 3,4/1,2/1,4 repeat units in the ratio 59/41/0, and glass transition temperatures measured by differential scanning calorimetry were found to lie in the range 15-19°C regardless of cation. Specimens for testing were compression-molded for five minutes at 70°C into sheets approximately 0.5 mm thick, or disks approximately 2 mm thick. All samples were stored in a desiccator over $CaSO_4$ until use.

Instrumental Conditions. Dogbone samples for tensile testing were stamped out with a standard ASTM D1708 die. Measurements were performed in air, at 30°C, on an Instron TM at a crosshead speed of 0.5 in./min. All reported data are the average of three tests.

The small-angle x-ray scattering (SAXS) experiments were performed with an Elliot GX-21 rotating anode x-ray generator operated with a copper target at 40 kV accelerating potential and 15 mA emission current. Cu $K\alpha$ x-rays were selected by filtering with nickel foil and by pulse-height analysis at the detector. An Anton-Paar compact Kratky camera was used to collimate the x-rays into a line measuring 0.75 cm by 100 μm. The scattered x-rays were detected with a TEC 211 linear position sensitive detector, positioned at a sample-to-detector distance of 60 cm for a q range ($q=4\pi\sin\theta/\lambda$) of 0.15-5.4 nm^{-1}. The data were corrected for detector sensitivity, empty beam scattering, and sample absorption. A moderate amount of cubic spline smoothing was applied to the data, which were then desmeared by the iterative method of Lake[19] using an experimentally-determined weighting function. The data was placed on an absolute intensity scale by comparison with a calibrated Lupolen polyethylene standard[13]. To eliminate scattering from thermal fluctuations, the data in the range 3.0-5.0 nm^{-1} were fit to Porod's Law[20] plus a constant background term. This background was then subtracted from the curves.

The transmission extended x-ray absorption fine structure (EXAFS) spectra were collected on the A-2, C-1, and C-2 stations of the Cornell High Energy Synchrotron Source (CHESS). Data reduction followed a standard procedure of pre-edge and post-edge background removal, extraction of the EXAFS oscillations $\chi(k)$, taking the Fourier-transform of $\chi(k)$, and finally applying an inverse transform to isolate the EXAFS contribution from a selected region in real space[14-16].

Tensile Testing Results. Stress-strain curves are shown in Figure 1 for all five materials, where differences between the materials neutralized with different cations can be clearly seen. Note that all the cations used here are divalent, so that the cation charge is not a variable. Figure 1a shows the materials neutralized with the alkaline earth cations Ca and Sr, while those neutralized with the transition metal cations Ni, Zn, and Cd are shown in Figure 1b. Note that the Cd ionomer actually extended to approximately 1600%, and exhibited essentially no recovery upon breaking. The Ca and Ni materials, by contrast, snapped back to within 40% of their unstressed length after breaking. The Zn and Sr telechelics exhibited intermediate behavior. The zero-strain Young's moduli are listed in Table I.

Table I. Telechelic Tensile Testing and SAXS Modelling Results

Cation^{2+}	E (MPa)	R_1 (nm)	R_2 (nm)	v_p (nm^3)	$\rho_1 - \rho_0$ (nm^{-3})	n
Ca	4.4	1.06	2.01	262	163	34
Sr	5.6	1.04	1.95	250	139	33
Ni	3.7	0.95	2.02	205	208	27
Zn	3.2	0.92	2.10	194	169	25
Cd	3.2	0.90	2.06	197	185	26

The two alkaline earth materials have an average modulus nearly 50% higher than the average of the three transition metal materials, despite having identical ion contents and molecular weights. Moreover, these modulus values are higher than the values predicted by rubber elasticity theory, if we consider the effective network elements to be those polyisoprene chains terminated at both ends with metal carboxylate groups. In this case, the zero-strain tensile modulus E is given by[17]:

$$E = 3[(g-2)/g]\nu RT \qquad (1)$$

where g is the crosslink functionality, ν is the density of elastically effective chains, R is the gas constant, and T is the absolute temperature. The bracketed quantity reflects the mobility of the crosslinks within the matrix. Assuming that all chains with metal carboxylate groups at both ends are elastically effective, and that the 5% of all chain ends which are not carboxylate groups are randomly distributed among all the chain ends, the ν value for these telechelics is approximately 107 μmole/cm^3. If we further assume that the functionality of an ionic crosslink is large, such that the bracketed quantity in Equation 1 approaches unity, we calculate an upper bound on E equal to 0.81 MPa. The observed values lie in the range 3.2-5.6 MPa, so some factor must be acting to enhance the modulus. Because of the relatively low concentration of ionic groups, the ionic aggregates do not occupy a significant fraction of the materials' volumes, so the filler effect is small, as will be shown in the following section. Moreover, the volume fraction of ionic aggregates should be nearly the same for all five materials.

Strain-hardening behavior can also be observed in Figure 1 for the Ca telechelic beginning near 50% elongation, and to a lesser extent for the Ni ionomer near 100% elongation. This feature is absent in the other three materials. As these telechelics are copolymers of the two types of vinyl addition, stress-induced crystallization cannot be the source of the strain-hardening, nor is it due to the finite extensibility of the primary chains, of molecular weight 8000, which should occur above 300% extension.

Trapped entanglements often contribute appreciably to the moduli of networks formed by chemically crosslinking in bulk[18]. The situation for the telechelic ionomers is somewhat different, since the primary chains are of low molecular weight and would not be expected to be highly entangled. However, entanglements can still form when the ionic groups aggregate, and a large fraction of these are likely to be interlocking "loops". In addition, trapped entanglements between two linear chains which cross are possible (as in chemically-crosslinked networks) or, between a single loop and a linear chain, nonw of which would be present in the nonionic polyisoprene precursor. It has long been recognized[5] that a large fraction of the ionic sites would, upon ionic aggregation, coalesce into the same aggregate as their topological neighbor. For a linear telechelic ionomer, this corresponds to both ends residing in the same aggregate, creating a "loop" of polymer chain. In bulk, this loop will be entangled with other loops, and as long as the ionic aggregates are not disrupted under stress, these interlocking loops are elastically-effective entanglements. These loops will pull taut at a much lower extension than would be required to fully extend the entire polymer chain.

When these interlocking loops pull taut, one of two effects may result. If the ionic aggregates are highly cohesive and do not rupture, then the polymer chains must break, and the material will snap back to approximately its unstressed length. If the aggregates are weakly cohesive, the stressed entanglements can relax by "ion-hopping", or pulling the ionic groups out of the aggregates.

Two questions remain, however: why is the modulus enhancement, attributed primarily to these interlocking loops, greater for the Ca and Sr ionomers than for the Ni, Zn, and Cd telechelics, and why is the stress-hardening behavior exhibited only by the telechelics neutralized with Ca and Ni? Small-angle x-ray scattering (SAXS) and extended x-ray absorption fine structure (EXAFS) spectroscopy were employed to address these questions.

SAXS Results. The SAXS patterns for the telechelics are shown in Figure 2; the curves have each been offset by 1200 intensity units for clarity. Due to the low concentration of ions in these materials, they are poor scatterers compared with many ionomers, but these curves do show two typical features: a peak, seen here from 1.2-1.5 nm^{-1}, and a steep upturn at very low angle. To quantify the morphological differences between the materials, the SAXS patterns were fit to the Yarusso liquid-like model[9], which attributes the peak to interparticle scattering. A schematic diagram of the Yarusso model is shown in Figure 3. The ionic aggregates have a core radius R_1, but are coated with an impenetrable sheath of polymer due to the

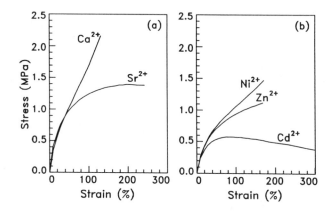

Figure 1. Stress-strain behavior of carboxy-telechelic polyisoprenes. (a) Ca^{2+} and Sr^{2+}, (b) Ni^{2+}, Zn^{2+}, and Cd^{2+}. (Reprinted from ref. 11. Copyright 1988 American Chemical Society.)

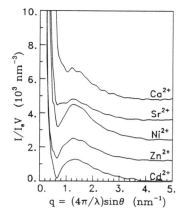

Figure 2. Desmeared, background-subtracted SAXS patterns for carboxy-telechelic polyisoprenes. Curves are offset 1200 intensity units for clarity.

chain units near the ionic groups. As a result, the radius of closest approach, R_2, is greater than R_1. The ionic aggregates are assumed to have a uniform electron density ρ_1, while the matrix has an electron density ρ_0. The equation for the scattered intensity is given by:

$$I/I_e V = (1/v_p)(4\pi R_1^3/3)^2 (\rho_1-\rho_0)^2 \Phi(qR_1)S(q,R_2,v_p) \qquad (2)$$

where v_p is the total volume of material per ionic aggregate (reciprocal of aggregate number density), $\Phi(x) = 3(\sin x - x\cos x)/x^3$, and $S(q,R_2,v_p)$ is the interference function. The Percus-Yevick[21] total correlation function was employed for S, as suggested by Kinning and Thomas[22]. The Yarusso model produces good fits of the ionomer peak when applied to sulfonated polystyrene[9] and polyurethane[6] ionomers, with physically satisfying parameters. However, it does not predict the observed intensity upturn at very low angle, so only data above 0.65 nm^{-1} was considered in the fits. The best-fit parameters of the Yarusso model to the SAXS data are given in Table I. Note that these values are slightly different from those reported previously[11]; this is due to corrections to our desmearing procedure and absolute intensity calibration. However, since the discussion was originally based on trends rather than absolute values, the discussion here will parallel the original[11]. The volume fraction of ionic aggregates, equal to $(4\pi R_1^3/3v_p)$, is less than 2.0% in all cases. Therefore, the filler effect[17] in these materials leads to less than 5% modulus enhancement, and thus cannot be the source of the 300-600% enhancement observed here.

The R_1 values for the Ca and Sr ionomers are larger than those for the Ni, Zn, and Cd ionomers, suggesting that the aggregates in the former two materials contain more ionic groups. However, the size of the ionic aggregates is also influenced by the size of the neutralizing cation, any hydrocarbon incorporated into the aggregates, and any water absorbed from the atmosphere. These telechelic ionomers should be highly phase-separated[8], due to the regularity of the chain architecture. Assuming that all the ions reside in aggregates, a more reliable indicator of the number of cations per ionic aggregate is v_p. Based on the molecular weight, functionality, and neutralization level of these telechelics, there is an average of 0.13 ions/nm^3. Multiplying this factor by the v_p value yields the number of ions per aggregate, n, listed in Table I. There is a clear division in n between the Ca and Sr materials (n=33-34) and the Ni, Zn, and Cd materials (n=25-27). Larger values of v_p and n indicate that the aggregation process has proceeded further for the Ca and Sr telechelics, forming larger aggregates. This will, in turn, lead to more trapped entanglements. It can be concluded that the higher small-strain tensile moduli observed for the Ca and Sr materials is due to a greater density of interlocking loops as a result of the larger aggregates in these materials.

EXAFS Results.   In order to estimate the cohesiveness of the ionic aggregates, we applied EXAFS spectroscopy to examine the coordination structure about the cation. The EXAFS signal is a modulation of the x-ray absorption coefficient on the high-energy side of an elemental absorption edge. The photoelectrons that are ejected by the absorbed x-rays can be backscattered by atoms coordinated to the

absorbing atom;   superposition   of   the outgoing and backscattered
electron waves gives rise to an   interference   pattern.   The EXAFS
signal $\chi(k)$, where k is the photoelectron wavevector, contains information on the number N_j and type of atoms in coordination shell
j, the distance R_j to this shell, and the static and vibrational
disorder of the shell, measured as the Debye-Waller factor σ_j.   To
convert the EXAFS signal from wavevector to real space, it is
Fourier-transformed to yield the radial structure function, or RSF.
Each non-artifactual peak in the RSF represents a distinct coordination shell;   the peak positions are shifted slightly from the true
shell distances by a phase shift ϕ_j that the photoelectron experi-
ences in backscattering. Previous applications of EXAFS to ionomers
have revealed varying degrees of order within the ionic domains,
ranging from   highly   ordered   domains   in ferric-neutralized
carboxy-telechelic polybutadienes[23] to essentially no local order
in domains of rubidium-neutralized sulfonated polystyrene[24].
     The   EXAFS   data   were   analyzed   with   single-electron
single-scattering theory[15]:

$$\chi(k) = \sum_{j} \frac{N_j \gamma_j}{kR_j^2} F_j(k)\sin[2kR_j+\phi_j(k)]\exp(-2k^2\sigma_j^2) \tag{3}$$

where k is the wavevector defined as k = $(2\pi/h)[2m(E-E_o)]^{1/2}$, where
E is the incident x-ray energy, h is Planck's constant, and m is the
mass of an electron. E_o is approximately equal to the edge energy,
but is allowed to vary slightly to provide the best fit to the data
and to correct for any errors in energy calibration[25]. γ_j accounts for the amplitude reduction due to inelastic scattering, and
is approximated as $\exp(-2R_j/\lambda)$, with λ defining a mean-free-path
parameter. λ was obtained by analyzing model compounds of known
structure and is assumed to be transferable to other structures having the same coordination pairs. The model compounds used here were
the 1:1 oxides CaO, SrO, NiO, ZnO, and CdO, all of which have the
sodium chloride crystal structure[26], except for ZnO, which has a
pseudo-hexagonal structure ("zincite" structure)[27]. These made
particularly useful model compounds, as the first coordination shell
consists of M-O pairs and the second of M-M pairs, the two types
needed for the analysis. $F_j(k)$ and $\phi_j(k)$ are the backscattering amplitude and phase-shift functions, respectively, which are characteristic of the types of atoms in shell j and the absorbing atom.
The calculations of Teo and Lee[25] were used throughout for these
functions, except for certain modifications to accomodate previous
experimental data taken on model compounds, as discussed by Pan[46].
To correct for minor errors in the calculated phase shifts, the M-O
and M-M distances determined by EXAFS for the model compounds were
compared with the known crystallographic distances and the difference between the crystallographic and EXAFS values added to the
values determined for the telechelics. Any errors in the amplitude
functions would be largely cancelled out by using model compounds
with the same coordination pairs.
     Typical EXAFS data, both $\chi(k)$ and the RSF, are shown in Figure
4 for the Ni ionomer. Note that peaks below 1.5 Å in the RSF are
artifacts of imperfect background subtraction and do not represent

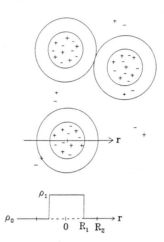

Figure 3. Schematic representation of Yarusso interparticle model. Top: physical picture; bottom: electron density profile. (Reprinted from ref. 9. Copyright 1983 American Chemical Society.)

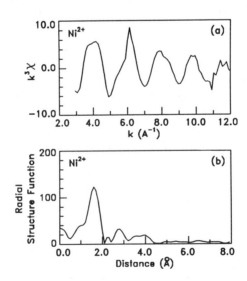

Figure 4. EXAFS data for the Ni^{2+} carboxy-telechelic polyisoprene. (a) $k^3\chi$ vs. k, (b) radial structure function. (Reprinted from ref. 11. Copyright 1988 American Chemical Society.)

coordination shells. The Ca, Ni, and Zn telechelics have more than
one coordination shell visible in the RSF, while the other two
telechelics have only one visible shell. To quantify these results,
the parameters for each of the first two shells, where visible, were
computed; the results are listed in Table II. The R_j values are
generally considered to be accurate to within 1%, while the N_j
values are correct to within about 20%[28,29].

Table II. EXAFS Structural Parameters for Telechelics

Cation^{2+}	Shell Number (j)	Shell Element	R_j (Å)	N_j	σ_j (Å)	Q (%)^a
Ca	1	O	2.40	5.5	0.05^b	26.69^c
	2	Ca	3.48	3.6	0.07^b	
Sr	1	O	2.58	5.1	0.082	3.63
Ni	1	O	2.08	7.1	0.071	4.95
	2	Ni	3.09	1.4	0.086	2.82
Zn	1	O	1.96	4.6	0.060	1.72
	2	Zn	2.97	0.9	0.098	3.94
Cd	1	O	2.29	7.0	0.086	3.02

^{a}Square root of sum-squared residuals divided by sum-squared data.
^{b}Held fixed during regression. ^{c}Simultaneous two-shell fit.

      In all cases, the first peak arises from oxygen atoms, while
the second peak (where present) is due to metal atoms.   This
strongly suggests that the first shell is due to water molecules absorbed by the ionomer in handling, as hydrogen atoms are poor electron scatterers and not visible by EXAFS. While some of the oxygen
atoms could be from carboxylate groups, the absence of a discernable
carbon shell[30] suggests that such species, if present, are in the
minority. The coordination number for the first shell in each of
the Ca, Sr, Ni, and Cd ionomers is within 20% of six, the value for
the oxide, so all four are likely to be octahedrally coordinated by
oxygen. The value of 4.6 for the Zn telechelic could reflect either
four- or fivefold coordination; however, the Zn-O distance of 1.96
Å strongly implies tetrahedral coordination, as found by Pan et
al.[33] in a study of model compounds.
      As for the second shell, such metal-metal distances have been
observed previously by EXAFS in nickel[32] and iron[33] neutralized
Nafion ionomers.   For such a well-defined M-M distance to be
present, the two metal atoms must be somehow connected, such as by
bridging through hydroxyl groups[33], wherein the two metal atoms
share oxygen atoms in their first coordination shells.   The nature
of the bridging oxygen is difficult to determine by EXAFS, however,
since hydrogen atoms are not visible. Regardless of the nature of
the bridge, such ionic "oligomers" form a more cohesive ionic aggregate, since pulling an ionic group from the aggregate would require
disruption of these M-M distances. The coordination number of the
second shell decreases in the order Ca(3.6) > Ni(1.4) > Zn(0.9) >
Sr,Cd(0). The same order is observed in the strain-hardening
behavior; the Ca telechelic exhibits marked strain-hardening at

approximately 50% elongation, the Ni telechelic begins to harden near 100% elongation, and the other three ionomers show no strain-hardening behavior. This remarkable correlation strongly supports the hypothesis set forth earlier: stressed entanglements cannot relax in a material with tightly-bound aggregates, causing strain-hardening and subsequent sample failure, while those in materials with weakly-bound aggregates can relax by ion-hopping.

As a final note, because of the presence of water in the ionic aggregates, we cannot state conclusively which cation would produce an "inherently" tougher ionomer. It is possible that in the complete absence of water, a different ordering in the mechanical properties would prevail, as the relative size and cohesiveness of the aggregates composed of different cations may be altered in the dry state. However, the general relationship between tensile behavior and aggregate size and cohesion should remain.

Anomalous Small-Angle X-Ray Scattering from NiSPS

Experimental. The sulfonated polystyrene (SPS) was obtained from Dr. Robert Lundberg of the Exxon Research and Engineering Company, and was prepared by procedures described elsewhere[34]. The sample used in this study had 5.6 percent of the styrene repeat units sulfonated. The acid SPS was neutralized with 95% of the stoichiometric amount of nickel acetate in tetrahydrofuran/water solution, precipitated into water, then dried under vacuum at $120^{\circ}C$. A suitable sample for ASAXS experiments was prepared by compression-molding the nickel ionomer (NiSPS) into a 1 mm thick disc.

ASAXS experiments were conducted on Beamline II-2 at the Stanford Synchrotron Radiation Laboratory (SSRL). X-ray energies 5 and 100 eV below the measured K edge for Ni in the ionomer (8333 eV in Ni metal) were selected by a double-crystal monochromator using planar Si(111) crystals with a nominal resolution of ± 4 eV. The scattering apparatus has been described in detail elsewhere[35,36]. The data were corrected for transmittance of the sample and fluctuations in incident beam intensity and are shown herein on a consistent relative intensity scale.

Anomalous Scattering Background. The intensity of x-rays scattered from an object can be expressed[37] as:

$$I/I_e = \sum_i \sum_j f_i f_j^* e^{-i\mathbf{q}\cdot(\mathbf{r}_i - \mathbf{r}_j)} \qquad (4)$$

where i and j are any two atoms, and $\mathbf{r}_i - \mathbf{r}_j$ is the vector connecting these two atoms. In general, the atomic form factors f are both energy-dependent and complex, as expressed below:

$$f = f_o + f'(E) + if''(E) \qquad (5)$$

where f_o is energy-independent. The anomalous dispersion terms f' and f" are small compared with f_o at most energies, but close to an atomic absorption edge, they can become as much as 30% of f_o. The f' term becomes increasingly negative as an absorption edge is approached from lower energy, which causes the scattered x-ray intensity to diminish near the edge. The f" term is equal to $E\mu(E)$, where $\mu(E)$ is the x-ray absorption coefficient[38]; as such, f" is small at energies below the edge, but at the edge it jumps sharply and then decreases as the energy is further increased. As all our measurements were conducted below the absorption edge, the only effect of f" is to produce a change in the background due to fluorescence, as described below.

Scattered intensity for an ionomer can be expressed as:

$$I = I_{cations} + I_{background} \qquad (6)$$

where $I_{cations}$ represents scattering from all cation-containing sources, such as scattering between the ionic aggregates, while $I_{background}$ includes all scattering sources not involving cations, such as crystallites (though not for SPS, which is amorphous), voids, impurities, the low-angle tail of the amorphous halo, and thermal density fluctuations. The separation into cation and background scattering is valid when there is no positional correlation between the different entities.

Since the background scattering does not contain a cation contribution, its intensity is independent of energy. In contrast, scattering centers containing the cation will have their scattering power changed as the x-ray energy is tuned near the cation's absorption edge. Therefore, if two scattering patterns are collected, one at an energy E_1 far from the cation's absorption edge and one at an energy E_2 close to the edge, subtraction of the two patterns will yield a "difference pattern" reflecting only the scattering involving cation-containing entities:

$$I(E_1) - I(E_2) = \Delta I_{experimental} = \Delta I_{cations} \qquad (7)$$

When working near the absorption edge, however, fluorescent x-rays will also be generated in the sample and contribute to the observed scattering intensity. The fluorescent x-rays arise from the finite energy bandpass of the monochromator as well as from the phenomenon known as resonant Raman scattering[39]. However, the intensity of the fluorescent radiation[40] at these energies is independent of angle to within 0.03% over the small angular range used. Therefore, fluorescence merely appears as a constant upward shift in the scattering pattern collected close to the edge. The fluorescent contribution was estimated by matching the high-q portions of the scattering patterns, beyond the ionomer peak, where the anomalous effects should be small. This contribution was then subtracted from the data. The final equation of interest is:

$$\Delta I_{experimental} - \Delta I_{fluorescence} = \Delta I_{cations}(f') \qquad (8)$$

where the term on the right represents the scattering pattern arising from interactions involving only the cation.

Results and Discussion. The scattering patterns of the NiSPS sample, taken 5 and 100 eV below the measured cation absorption edge, are shown in Figure 5. The constant shift due to fluorescence is apparent by looking at the difference between the two curves on either side of the ionomer peak. Once this component is subtracted, the ionomer peak has a lower peak intensity at 5 eV below the edge than at 100 eV below the edge. This demonstrates the anomalous effect discussed above. The difference pattern is shown in Figure 6, as well as the pattern 100 eV below the edge. To make comparison easier, the difference pattern has been scaled to equal the raw data in integrated intensity, which entailed multiplication by a factor of 12.5. Therefore, the anomalous effect is 8% of the total intensity for the NiSPS, as compared with 23% for pure Ni metal[38]. Given the expected compositional differences between the aggregates and pure Ni metal, such a value is reasonable.

An important fact can be gleaned simply from observation of Figure 6: the upturn in scattered intensity near zero angle is still present, indicating that it must be related to the neutralizing cation. If it were from a non-cation-containing source, it would have cancelled out when the difference pattern was computed. Note also that the ionomer peak is present in the difference pattern, and appears similar to that in the original data. This simply confirms the generally-accepted postulate that the ionomer peak reflects scattering from the ionic aggregates, which necessarily involves the cation.

As discussed in the previous section, the Yarusso model provides a good fit to the ionomer "peak", but does not produce an upturn at low angle. To reproduce the entire scattering pattern, another type of scattering entity must be postulated. Yarusso and Cooper[9] have estimated that approximately half the ions in zinc-neutralized sulfonated polystyrene remain outside the ionic aggregates, as contact pairs or small multiplets. (The use of the term "dissolved cations" below is merely meant to reflect that it is the cations which are responsible for the anomalous effect. We expect that these cations remain coordinated to anions such that charge neutrality is maintained.) If these dissolved ions are not uniformly distributed throughout the material, then this inhomogeneity would also give rise to scattering. To model this possibility, we have employed the Debye-Bueche random-two-phase model[41,42]. Here, the heterogeneities follow the correlation function:

$$\gamma(r) = \exp(-r/c) \tag{9}$$

The constant c is a correlation length, reflecting the average size of the heterogeneities, while r is the radial distance from any point in the material. The scattered intensity predicted by this model is:

$$I/I_e V = 8\pi\phi_1\phi_2(\Delta\rho)^2/c(q^2+c^{-2})^2 \tag{10}$$

Here, $\Delta\rho$ is the electron density difference between the ion-rich and ion-poor regions in the material, and ϕ_1 and ϕ_2 are the volume fractions of the two types of region.

The combined Yarusso/Debye-Bueche model provides an excellent fit to the full range of the ASAXS data, as shown in Figure 7 for the difference pattern. The fit parameters are listed in Table III.

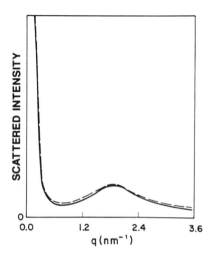

Figure 5. SAXS patterns for nickel-neutralized sulfonated polystyrene at 100 eV (———) and 5 eV (----) below the measured nickel K-edge. Relative intensity scale. (Reprinted from ref. 12. Copyright 1988 American Chemical Society.)

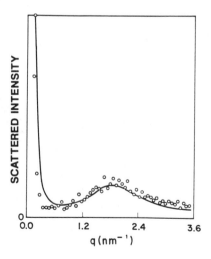

Figure 6. SAXS patterns for nickel-neutralized sulfonated polystyrene at 100 eV (———) below the edge and difference pattern of curves shown in Figure 5 (○). Difference pattern has been scaled to equal the solid curve in integrated intensity. (Reprinted from ref. 12. Copyright 1988 American Chemical Society.)

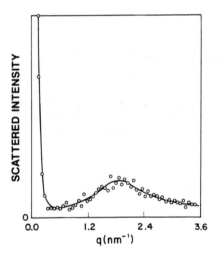

Figure 7. Yarusso/Debye-Bueche model fit to the difference pattern. Solid line is the fit; circles are data. (Reprinted from ref. 12. Copyright 1988 American Chemical Society.)

Table III. Yarusso/Debye-Bueche Model Fit Parameters for NiSPS[d]

Pattern	R_1(nm)	R_2(nm)	v_p(nm^3)	$\rho_1 - \rho_0$	$\phi_1\phi_2(\Delta\rho)^2$	c (nm)
Original (-100 eV)[e]	0.76	1.57	76.	7.6	0.002	90
Difference[e]	0.67	1.57	76.	11.0	0.001	300

[d]Relative values for $\rho_1 - \rho_0$ and $\Delta\rho$.
[e]Scaled to match original pattern intensity.

Considering the noise level in the difference pattern, the Yarusso model parameters are essentially the same for the two patterns, while the values of the correlation length differ considerably between the original and difference patterns. However, an accurate determination of $\Delta\rho$ and c is complicated by the fact that the upturn is very steep, and thus it is difficult to reliably regress these two correlated parameters from the data. It is worth noting that a similar analysis of small-angle neutron scattering data from ampholytic polystyrene ionomers by Clough et al.[43] yielded a good fit with correlation lengths of 30-40 nm.

Since we do not know the volume fractions of the two types of heterogeneity, a definitive value of $\Delta\rho$ cannot be obtained. However, the $\phi_1\phi_2$ product has its maximum value when the two volume fractions are equal, so the lower bound on $\Delta\rho$ is less than 1% of $\rho_1 - \rho_0$, the electron density difference between the ionic aggregates and the matrix. Therefore, the electron density difference required to produce the observed low-angle scattering is very small, while the correlation length is very large (two orders of magnitude greater than the diameter of an ionic aggregate).

It should also be noted that assigning the upturn to dissolved cations is in good agreement with a recent study of manganese-neutralized SPS by Galambos et al.[44], using simultaneous SAXS and differential scanning calorimetry. This material, when quenched, exhibits no ionomer peak, although the upturn is quite pronounced. As the material is heated, the SAXS peak develops, while the upturn diminishes in intensity. This observation is consistent with attributing the ionomer peak to ionic aggregates and the upturn to dissolved cations.

Future Directions

These recent investigations demonstrate the utility of x-ray analysis techniques in the study of ionomers, and this utility is certain to continue undiminished. The growing availability of synchrotron x-ray beamlines will certainly provide a major impetus in the development and use of these techniques. In the area of EXAFS, we are currently studying the effect of anion type on coordination structure and physical properties. Since the acid or base strength of commonly-used bound ions, such as sulfonate, carboxylate, phosphonate, or quaternary amine, varies widely, it is expected that the local coordination structure will also vary dramatically. To date, such experiments have only been performed on sulfonated and carboxylated polystyrene[24]. As shown in the investigation described above on the carboxy-telechelic ionomers, however, the cohesiveness of the ionic aggregates plays a major role in the materials' mechanical behavior, so by changing the bound ion type we expect to effect large changes in mechanical properties as well.

As for ASAXS, we are currently using this technique to separate the crystallite and ionic "peak" scattering in the semicrystalline ionomers Nafion and Surlyn. Since these two scattering sources overlap in q-space, it has in the past proved difficult to separate and model these scattering sources, but with ASAXS this task becomes possible. Also, we are investigating the high-q shoulder we have observed in the SAXS patterns of sulfonated polyurethane ionomers[45], which may be evidence of structural order within the ionic aggregates.

It should be noted that EXAFS and ASAXS are relatively new techniques, however, having emerged only within the last fifteen years. Therefore, it is reasonable to assume that their full potential is largely unexplored. Moreover, it is certain that with current and future synchrotron sources[47], new techniques will be developed, and that some of these will be profitably applied to the study of ionomers.

Acknowledgments

R.A.R. wishes to thank S.C. Johnson & Son and the Fannie and John Hertz Foundation for support while this work was performed, while M.F. and R.J. are very much indebted to Service de la Programmation de la Politique Scientifique (SPPS, Brussels). The assistance of James G. Homan in preparing this photo-ready manuscript is also greatly appreciated. The staff of the Cornell High Energy Synchrotron Source (CHESS) of the Wilson Synchrotron Laboratory at Cornell University were invaluable in carrying out the EXAFS experiments. Partial support of this research by the U.S. Department of Energy under Agreement No. DE-FG02-84-ER45111 is gratefully acknowledged. The Stanford Synchrotron Radiation Laboratory is supported by the Department of Energy, Office of Basic Energy Sciences and the National Institutes of Health, Biotechnology Resources Program, Division of Research Resources.

Literature Cited

1. Eisenberg, A.; King, M., Eds. *Ion Containing Polymers*; Halsted-Wiley: New York, 1975.
2. Bazuin, C.G.; Eisenberg, A. *Ind. Eng. Chem. Prod. Res. Dev.* 1981, *20*, 271.
3. MacKnight, W.J.; Earnest, T.R. *J. Polym. Sci.: Macromol. Rev.* 1981, *16*, 41.
4. Wilson, F.C.; Longworth, R.; Vaughan, D.J. *Polym. Prepr. (Am. Chem. Soc. Div. Polym. Chem.)* 1968, *9*, 505.
5. Eisenberg, A. *Macromolecules* 1970, *3*, 147.
6. Ding, Y.S. Ph.D. Thesis, University of Wisconsin - Madison, 1986.
7. Broze, G.; Jérôme, R.; Teyssié, P. *Macromolecules* 1982, *15*, 920.

8.   Williams, C.E.;   Russell, T.P.;   Jérôme, R.;   Horrion, J. *Macromolecules* 1986, 19, 2877.
9.   Yarusso, D.J.;   Cooper, S.L.   *Macromolecules* 1983, 16, 1871.
10. Handlin, D.L.;   MacKnight, W.J.;   Thomas, E.L.   *Macromolecules* 1981, 14, 795.
11. Register, R.A.;   Foucart, M.;   Jérôme, R.;   Ding, Y.S.; Cooper, S.L.   *Macromolecules* 1988, 21, 1009;   correction, 1988, 21, 2652.
12. Ding, Y.S.;   Hubbard, S.R.;   Hodgson, K.O.;   Register, R.A.; Cooper, S.L.   *Macromolecules* 1988, 21, 1698.
13. Pilz, I.;   Kratky, O.   *J. Colloid Interface Sci.* 1967, 24, 211.
14. Sayers, D.E.;   Stern, E.A.;   Lytle, F.W.   *Phys. Rev. Lett.* 1971, 27, 1024.
15. Stern, E.A.;   Sayers, D.E.;   Lytle, F.W.   *Phys. Rev. B* 1975, 11, 4836.
16. Lee, P.A.;   Citrin, P.H.;   Eisenberger, P.;   Kincaid, B.M. *Rev. Mod. Phys.* 1981, 53, 769.
17. Ferry, J.D.   *Viscoelastic Properties of Polymers*, 3rd ed.; Wiley: New York, 1980.
18. Mark, J.E.   *Rubber Chem. Technol.* 1975, 48, 495.
19. Lake, J.A.   *Acta Crystallogr.* 1967, 23, 191.
20. Porod, G.   *Kolloid Z.* 1951, 124, 83.
21. Percus, J.K.;   Yevick, G.   *Phys. Rev.* 1958, 110, 1.
22. Kinning, D.J.;   Thomas, E.L.   *Macromolecules* 1984, 17, 1712.
23. Meagher, A.;   Coey, J.M.D.;   Belakhovsky, M.;   Pinéri, M.; Jérôme, R.;   Vlaic, G.;   Williams, C.;   Dang, N.V.   *Polymer* 1986, 27, 979.
24. Yarusso, D.J.;   Ding, Y.S.;   Pan, H.K.;   Cooper, S.L.   *J. Polym. Sci.: Polym. Phys. Ed.* 1984, 22, 2073.
25. Teo, B.K.;   Lee, P.A.   *J. Am. Chem. Soc.* 1979, 101, 2815.
26. Donnay, J.D.H.;   Ondik, H.M., Eds.   *Crystal Data: Determinative Tables*, Vol. II, 3rd. ed.;   National Bureau of Standards: Washington, D.C., 1973.
27. Abrahams, S.C.;   Bernstein, J.L.   *Acta Cryst. B* 1969, 25, 1233.
28. Lengler, B.;   Eisenberger, P.   *Phys. Rev. B* 1980, 22, 4507.
29. Stern, E.A.;   Kim, K.   *Phys. Rev. B* 1981, 23, 781.
30. Ding, Y.S.;   Register, R.A.;   Nagarajan, M.R.;   Pan, H.K.; Cooper, S.L.   *J. Polym. Sci. B: Polym. Phys.* 1988, 26, 289.
31. Pan, H.K.;   Knapp, G.S.;   Cooper, S.L.   *Colloid Polym. Sci.* 1984, 262, 734.
32. Pan, H.K.;   Yarusso, D.J.;   Knapp, G.S.;   Cooper, S.L.   *J. Polym. Sci.: Polym. Phys. Ed.* 1983, 21, 1389.
33. Pan, H.K.;   Yarusso, D.J.;   Knapp, G.S.;   Pinéri, M.;   Meagher, A.;   Coey, J.M.D.;   Cooper, S.L.   *J. Chem. Phys.* 1983, 79, 4736.
34. Makowski, H.S.;   Lundberg, R.D.;   Singhal, G.H.:   U.S. Patent 3,870,841 (1975), to Exxon Research and Engineering Company.
35. Fairclough, R.H.;   Miake-Lye, R.C.;   Stroud, R.M.;   Hodgson, K.O.;   Doniach, S.;   *J. Mol. Biol.* 1986, 189, 673.
36. Hubbard, S.R.   Ph.D. Thesis, Stanford University, 1987.
37. James, R.W.   *The Optical Principles of the Diffraction of X-Rays*;   Cornell University Press: Ithaca, 1965.

38. Bonse, U.; Hartmann-Lotsch, I. *Nucl. Inst. Meth.* A 1984, 222, 185.
39. Eisenberger, P.; Platzman, P.M.; Winick, H. *Phys. Rev.* B 1976, 13, 2377.
40. Aur, S.; Kofalt, D.; Waseda, Y.; Egami, T.; Chen, H.S.; Teo, B.-K.; Wang, R. *Nucl. Inst. Meth.* A 1984, 222, 259.
41. Debye, P.; Bueche, A.M. *J. Appl. Phys.* 1949, 20, 518.
42. Debye, P.; Anderson, H.R.; Brumberger, H. *J. Appl. Phys.* 1957, 28, 679.
43. Clough, S.B.; Cortelek, D.; Nagabhushanam, T.; Salamone, J.C.; Watterson, A.C. *Polym. Eng. Sci.* 1984, 24, 385.
44. Galambos, A.F.; Stockton, W.B.; Koberstein, J.T.; Sen, A.; Weiss, R.A.; Russell, T.P. *Macromolecules* 1987, 20, 3091.
45. Lee, D.-c.; Register, R.A.; Yang, C.-z.; Cooper, S.L. *Macromolecules* 1988, 21, 998.
46. Pan, H.K. Ph.D. Thesis, University of Wisconsin - Madison, 1983.
47. Elsner, G. *Adv. Polym. Sci.* 67, 1 (1985).

RECEIVED November 7, 1988

Chapter 18

Scanning Transmission Electron Microscopy To Observe Ionic Domains in Model Ionomers

C. E. Williams[1], C. Colliex[2], J. Horrion[3], and R. Jérôme[3]

[1]LURE–Centre National de la Recherche Scientifique, University of Paris-Sud, 91405 Orsay, France
[2]Physique du Solide, University of Paris-Sud, 91405 Orsay, France
[3]Laboratory of Macromolecular Chemistry and Organic Catalysis, University of Liège, Sart Tilman B6, B–4000 Liège, Belgium

Small domains of high electronic density have been imaged in halato-telechelic ionomers by Scanning Transmission Electron Microscopy (STEM) using the technique of atomic number or Z–contrast. The possibility that these are ionic domains is evaluated and the morphology compared with that derived from recent SAXS experiments.

Although there is little doubt that the unique mechanical and transport properties of ionomers are due to a microphase separation between organic monomers and ionizable groups, their exact morphology is not yet understood.

Direct observation of the ionic domains by electron microscopy has yielded controversial results : large variations in size and shape of what was assumed were the domains have been found, and the observed structure has seldom agreed with that derived from X-ray scattering studies. A recent critical evaluation of electron microscopy investigations of ionomers (1) claims that the images are either plagued with artifacts related to the preparation (e.g. large scale phase separation in solvent cast films, surface features, microcrystals of neutralizing agent) or are too thick to allow a determination of size and shape of the domains. The conjunction of the recent development of sophisticated imaging techniques using scanning transmission electron microscopy (STEM) and a better understanding and control of the process of ion aggregation in model ionomers prompted us to attempt again to image directly the ionic domains in solvent-cast films of barium-neutralized carboxy telechelics (2).

0097–6156/89/0395–0439$06.00/0

Indeed recent structural studies (3,4,5) of α, ω-carboxylato-
polybutadiene and polyisoprene by small angle X-ray scattering
(SAXS) have shown that provided the chains between ionic groups
have a narrow distribution in length, there is microphase separation
of the ionic monomers into small domains with sharp interfacial
boundaries. The domains are distributed in the bulk of the polymer
at an average distance on the order of the end-to-end distance of
the chain between ionic groups. Few ionic groups are incorporated
in the organic matrix. So each small ionic aggregate can be viewed
as being surrounded by a larger volume on the order of the chain
length from which all other domains are excluded. Hence, for a Ba-
polybutadiene, it was found that the domains have a radius of
gyration of 7 ± 1 Å, assuming spherical shape, and are separated by
an average distance of 75 Å for a chain of $\overline{M}_n$ = 4600. Of importance
for our investigation are the facts that (i) there is large electronic
density difference between sharply localized ionic domains
(*multiplets*) and the organic matrix providing good amplitude
contrast and (ii) the distance between the domains can be increased
by varying the molecular weight of the polymer.

Experimental

For this electron microscopy examination, we used a di-
carboxylic polystyrene neutralized with Ba, (PS–Ba) of $\overline{M}_w$ = 60 000
and $\overline{M}_w/\overline{M}_n$ = 1.24. The end-to-end distance is about 198 Å. Note
that no SAXS analysis is possible for such a molecular weight
because the "ionomer peak" is very weak (low ionic content) and its
position is such that it is washed out by the small angle upturn
observed in all ionomers.

The samples were cast directly on a 250 Å – thick carbon film
supported on a copper grid. A known amount of a dilute solution of
PS–Ba in toluene containing 0.1 % methanol was deposited on the
grid, the solvent was slowly evaporated under dry nitrogen, then
the films were dried at 25°C in vacuum. Assuming uniform spread of
the films over the entire grid, the nominal film thickness should be
between 250 Å and 500 Å. Three films were then observed without
staining with a VG–HB 501 STEM microscope operating at 100 kV. We
used the so-called *atomic number or Z-contrast* method in which
the annular dark field image intensity produced by the elastically
scattered electrons is divided by the inelastic component of the
bright field image intensity, simultaneously recorded (6). With this

technique, first introduced by Crewe and developed recently for biological applications (7), the ratio signal is primarily dependent on the atomic number of the elements in the sample ($I_{ratio} \sim Z$) and only weakly dependent on thickness and density (8). The spatial resolution was on the order of 5 to 6 Å. It is mostly governed by the size of the field emission probe which was near optimal size with a semi-angle of emission of 7.5 mrad (9).

Results and discussion

Figure 1 shows two typical images of one of the films ; similar images have been obtained with all films. Approximately spherical, high electronic density domains of an average diameter smaller than 20 Å are clearly seen evenly distributed throughout the sample. Note that a carbon film observed in the same experimental conditions was featureless, except for a "salt and pepper" texture which appeared on defocussing. In order to ascertain that the high electronic density domains were not due to some artifacts or impurities in the solvent, a film of the non-neutralized diacid polystyrene (PS-H) was also prepared in similar conditions. Only large spots appeared, similar to those marked by an arrow on the figure. None of the high density small domains could be seen. It is believed that the large features are related to impurities in the solvent, which was the same for both films.

The question is, are the high electronic density features the ionic domains ? Let us emphasize a few important points.

First the presence of barium and oxygen in the film was checked by X-ray analysis and electron energy loss spectroscopy respectively. It was taken as a signature of the ionomer films which were found to cover the whole grid. Silicon also appeared as a trace impurity of unknown origin.

Solvent casting has been used to obtain sufficiently thin films. To retard the formation of a gel, methanol has been added to the solvent ; its effect on the size of individual ionic domains is unknown. Although the film thickness is much larger than the domain size, it is only slightly larger than the average distance between the domains, determined by the chain length (about 200 Å), - so that the images of very few layers are projected onto the micrograph. However it must be emphasized that we did not attempt a direct measurement of the films thickness for this preliminary investigation. Rather we observed only those parts of the films which appeared the thinnest in the bright field images and were believed to be close to the nominal thickness.

(a)

(b)

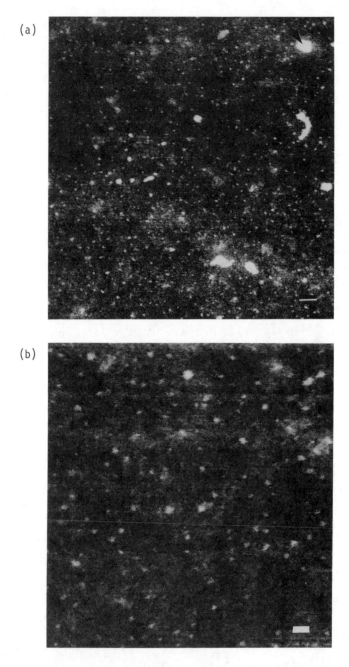

Figure 1. Z-constrast images of a solvent cast film of barium-neutralized dicarboxy-polystyrene of molecular weight 60,000. (a) bar = 150 Å, (b) = 60 Å.

It has been claimed (1) that a slight excess of neutralizing agent could be present in the material and could appear as microcrystals dispersed in the sample. However it has not been in evidence otherwise in the bulk samples quantitatively neutralized in the same conditions, using for instance EXAFS spectroscopy (5).

In the PS-Ba films the heavy atom domains remained stable under the electron beam up to a magnification of X500,000 when the dose received was 1 coulomb/cm^2. As pointed out in (1), although the macromolecules might be subjected to radiation damage, the ionic multiplets are likely to remain unchanged. However, the polybutadiene based films, which would have allowed a direct comparison with SAXS data, were unstable under the electron beam, and the domains drifted. Furthermore, since the film thickness was not uniform and the domains too close, we did not attempt to interpret the images.

Conclusion

The STEM images of solvent cast films of barium-dicarboxylato-polystyrene have shown the presence of small domains of high electronic density, less than 20 Å wide, dispersed in a low density matrix. The technique of Z-contrast insures that the observed features are not related to variation in film thickness. In situ composition analysis shows the presence of barium and oxygen in the film. The very large features (>100 Å) observed on some micrographs could be due to impurities present in the solution (silicon was detected by X-ray analysis). A slight excess of neutralizing agent could also be present, although it has not been in evidence otherwise in the bulk samples. The favorable experimental data on samples prepared in similar conditions lead us to believe that ionic domains have been visualized in the telechelic ionomers where microphase separation is known to occur. However more experiments with films of different nature, thickness, and polymer molecular weight are necessary to ascertain the nature of these domains.

Rather than a detailed study of the morphology of halatotelechelics, this note is intended to show that with the improved resolution and capabilities of electron microscopy and with the better control of the molecular characteristics of ionomers, more research should be devoted to a direct characterization of the morphology of these materials. Techniques developed for other fields such as biology could be advantageously utilized.

Acknowledgments

Thanks are due to Marcel Tencé and Pascal Delzenne for help with the experiment and to E.L. Thomas and D.L. Handlin, Jr. for their comments.

Literature cited

1 – Handlin, D.L. ; Macknight, W.J. ; Thomas, E.L. Macromolecules, 1981, 14, 795
2 – Broze, G. ; Jerome, R. ; Teyssie, Ph. ;   Marco, C. Macromolecules, 1985 15, 1376 , and other references therein.
3 – Williams, C.E. ; Russell, T.P. ; Jerome, R. ; Horrion, J. Macromolecules, 1986 19, 287
4 – Ledent, J. ; Fontaine, F. ; Reynars, H. ; Jerome, R.Polym. Bull. 1985 14, 461
5 – Williams, C.E. ; "Contemporary Topics in Polymer Science", Vol. 6, W.H. Culbertson, ed., Plenum Publ. 1989
6 – It is worth noting here that in catalysis studies, small metallic clusters, of dimension down to 10 Å, have been clearly visualized by annular dark field imaging and ratio contrast. See for instance Treacy, M. ; Howie, A. ; Wilson, C.J. ; Phil. Mag. 1978 A38, 5698 and Delcourt, M.O. ; Belloni, J. ; Marigner, J.L. Mory, C. ; Colliex, C. ; Rad. Phys. Chem. 1984 23, 4785
7 – Crewe, A.V. ; Langmore E, J.P. ; Isaacson, M.S. "Physical Aspects of Electron Microscopy and Analysis", Ed. Siegel B.M. and Beaman D.R., Wiley 1975
8 – Colliex, C. ; Jeanguillaume, C. ; Mory, C., J. Ultrastructure Res.1984 88, 177
9 – Mory, C. ; Ph. D Thesis, University of Paris–Sud, Orsay,1985

RECEIVED February 22, 1989

SOLUTIONS AND PLASTICIZED SYSTEMS

Chapter 19

Light-Scattering Study of Ionomer Solutions

M. Hara and J. Wu

Department of Mechanics and Materials Science, Rutgers, The State University of New Jersey, Piscataway, NJ 08855-0909

Low-angle light scattering and dynamic light scattering experiments were conducted for partially sulfonated polystyrene ionomers (sodium salt) in a polar and a low polarity solvent. In the polar solvent (DMF), the concept of effective diameter of ionomers (macroions) was adopted to discuss the effect of counterion and ion content on the solution properties of ionomers. Also, preliminary dynamic scattering data suggested the existence of three different concentration regions for ionomers in the polar solvent. In the low-polarity solvent (THF), a small degree of aggregation of ionomers due to dipolar attractions was observed over the concentration range studied (10^{-4}-10^{-3} g/cm^3). The dynamic scattering data suggested the polydispersity of the molecular aggregates.

Ion-containing polymers, such as polyelectrolytes, biopolymers, and glasses, have attracted the attention of researchers for many years. A recent addition to these ion-containing polymers is ionomers, which have a small number of ionic groups (up to 10-15 mol %) along nonionic backbone chains (1-3). To understand the significant changes in structure and properties caused by strong ionic interactions, various studies have been conducted in the solid state. However, during the past several years, the study of ionomer solution properties has become active and interesting results have been reported (1-18). It has now been widely recognized that ionomers show two types of behavior depending on the polarity of the solvents (6): (1) polyelectrolyte behavior due to Coulombic interactions between ions in polar solvents; and (2) aggregation behavior due to dipolar attractions of ion pairs in low-polarity or nonpolar solvents.

Various techniques have been used to study the solution properties of ionomers. These include viscosity (4-8,16), static and dynamic light scattering (12,13,15-18), small-angle neutron scattering (11,14), and spectroscopy (10). Here, we use (static and

0097–6156/89/0395–0446$06.00/0

dynamic) light scattering techniques to study the structure of ionomers in both polar and low-polarity solvents. The light scattering technique is one of the direct methods for obtaining information on the structure of polymers in solution (19,20).

Experimental

Lightly sulfonated polystyrenes were prepared by solution sulfonation of standard polystyrenes ($M_w = 5.0 \times 10^4$, 2.0×10^5, 4.0×10^5, $M_w/M_n < 1.06$). Details about the preparation of lightly sulfonated polystyrenes and their characterization were described elsewhere (16). In this work, sodium ionomers were used. The following designation of ionomers is used (2): S-xSSA-Na (Mw: 50,000) means the copolymer of styrene (S) with styrene sulfonic acid (SSA) whose mole fraction is x, followed by the conversion to sodium (Na) salt (i.e., ionomer); also, the weight-average molecular weight is 50,000.

Polymer solutions were prepared by dissolving the freeze-dried ionomer samples in a proper solvent (DMF, THF) under stirring for a day at room temperature. Small-angle light scattering experiments were conducted with a KMX-6 low-angle light scattering photometer (Chromatix) at a wavelength of 633 nm at 25 ± 0.5°C. The effect of large dust particles was easily removed by using a flow cell. Dynamic light scattering measurements were conducted with a BI-200SM photogoniometer (Brookhaven) and a BI-2030 digital correlator (Brookhaven) at a wavelength of 633 nm. Measurements were conducted at 9 angles from 30 to 150° at 25 ± 0.1°C. The light scattering cells were cleaned for 2 hours with condensing acetone vapors using a Thurmond-type apparatus (21). The optical clarification of the solution for both static and dynamic scattering was carried out by passing the solution through two membranes, whose pore sizes were 0.5 and 0.2 (or 0.1) μm in succession. The specific refractive index increment, dn/dc, was measured at 25 ± 0.1°C by using a KMX-16 differential refractometer (Chromatix). Details concerning the small-angle light scattering experiments are described elsewhere (18).

Light Scattering Analyses

The analyses of static light scattering data are presented elsewhere (18); here, only the essential equations are shown.

When the interaction is weak, as is the case of ionomers in THF, the interference effect is small and expressed in terms of a virial expansion (22)

$$Kc/R_\theta = 1/MP(\theta) + 2A_2c + \ldots \qquad (1)$$

where R_θ is the excess reduced scattered intensity (at an angle θ) from the solution, c represents polymer concentration, M (weight-average) molecular weight, and, $P(\theta)$ the particle scattering factor. K is the optical constant, i.e., $K = 4\pi^2 n^2 (dn/dc)^2/N_0\lambda^4$, where n is the refractive index of the solvent, λ the wavelength in vacuo, and N_0 is Avogadro's number. Especially for the low-angle data,

$$Kc/R_0 = 1/M + 2A_2c + \ldots \qquad (2)$$

When the molecular interaction is strong, as is the case of ionomers in DMF, a virial expansion is not sufficient to describe the interference effect. We use a simple effective potential model with an effective diameter. This model, which treats the macroions (ionomers) as if they were neutral but have an effective size, was originally used by Doty and Steiner (23) to analyze light scattering data from protein solutions. The equation obtained is

$$\frac{Kc}{R_\theta} = \frac{1}{MP(\theta)} \left[1 + \left(\frac{2B'}{M}\right) c\Phi(hD) \right] \tag{3}$$

Here, $\Phi(x) = (3/x^3)(\sin x - x \cos x)$ and $B' = (2\pi/3) D^3 N_o$. D is the effective diameter of the macroions, which is a measure of the range of interaction of ionomers with other ionomers and is a decreasing function of concentration. Since the function of D(c) is not known, we analyze the initial slope of the Kc/R_o vs. c curves. From Equation 3,

$$(\text{initial slope}) = \left[\frac{d}{dc}\left(\frac{Kc}{R_o}\right)\right]_{c=0} = \frac{4\pi N_o}{3M^2} D_o^3 \tag{4}$$

where D_o represents the effective diameter at zero polymer concentration. The initial slope is obtained either by a simple graphical method (18) or by curve fitting (15).

Dynamic light scattering data were analyzed by the method of cumulants (24).

$$\ln[C(\tau)/B-1]^{1/2} = \ell nb^{1/2} - \bar{\Gamma} \tau + \mu_2 \tau^2/2 + \dots \tag{5}$$

where $C(\tau)$ is the measured correlation function, B is the baseline, and b is an optical constant. $\bar{\Gamma}$ represents the first cumulant and μ_2 is the second cumulant. An effective diffusion coefficient, D_{eff}, was obtained from the following equation,

$$\bar{\Gamma} = D_{eff} q^2 \tag{6}$$

where $q = \frac{4\pi}{\lambda} \sin(\theta/2)$ is the scattering vector.

When the higher terms in Equation 5 are negligible, the correlation function is expressed as a single exponential function,

$$C(\tau) = B(1 + b \ e^{-2\bar{\Gamma}\tau}) \tag{7}$$

Our experiments showed that the PS samples with sharp molecular weight distributions and ionomers with a large amount of salt followed this equation. The deviation from single exponentiality is caused by various factors, such as (i) polydispersity of particles in solution (ex. ionomers in THF) and (ii) strong electrostatic interactions (ex. ionomers in DMF).

Diffusion coefficients may be converted to the molecular parameters (25). In dilute solution where each molecule is separated, the diffusion coefficient is the translational diffusion coefficient of individual molecules and is related to the hydrodynamic radius, R_H, by the well-known Stokes-Einstein relationship (20)

$$D = \frac{kT}{6\pi\eta R_H}$$ (8)

Here, η is the viscosity of the solvent.

In the semi-dilute solution where the molecules overlap, the diffusion coefficient may be the cooperative diffusion coefficient of entangled chain and may be related to the correlation length, ξ, by

$$D = \frac{kT}{6\pi\eta\xi}$$ (9)

Here, ξ is a measure of the average "mesh" size of the entangled network. Understanding of these molecular parameters, for example, the concentration dependence of the diffusion coefficient is well developed for neutral polymer solutions (25) and polyelectrolytes with a large amount of simple salts. However, for salt-free polyelectrolytes, reliable data are lacking and understanding is rather poor.

Results and Discussion

Polyelectrolyte Behavior. Figure 1 shows the characteristic Kc/R_o vs. c plot of ionomers in polar solvents: the reciprocal reduced scattered intensity rises steeply from the intercept at zero polymer concentration, bends over, and becomes nearly horizontal at higher concentration. This type of behavior was reported for some salt-free polyelectrolytes in aqueous solution (23,26), although the reliability of these early measurements is rather poor because of very small scattered intensity from polyelectrolyte/aqueous solution systems. For example, the excess scattered intensity from salt-free sodium poly(methacrylate) in aqueous solution over that of water was only 10 to 100% (26): $R_\theta = (0.1-1) \times 10^{-6}$. In ionomer solutions, the excess scattered intensity is twenty to several hundred percent of that from solvent (ex. DMF): $R_\theta = (1-20) \times 10^{-6}$. This is one of the advantages of using ionomers to study the characteristic behavior of salt-free polyelectrolytes.

By using Equation 4, the effective ionic diameters, D_o, can be estimated. The initial slope of each curve in Figure 1 may be obtained by either a simple graphical method or a curve fitting method. The effective diameter is a measure of the distance of closest approach of the centers of the macroions and reflects the range of interaction of ionomer molecules with other ionomer molecules.

Effective diameters of ionomers can be used to discuss quantitatively the effects of various molecular parameters, such as ion content and counterions, on the structure of ionomers. For example, for the ion content effect it is found (18) that D_o increases linearly from 420 Å for PS, which is close to the root-mean-square end-to-end distance of PS, to ca. 3100 Å at 6 mol % ion content. The linear relationship between D_o and ion content, f, is expressed as

$$D_o = 45{,}000\ f + 420$$ (10)

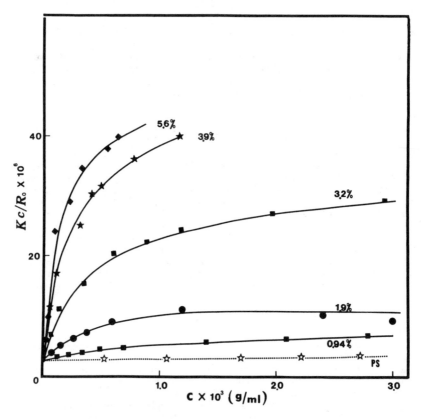

Figure 1. Reciprocal reduced scattered intensity at zero angle,
Kc/R_0, against c for S-xSSA-Na (M_w: 400,000) of various ion
contents as well as PS in DMF. (Reprinted from ref 15. Copyright
1986 American Chemical Society.)

Also, for the counterion effect on the effective diameters it is found ($\underline{18}$) that D_0 decreases with increasing ionic radius of the counterions: 1800 Å for Na ion (0.96 Å), 1600 Å for K ion (1.33 Å), and 1500 Å for Cs ion (1.69 Å). Original Kc/R_0 vs. c curves were reported elsewhere ($\underline{16}$). We also pointed out by viscosity measurements ($\underline{16}$) that for sulfonated ionomers, the effect of counterion (i.e., counterion binding) increased in the order of Li < Na < K < Cs, since the solvated ion size decreased in the order of Li > Na > K > Cs. Light scattering results show this tendency more quantitatively.

In addition to the static scattering experiments described above, some preliminary dynamic scattering experiments were conducted for the same ionomer systems. Except for the case of PS and ionomers with a large amount of simple salts, the cumulant fit was better than the single-exponential fit (e.g. χ^2-test showed smaller values). This discrepancy from a single-exponential function is due to the long-range electrostatic interactions of the ionomer/polar solvent system, as is the case of polyelectrolyte/water system ($\underline{27}$). Therefore, we used the cumulant analysis for all samples. The linear relation between the first cumulant, $\bar{\Gamma}$, and $\sin^2(\theta/2)$ was obtained for all concentrations studied. By using Equation 6, the effective diffusion coefficient, D_{eff}, was calculated from the slopes of each $\bar{\Gamma}$ vs. $\sin^2(\theta/2)$ lines.

Figure 2 shows the relation between D_{eff} and polymer concentration for ionomers with higher molecular weight ($M_w = 4 \times 10^5$). It is seen that the higher the ion content, the larger the value of D_{eff} for most of the concentration range studied. This tendency is opposite to that observed for ionomers with lower molecular weight ($M_w = 5 \times 10^4$): Figure 3 shows that the values of D_{eff} decrease with increasing ion content. A similar tendency was reported for sulfonated polystyrene ionomers with similar molecular weight ($M_w = 1.0 \times 10^5$) by Lantman et al. ($\underline{13}$).

In order to find the cause of this seemingly opposite tendency, the log D_{eff} is plotted as a function of log c (Figure 4). It is seen that apparently two regions can be distinguished; the region where D_{eff} increases linearly with polymer concentration and the region where D_{eff} is almost constant. It is of interest to notice that the two lines in the first region intersect at low concentration. Grüner et al. ($\underline{28}$) reported that salt-free polyelectrolytes in water showed three regions according to the concentration; in addition to two regions which are similar to our results on ionomers with higher molecular weight ($M_w = 4 \times 10^5$), the lowest concentration region where D_{eff} is almost constant existed, which they attributed to the dilute solution region of polyelectrolytes. They attributed three different concentration regions to dilute, semi-dilute, and concentrated regimes by using a scaling approach ($\underline{25,29,30}$).

Since the applicability of scaling approach to salt-free ionomer/polar solvent system has not been established, only some tentative speculations may be given. At least it appears that different regions are separated experimentally according to concentration. One measure of concentration is so-called overlap concentration where molecules begin to overlap by assuming the homogeneous entanglements. Again, although we are not sure whether

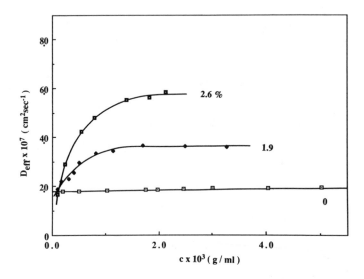

Figure 2. The effective diffusion coefficient, D_{eff}, against polymer concentration for S-xSSA-Na (M_w^{eff}: 400,000) in DMF.

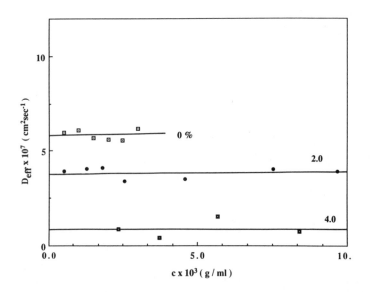

Figure 3. The effective diffusion coefficient, D_{eff}, against polymer concentration for S-xSSA-Na (M_w^{eff}: 50,000) in DMF.

the scaling argument can be applied to ionomer system, we may use calculated overlap concentrations as a rough measure of concentration region. The values of overlap concentrations (C^*_{coil} and C^*_{rod})for polystyrene with different molecular weights are shown in Table I. Since the shape of ionomer molecules is expected to be intermediate between extended rod and random coil, C* value of ionomers may be between these two extreme cases, depending on ion content. It may be that the concentration studied for ionomers of lower molecular weight (M_w = 5 x 10^4) is in the dilute region; D_{eff} may reflect the translational diffusion coefficient of individual ionomer chains and decrease with ion content, since the size of ionomer chains may increase with ion content (Equation 8). The data for the higher molecular weight samples (4.0 x 10^5) may reflect the higher concentration regions.

In order to study further this point, we conducted dynamic scattering experiments for ionomers with medium molecular weight (M_w = 2 x 10^5). These ionomers are expected to have a higher critical concentration than the ionomer with higher molecular weight (M_w = 4 x 10^5), therefore, we may observe the dilute region in addition to the two regions observed for ionomers with higher molecular weight (Figure 4). Figure 5 shows preliminary results for the system. It seems that three different concentration regions are apparently distinguished: a most dilute region where D_{eff} increases slightly with polymer concentration, a middle region where log D_{eff} increases linearly with log c, and a last region where D_{eff} is almost independent of polymer concentration.

Because of the preliminary nature of the data, we do not intend to analyze the data according to scaling approach at this time. Important questions to be answered are: whether the existence of different concentration regions is general for ionomer systems, if so, how to relate the critical concentrations to molecular conformation of ionomers; whether the scaling relation derived for salt-free polyelectrolytes can be applied to ionomer systems; why the effective diffusion coefficient in the last concentration region stays the same. Also, another model (31) on salt-free polyelectrolytes, which does not consider the homogeneous overlapping of polymer chains, could be used to interpret the data on ionomers. These discussions will be presented in forthcoming papers.

Aggregation Behavior. The aggregation of ionomers in a low-polarity solvent (THF) was studied by low-angle light scattering experiments. Figure 6 shows the reciprocal reduced scattered intensity at zero angle, Kc/R_0, against polymer concentration for ionomers (M_w = 4 x 10^5) as well as the PS in THF. These data are analyzed using Equation 2. It is seen that the weight-average molecular weight increases and the second virial coefficient decreases with increasing ion content. These values are plotted in Figures 7 and 8.

Figure 8 shows that the interaction parameter, A_2, decreases rather sharply with ion content, reaching zero at ca. 2 mol % ion content level. By using low molecular weight samples (M_w = 3,500 and 9,000), we could measure light scattering from the samples with higher ion content (18). These samples show that A_2 decreases rather sharply, passes the line of A_2 = 0 and becomes negative.

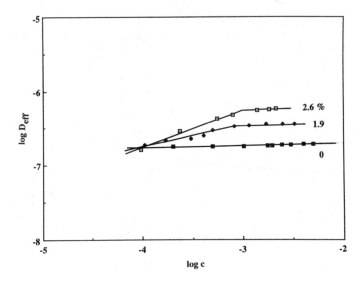

Figure 4.   The log-log plot of Fig. 2.

Table I.   Overlap Concentration of Polystyrene

M.W.	C^*_{rod}	C^*_{coil}
50,000	4.7×10^{-5} (g/cm^3)	2.4×10^{-2} (g/cm^3)
200,000	2.9×10^{-6}	1.2×10^{-2}
400,000	7.3×10^{-7}	8.6×10^{-3}

$c^*_{rod} = M/(No\ell^3)$ (ℓ: extended chain length of a rod)

$c^*_{coil} = M/(NoR_F^3)$ (R_F: root-mean-square end-to-end distance of a coil)

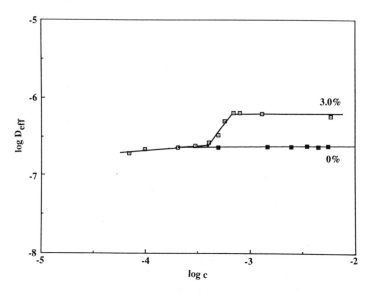

Figure 5. The log-log plot of the effective diffusion coefficient, D_{eff}, against polymer concentration for S-xSSA-Na (M_w: 200,000) in DMF.

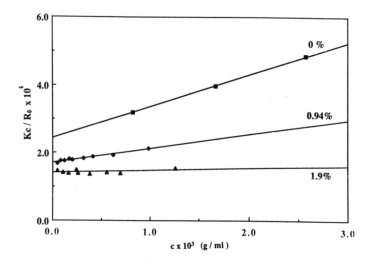

Figure 6. Reciprocal reduced scattered intensity at zero angle, Kc/R_o, against polymer concentration for S-xSSA-Na (M_w: 400,000) in THF. (Reprinted from ref. 18. Copyright 1988 American Chemical Society.)

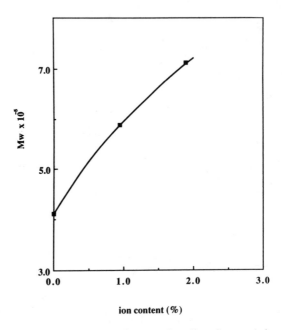

Figure 7.   Ion-content dependence of molecular weight of S-xSSA-Na (M_w: 400,000) in THF.

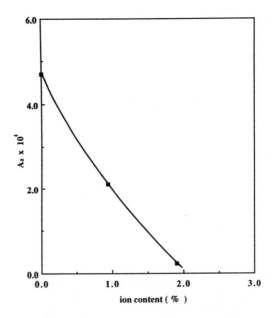

Figure 8.   Ion-content dependence of second virial coefficient, A_2, of S-xSSA-Na (M_w: 400,000).

The decrease in A_2 is attributed to two factors: (1) poor solubility of solvent, since THF is not a good solvent for ionic groups; (2) attraction between ion pairs in low dielectric constant medium. If the former is the only cause, we should obtain the same molecular weight irrespective of the ion content. However, as is shown in Figure 7, the molecular weight increases with ion content. Therefore, the decrease in A_2 definitely reflects the attractions between ion pairs in THF. The second virial coefficient of ionomer solutions is composed of two parts (32,33); i.e., $A_2 = A_2' + f^2 A_2''$, where A_2 is the overall second virial coefficient, A_2' is the second virial coefficient due to the excluded volume of the backbone chains, A_2'' is the second virial coefficient due to attractions between dipoles (therefore, A_2'' is always negative), and f is the ion content. Although, in the original treatment (32), A_2' reflects only the excluded volume of homopolymer chains (e.g., polystyrene), it should include the interaction between ionic groups and solvent (i.e., solvent quality); A_2' is actually a decreasing function of ion content. Thus, the overall second virial coefficient, A_2, decreases with increasing ion content, since A_2' decreases and f^2 increases with ion content.

It seems obvious that small molecular aggregates are formed mainly due to attraction between ion pairs. However, it should be pointed out that only a small percentage of ion pairs are participating in the intermolecular association at very dilute concentration. For example, the ionomer whose ion content is 2 mol % and molecular weight is 400,000 has about 80 ion pairs per chain. If one out of 80 ion pairs participates in the intermolecular association, the dimer whose molecular weight is 800,000 can be found on average. Our experiments show that the weight-average molecular weight is about 700,000; therefore, the number of ion pairs participating in intermolecular associations at very dilute concentration is only about 1% of the total number of ion pairs.

Preliminary dynamic light scattering experiments for ionomers in THF show that the measured correlation function deviates from a single-exponential function. This may be due to the polydispersity of molecular aggregates of ionomers, since the aggregation occurs randomly.

Acknowledgments

We thank Drs. J. Scheinbeim and U.P. Strauss for useful discussions. Acknowledgment is also made to the donors of the Petroleum Research Fund, administered by the American Chemical Society, for partial support of this research. Research support by NSF (DMR-8513893) is also gratefully acknowledged.

Literature Cited

1. Holliday, L., Ed. *Ionic Polymers*; Applied Science:London, 1975.
2. Eisenberg, A.; King, M. *Ion-Containing Polymers*; Academic:New York, 1977.
3. MacKnight, W.J.; Earnest, T.R. *J. Polym. Sci., Macromol. Rev.* 1981, 16, 41.
4. Rochas, C.; Domard, A.; Rinaudo, M. *Polymer* 1979, 20, 1979.

5. Lundberg, R.D.; Makowski, H.S. J. Polym. Sci., Polym. Phys. Ed. 1980, 18, 1821.
6. Lundberg, R.D.; Phillips, R.R. J. Polym. Sci., Polym. Phys. Ed. 1982, 20, 1143.
7. Broze, G.; Jerome, R.; Teyssie, Ph. Macromolecules 1981, 14, 224; 1982, 15, 920; 1982, 15, 1300.
8. Niezette, J.; Vanderschueren, J.; Aras, L. J. Polym. Sci., Polym. Phys. Ed. 1984, 22, 1845.
9. Tant, M.R.; Wilkes, G.L.; Storey, R.F.; Kennedy, J.P. Polym. Prepr. (Am. Chem. Soc., Div. Polym. Chem.) 1984, 25, 118.
10. Fitzgerald, J.J.; Weiss, R.A. ACS Symp. Ser. 1986, No. 302, 35.
11. MacKnight, W.J.; Lantman, C.W.; Lundberg, R.D.; Sinha, S.K.; Peiffer, D.G.; Polym. Prepr. (Am. Chem. Soc., Div. Polym. Chem.) 1986, 27(1), 327.
12. Lantman, C.W.; MacKnight, W.J.; Lundberg, R.D.; Peiffer, D.G.; Sinha, S.K. Polym. Prepr. (Am. Chem. Soc., Div. Polym. Chem.) 1986, 27(2), 292.
13. Lantman, C.W.; MacKnight, W.J.; Peiffer, D.G.; Sinha, S.K.; Lundberg, R.D. Macromolecules 1987, 20, 1096.
14. Aldebert, P.; Dreyfus, B.; Pineri, M. Macromolecules 1986, 19, 2651.
15. Hara, M.; Wu, J. Macromolecules 1986, 19, 2887.
16. Hara, M.; Lee, A.H.; Wu, J. J. Polym. Sci., Polym. Phys. Ed. 1987, 25, 1407.
17. Hara, M.; Lee, A.H.; Wu, J. Polym. Prepr. (Am. Chem. Soc., Div. Polym. Chem.) 1985, 26(2), 257; 1986 27(1), 335; 1986, 27(2), 177; 1987, 28(1), 198.
18. Hara, M.; Wu, J. Macromolecules 1988, 21, 402.
19. Huglin, M.B., Ed. Light Scattering from Polymer Solutions; Academic:New York, 1972.
20. Pecora, R., Ed. Dynamic Light Scattering; Plenum:New York, 1985.
21. Thurmond, C.D. J. Polym. Sci. 1952, 8, 607.
22. Zimm, B.H. J. Chem. Phys. 1948, 16, 1093; 1948, 16, 1099.
23. Doty, P.; Steiner, R.F. J. Chem. Phys. 1952, 20, 85.
24. Koppel, D.D. J. Chem. Phys. 1972, 57, 4814.
25. de Gennes, P.-G. Scaling Concepts in Polymer Physics; Cornell University Press:Ithaca, 1979.
26. Oth, A.; Doty, P. J. Phys. Chem. 1952, 56, 43.
27. Koene, R.S.; Mandel, M. Macromolecules 1983, 16, 220.
28. Grüner, F.; Lehmann, W.P.; Fahlbusch, H.; Weber, R. J. Phys. 1981, A14, L307.
29. de Gennes, P.-G.; Pincus, P.; Velasco, R.M.; Brochard, J. J. Phys. (Paris) 1976, 37, 1461.
30. Odijk, T. Macromolecules 1979, 12, 688.
31. Ise, N. Angew Chem. 1986, 25, 323.
32. Joanny, J.F. Polymer 1980, 21, 71.
33. Gates, M.E.; Witten, T.A. Macromolecules 1986, 19, 732.

RECEIVED December 21, 1988

Chapter 20

Characterization of Sulfonate Ionomers in a Nonionizing Solvent

C. W. Lantman[1], W. J. MacKnight, J. S. Higgins[2], D. G. Peiffer[3], S. K. Sinha[3], and R. D. Lundberg[4]

Polymer Science and Engineering Department, University of Massachusetts, Amherst, MA 01003

Rheological properties of flexible long chain polymers can be dramatically altered by introducing a very small fraction of ionic groups along the hydrocarbon backbone. The molecular parameters responsible for the solution properties of lightly sulfonated polystyrene ionomers have been studied in a non-ionizing solvent. A combination of solution viscosity measurements, quasi-elastic light scattering, static small-angle light scattering and small-angle neutron scattering studies have been performed. The results yield values for the effective diffusion coefficient, radius of gyration and apparent molecular weight. From this combination of scattering and rheological information, a more detailed description emerges of ionomer solution behavior over a range of polymeric molecular weights, ionic contents and concentrations.

Solution properties of flexible long-chain polymers can be dramatically altered by introducing a very small fraction of ionic groups along the hydrocarbon backbone. Polymers containing up to 10 mol% of such groups are commonly called ionomers. Recent solution viscosity studies (1-5) have revealed several unusual phenomena in dilute and semidilute ionomer solutions. These phenomena are due to the presence of ionic groups in the polymer and hence are highly dependent upon solvent polarity. This discussion focuses on sulfonated polystyrene ionomers dissolved in a relatively nonpolar solvent. In such an environment, a majority of the metal counterions remain attached to or in the near vicinity of the sulfonate groups. These metal sulfonate species attempt to escape the non-ionizing solvent

[1]Current address: Mobay Corporation, Pittsburgh, PA 15205
[2]Current address: Imperial College, London, England
[3]Current address: Exxon Research and Engineering Company, Clinton Township, Annandale, NJ 08801
[4]Current address: Exxon Chemical Company, Linden, NJ 07036

environment by forming ionic associations. This process is however hindered by the chemical attachment of the sulfonate groups to the hydrocarbon chain.

Ionomers dissolved in these non-ionizing solvents exhibit a dramatic reduction in solution viscosity as the dilute concentration regime is approached. As polymer concentration is increased, the solution viscosity increases greatly and gelation eventually occurs due to extensive intercoil associations. This behavior is typical of associating polymers.

The determination of an actual molecular basis for these viscometric phenomena is clearly of interest. To this end, a combination of small-angle light scattering (SALS), quasi-elastic light scattering (QELS) and small-angle neutron scattering (SANS) measurements have been performed on this model ionomer system. The solvent of choice was tetrahydrofuran (THF). This non-ionizing solvent ($\epsilon=7.6$) was used in its perdeuterated form for the SANS studies to reduce incoherent background scattering. The inclusion of known amounts of deuterated chains in the polystyrene ionomers made it possible to measure the radius of gyration of both an associated ionomer aggregate and a single chain within such an aggregate. Measurements were made on materials with a variety of sulfonation levels, molecular weights and counterions and are compared with the unmodified polystyrene precursor results. The ionomers used are summarized in Table I. Materials of narrow molecular weight distribution were chosen so that the effects of association could be seen more clearly in the absence of a large distribution of individual ionomer chain lengths. The results yield values for the polymer solvent interaction parameter A_2 as well as the molecular weight and effective size of a single scattering entity. The interested reader is referred to references 6 and 7 for a detailed description of sample preparation, experimental procedure and scattering theory.

Table I. Materials

mol wt	sulfonation level		counterion
	wt%	mol%	
100 000	0	0	
100 000	0.35	1.15	Na
100 000	0.91	3.05	Na
100 000	1.39	4.75	Na
100 000(H)		4.2	Na
100 000(D)		4.2	Na
900 000	0	0	
900 000	0.15	0.5	Na
900 000	0.52	1.72	Na
900 000	0.95	3.20	Na
1800 000	0	0	
1800 000	0.21	0.68	Na
1800 000	0.70	2.30	Na
1800 000	1.14	3.80	Na
115 000	0.75	2.49	H, Na, Zn

Quasielastic Light Scattering.

Typical solution viscosities for SPS ionomers dissolved in THF are shown in Figure 1. As mentioned above, the reduced viscosity of the ionomer solution at low concentrations is less than that of unmodified polystyrene. It is of interest to focus on this low concentration limit and to determine the molecular basis for the lowered viscosity.

In order to estimate the size of the individual polymer species in solution, the diffusion coefficient D_T was measured by QELS. The time correlation function was measured and used to calculate the effective diffusion coefficient D_Z. Use of the Stokes-Einstein relation allows the calculation of an effective hydrodynamic radius R_H for the dissolved species. These results are shown in Figure 2. From the results, it is readily apparent that aggregation occurs at the higher sulfonation levels. This aggregation is evidenced by the several-fold increase in effective size with increasing sulfonate content. It should be noted that the data for the highest sulfonation level of the 1.80×10^6 g mol^{-1} series is indicative of this trend but is of lower accuracy than the other data points since the aggregated species are too large to filter.

It is interesting to note that these aggregates can be "broken up" by dilution. This is most clearly seen in the case of 1.00×10^5 g mol^{-1} polystyrene with 4.73 mol% sulfonation as shown in Figure 2. The scattering from this solution was measured at concentrations as low as 1×10^{-6} g dl^{-1}, the experimental limit of our apparatus. The continuous change in effective hydrodynamic radius below concentrations of 1×10^{-5} g dl^{-1} is evidence of the dynamic nature of the aggregation process. At lower concentrations, less aggregation occurs. Similar trends are seen in the 9.00×10^{-5} and 1.80×10^6 g mol^{-1} series as well.

Small-Angle Light Scattering.

Having determined that the effective size of these aggregates is relatively constant throughout the limited concentration range of 0.1 to 0.5 g dl^{-1}, it is now possible to determine the quality of solution as measured by SALS. Light scattering from polymer solutions is usually interpreted according to the Debye equation (8). K is an optical constant incorporating the refractive index of the

$$\frac{Kc}{R_\theta} = \frac{1}{M_W} + 2A_2c \qquad (1)$$

solvent and the specific refractive index increment. M_W is the weight-average molecular weight of the dissolved polymer, and A_2 is directly related to the slope. Typical results from this series of lightly sulfonated polystyrenes dissolved in THF are shown in Figure 3 and Table II. The ionomer results are directly compared with those of the unmodified polystyrene precursor.

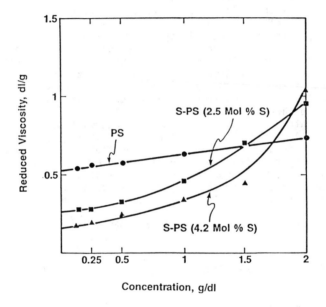

Figure 1. Reduced viscosity-concentration profiles of polystyrene and sulfonated sodium salt in tetrahydrofuran. (Reproduced from ref. 6. Copyright 1987 American Chemical Society.)

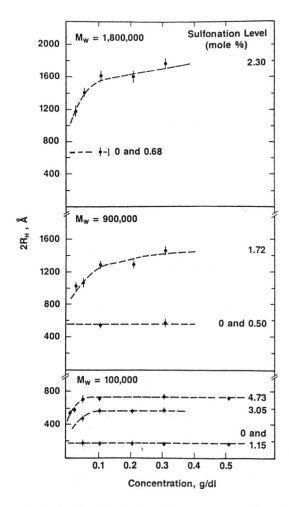

Figure 2. Effective hydrodynamic radii as a function of polymer concentration for a series of sodium sulfonated polystyrenes in tetrahydrofuran. (Reproduced from ref. 6. Copyright 1987 American Chemical Society.)

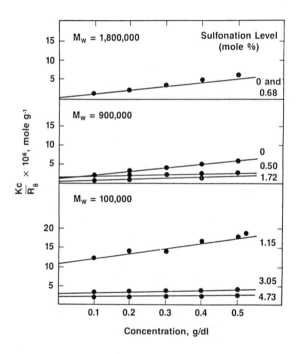

Figure 3. Kc/R$_\theta$ as a function of polymer concentration for a series of monodisperse sodium sulfonated polystyrenes in tetrahydrofuran. (Reproduced from ref. 6. Copyright 1987 American Chemical Society.)

Table II. Static Light Scattering Results: Na Salts/THF

mol wt	sulfonation level, mol%	measd M_w	$A_2 \times 10^6$, dL mol g^{-2}
100 000	1.15	91 000 ± 9 000	6.5 ± 0.7
100 000	3.05	284 000 ± 28 000	0.0 ± 0.7
100 000	4.73	543 000 ± 54 000	0.7 ± 0.7
900 000	0	1 023 000 ± 100 000	5.0 ± 0.4
900 000	1.72	5 208 000 ± 520 000	3.2 ± 0.7
900 000	3.20	insoluble	
1800 000	0	2 000 000 ± 200 000	5.0 ± 0.4
1800 000	0.68	2 000 000 ± 200 000	5.0 ± 0.4
1800 000	2.30	bluish, not filterable	
1800 000	3.80	insoluble	

A closer examination of Figure 3 shows that the second virial coefficient of the solution is rapidly reduced as the level of sulfonation is increased. This decrease in the value of A_2 corresponds to a decrease in solution quality. Theta conditions (where $A_2 = 0$) are approached at very low sulfonation levels. In general, the higher the molecular weight, the lower the charge density required to approach theta conditions. This effect can also be seen in Table II where the higher molecular weight materials are soluble over a more limited range of sulfonation levels. The measured decrease in second virial coefficient with increasing ionic content is consistent with recent theoretical predictions of Cates and Witten (9). Conditions similar to those found for ordinary chains near the Flory theta temperature have been predicted for idealized materials where the number of monomers between associating groups is large and the associating interaction is both local and very strong. A wider range of molecular weights and ionic contents must be studied to evaluate the scaling predictions of this model.

In this present study, the molecular weight of the scattering entity increases with increasing sulfonation level. It should be noted that the data for the highest molecular weight materials in Figures 2 and 3 appear to be insensitive to very low levels of sulfonation. This is probably erroneous and caused by the filtering process. The larger aggregates (>0.2μm) will not pass through the filter. In the lower molecular weight series, these aggregates are able to pass through the filter and therefore contribute to the scattering measurements. This aggregation phenomenon is apparently due to the inability of the metal sulfonate groups to interact in a purely intra-molecular manner. As a result, a small number interact intermolecularly. At sufficiently high sulfonation levels, these intermolecular associations lead to insolubility.

The values of A_2 obtained from static low-angle light scattering are presented in Table II. The second virial coefficient has previously been measured (10) for unmodified polystyrene dissolved in THF and is in reasonable agreement with the values determined in this study. In addition, calculation of $K'c_p/R_\theta$ requires the measurement of the specific refractive index increment, dn/dc. This quantity was measured for a number of the ionomer solutions and the results are compiled in Table III.

An additional aspect of this study of lightly sulfonated polystyrene ionomers in a non-ionizing solvent is the variation of solution properties with counterion. Hara and coworkers ($\underline{11}$) have correlated the degree of aggregation with counterion binding for a series of monovalent salts. The effects of counterion valency are considered here. To this end, the static and quasi-elastic scattering from the sulfonic acid and its sodium and zinc salt forms are compared at a fixed molecular weight of 1.15×10^5 g mol^{-1}. The variation in static scattering is shown in Figure 4 and the results are given in Table IV. As observed earlier, the sodium salt has a lower value of A_2 indicating poorer solution quality compared to unmodified polystyrene dissolved in THF. This effect is even more dramatic in the case of the zinc neutralized ionomer. Neither of the salt forms nor the free acid is significantly aggregated in solution at this level of sulfonation. This is shown by the molecular weight measured via SALS and by the effective hydrodynamic radius determined from QELS.

Table III. Refractive Index Increments: SPS Ionomers

mol wt	sulfonation level, mol%	solvent	dn/dc, mL g^{-1}
1 800 000	0	THF	0.20 ± 0.01
900 000	1.72	THF	0.20 ± 0.01
900 000	1.72	DMF	0.15 ± 0.01

Table IV. Variation of Counterion in THF for the Polymer with a Molecular Weight of 115 000 and a Sulfonation Level of 2.49 mol %

counterion	measd M_w	$A_2 \times 10^6$, dL mol g^{-2}	$D_T \times 10^7$ cm^2 s^{-1}
Na	190 000±19 000	2.2 ± 0.7	5.4 ± 0.2
Zn	116 000±12 000	0.0 ± 0.7	4.7 ± 0.2
H	123 000±12 000	7.3 ± 0.7	5.6 ± 0.2

Small-Angle Neutron Scattering - Aggregate Sizes.

Further study of this ionomer system was undertaken by small-angle neutron scattering. Due to the strong interactions present in this system, a conventional SANS experiment will not yield single-chain conformations. We may, however, consider the solution as a dilute or semidilute solution of aggregated clusters of chains and use conventional SANS data to extract the average values of the molecular weight and radius of gyration of these clusters. A second set of experiments including known amounts of perdeuterated chains in the polystyrene ionomers then makes it possible to extract M_w and R_g values for single chains within the clusters. Measurements were made on materials of fixed sulfonation level and molecular weight and are compared with the unmodified polystyrene precursor results. These SANS experiments are the first of their kind on ionomer solutions.

Solutions of lightly sulfonated polystyrene (SPS) sodium salts with 0 and 19% perdeuterated chains were studied throughout a concentration range of 0.5-4.0 g dL^{-1} in perdeuterated tetrahydrofuran (THF-d_6). A 0.5 g dL^{-1} solution of unmodified polystyrene (PS) was also measured for comparison. The scattered intensity I(q) can be interpreted using the well-known Zimm equation (12)

$$\frac{K'c}{I(q)} = \frac{1}{M} \left[\frac{1+q^2 <R_g^2>}{3} \right] + 2A_2 c \qquad (2)$$

where $<R_g^2>$ is the mean squared radius of gyration and q is the scattering vector.

The scattering from the solution of fully hydrogenous PS in THF-d_6 was used to scale the scattered intensity to absolute values. In order to determine molecular weight, corrections must be made for the effects of the second virial coefficient A_2. The literature value of A_2 was used for the PS/THF-d_6 solutions. The previously discussed light scattering experiments show that the apparent second virial coefficient for SPS/THF solutions is zero under these conditions.

The scattering from fully hydrogenous SPS in THF-d_6 over a range of concentrations is shown in Figure 5. The presence of aggregates larger than single chains is immediately evident from the upturn in intensity at low angles. When Equation 2 is used to analyze the scattering, the plots shown in Figure 6 are obtained. The values of radius of gyration and molecular weight that result from the low q limit are listed in Table V.

The molecular weight values listed in Table V indicate that interchain association is present even at concentrations as low as 0.5g dL^{-1}. As concentration is increased to 4.0 g dL^{-1}, the weight-average number of chains per aggregate rises from 3 to 12. It is important to emphasize that this is an average degree of aggregation. Though the individual ionomer chains make up a narrow molecular weight distribution ensemble, there is no reason to anticipate a narrow distribution of aggregate sizes. In fact, the strong downward curvature present in the plots of Figure 6 is consistent with a polydispersity greater than 2 (13,14). Curvature is also expected when the low-angle limit is exceeded. For this reason, only the initial portions of K'c/I(q) plots can be used.

The range of q values over which eq 2 has been used does not always satisfy the required criterion qR_g<<1 for the obtained R_g. This is a common occurrence in the field of SANS studies of polymer conformations. For values of qR_g up to ~4, the values for the molecular weight and R_g obtained from Equation 2 may be corrected using the method of Ullman(15) which also takes polydispersity into account. As is well-known, and as Ullman's calculations show, increasing the polydispersity reduces the error due to qR_g not being <<1. Since, as discussed above, the aggregates are likely to be quite polydisperse, we have not corrected the R_g listed in Table V

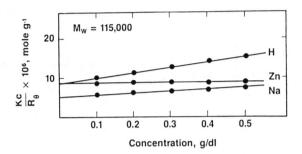

Figure 4. Kc/R$_\theta$ as a function of polymer concentration for acid and salt forms of monodisperse 2.49 mol % sulfonated polystyrene in tetrahydrofuran. (Reproduced from ref. 6. Copyright 1987 American Chemical Society.)

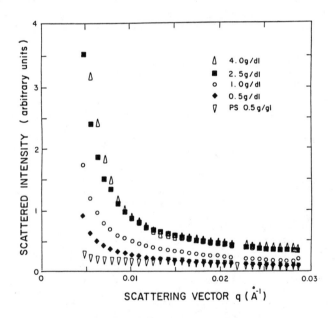

Figure 5. Scattered intensity for 4.2 mol % sulfonated polystyrene in perdeuteriated tetrahydrofuran. (Reproduced from ref. 7. Copyright 1988 American Chemical Society.)

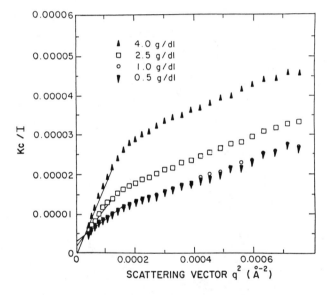

Figure 6. Zimm plot of data from Figure 5. (Reproduced from ref. 7. Copyright 1988 American Chemical Society.)

Table V

Molecular Parameters Determined from Zimm Analysis

system	concn, g dL⁻¹	total scattering			single-chain scattering	
		R_g, Å	R_g^C Å	M_w, g mol⁻¹	R_g, Å	R_g^C	M_w, g mol⁻¹
SPS(4.2 mol %)	4.0	790 ± 120		1164000±233000	104±10	91± 10	135000±20000
	2.5	390 ± 40		471000± 71000	124±12	109± 12	190000±29000
	1.0	226 ± 23		299000± 45000	110±11	97± 10	184000±28000
	0.5	230 ± 23		317000± 48000	128±13	112± 13	
PS	0.5	134 ± 13	118±13	assumed 100000	113±11	99± 11	104000±15000

The values of R_g^C have been corrected for the finite range of qR_g according to ref 15, as discussed in the text.

for the "finite qR_g" effect as we expect them to be accurate to 10%, except for one case. This is the case of the highest polymer concentration (4.0g dL^{-1}) where $3.2 < qR_g < 8.7$. For this concentration the R_g and M_w obtained from the data are not reliable but may be taken only as semiquantitative estimates of these quantities. Note that the quantity M_w extracted from the data represents the weight-average molecular weight and R_g represents the z-averaged radius of gyration.

Concomitant with the increase in molecular weight, the radius of gyration of an aggregate increases with concentration. These values also appear in Table V. At the lowest concentration of this study, the results may be compared with the previous light scattering measurements. Although the identical system has not been studied, it is possible to interpolate the light scattering results and predict an approximate hydrodynamic radius of 380 Å and an approximate average aggregate molecular weight of 4.00×10^s mol^{-1} for the 4.2 mol % ionomer of Table V at 0.5 g/dL concentration. These results are in reasonable agreement with the present measurements.

At first sight these results are unexpected in light of the solution viscosity behavior shown in Figure 1. The reduced viscosity at 0.5 g dL^{-1} is markedly higher for PS than for its sulfonated analogue, and yet PS has a smaller radius than the sulfonated aggregate. However, the solution viscosity of a molecularly dispersed solution can be greater than that of an aggregated solution if the density of material within a viscometric particle changes.

For example, if the well-known Flory relation (16) is assumed and the intrinsic and reduced viscosities are considered comparable at low concentrations, then the solution viscosity may be roughly approximated by

$$\eta_{red} = K''(R_g)^{3/2}/M \qquad (3)$$

If the data for PS shown in Figure 1 are now used to calculate the proportionality constant K'' and the measured radii of gyration and molecular weights of SPS at the two lowest concentrations are inserted in Equation 3 then a reduction of up to 40% in η_{red} is predicted for the SPS solutions. Clearly, this naive calculation cannot be expected to predict solution viscosity precisely, but its implication is nonetheless of great importance: the presence of aggregation at low SPS concentrations in THF is not inconsistent with decreased solution viscosity.

<u>Small-Angle Neutron Scattering - Single Chain Information.</u>

It now remains to determine the dimensions of a single ionomer chain within an associated aggregate. To this end, the mixed-labeling technique described in reference 7 was used. As demonstrated (17) this technique can be used to isolate single-chain information without additional corrections for the second virial coefficient.

This subtraction technique was tested on solutions of unsulfonated polystyrene. The scattering from a 0.5 g dL^{-1} solution of 19% perdeuterated PS in THF-d$_6$ was measured, and the results were combined with those of the fully hydrogenous polymer to calculate the single-chain scattering form factor $f(q)$. Since the solution is dilute for PS at this concentration and molecular weight, the subtraction technique should yield the same results as the fully hydrogenous chains (after appropriate corrections for A_2 as discussed earlier). The resulting values appear in Table V. As can be seen, the two measurements agree within experimental error and the correct molecular weight is measured by the subtraction technique.

Solutions of SPS containing 19% perdeuterated chains were measured at the same concentrations as the completely hydrogenous ionomers. The results were combined to yield single chain information. An example of this combination is shown in Figure 7. The resulting single chain scattering patterns are shown in Figure 8. As with the overall scattering from the aggregates, Equation 2 was used to determine the R_g and M_w of the single chain from the low q data. The appropriate plots are shown in Figure 9. The R_g's thus obtained satisfy the relationship $0.8 < qR_g < 2$ for the q range used to analyze the data. Since the individual chains are fairly monodisperse as discussed previously, a correction (of roughly 14%) has been made to the data by using the method of Ullman (15) and the corrected values are also displayed in Table V. Since the M_w's are normalized relative to polystyrene chains assumed to have $M_w = 1.00 \times 10^5$ and with roughly the same R_g values, no correction is necessary for the values of M_w.

The molecular weight of a single chain should, of course, be constant with variation in concentration. As can be seen in Table V, the molecular weights are roughly constant within experimental error. The values are somewhat greater than expected, reflecting the uncertainty in determining absolute scattered intensity. In this case, different diaphragms were used for the polymer and ionomer solutions. The resulting attempts at cross-normalizing the two geometries led to an overcalculation of the ionomer molecular weight.

It is important to note that the radius of gyration of a single ionomer chain also remains constant within experimental error as concentration increases. There is no evidence for significant coil collapse throughout the concentration range studied. Coil collapse might be observed at even lower concentrations where light scattering studies suggest that these aggregates dissociate and intrachain associations should dominate. Significantly, there is no evidence for single coil expansion as concentration is increased. The resulting picture of ionomer solutions, schematically shown in Figure 10, is one where single coils associate into aggregates without significant changes in individual size. Interestingly, the radius of gyration of a single ionomer chain within such an aggregate is very comparable to dimensions measured in the bulk (18).

Conclusion

In summary, these solution studies of sodium salts of lightly sulfonated polystyrene in tetrahydrofuran verify the presence of associating polymer behavior in ionomer solutions with nonionizing solvents. The results provide a molecular basis for the understanding of solution viscosity behavior. Individual ionomer coils are observed to retain constant dimensions while associating

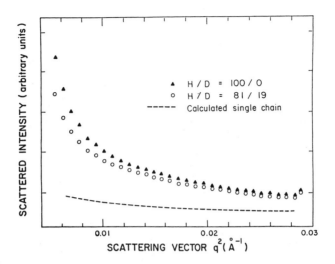

Figure 7. Mixed-labeling subtraction technique. (Reproduced from ref. 7. Copyright 1988 American Chemical Society.)

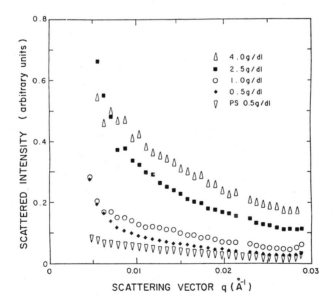

Figure 8. Calculated single-chain scattering for 4.2 mol % sulfonated polystyrene in perdeuteriated tetrahydrofuran. (Reproduced from ref. 7. Copyright 1988 American Chemical Society.)

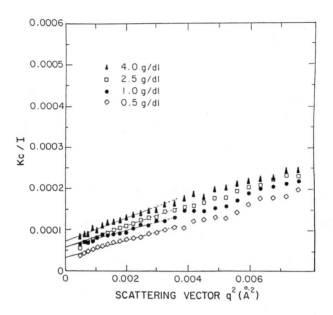

Figure 9. Zimm plot of data from Figure 8. (Reproduced from ref. 7. Copyright 1988 American Chemical Society.)

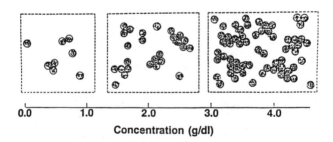

Figure 10. Association phenomenon in sodium sulfonated polystyrene solutions with tetrahydrofuran. (Reproduced from ref. 7. Copyright 1988 American Chemical Society.)

to form multicoil aggregates. At this low level of sulfonation, the single coil dimensions within an aggregate are not significantly different from those of unmodified polystyrene. Interchain associations are found to persist to unexpectedly low concentrations. The extent of aggregation observed by small-angle neutron scattering is very dependent on concentration in agreement with light scattering and solution viscosity studies. These measurements on SPS solutions are expected to apply in general to ionomer solutions in low polarity solvents with appropriate considerations for polymer architecture and counterion structure.

Acknowledgments

This work has been supported in part by the National Science Foundation, Grant DMR 8317590.

Literature Cited

1. Lundberg, R.D.; Makowski, H.S. J.Polym.Sci., Polym.Phys.Ed., 1980, 18, 1821.
2. Lundberg, R.D.; Phillips, R.R. J.Polym.Sci., Polym.Phys.Ed., 1982, 20, 1143.
3. Lundberg, R.D., J.Appl.Polym.Sci. 1982, 27, 4623.
4. Kim, M.W., Peiffer, D.G. J.Chem.Phys. 1985, 83, 4159.
5. Hara, M.; Tsao, I.; Lee, A.H.; Wu,J. ACS Polym.Prepr.,Polym.Chem.Div. 1985, 26 (2), 257.
6. Lantman, C.W.; MacKnight, W.J.; Peiffer, D.G.; Sinha, S.K.; Lundberg, R.D.; Macromolecules, 1987, 20, 1096.
7. Lantman, C.W.; MacKnight, W.J.; Higgins, J.S.; Peiffer, D.G.; Sinha, S.K.; Lundberg, R.D.; Macromolecules, 1988, 21, 1339.
8. Debye, P.J.; J.Phys. Colloid Chem., 1947, 51, 18.
9 Cates, M.E.; Witten, T.A.; Macromolecules, 1986, 19, 732.
10. Schultz, G.V.; Baumann, H. Die Makromolek.Chemie, 1968, 114, 122.
11. Hara, M.; Wu,J.; Lee, A.H.; ACS Polym.Prepr., Polym.Chem.Div., 1986, 27 (1), 335.
12. Zimm, B.H. J.Chem.Phys., 1948, 16, 1093.
13. Kirste, R.G.; Oberthur, R.C. in Small Angle X-ray Scattering; Glatter, O; Krathy, O., Eds.; Academic: London, 1982.
14. Oberthur, R.C., Makromol.Chem., 1978, 179, 2693.
15. Ullman, R. J.Polym.Sci.,Polym.Phys.Ed., 1985, 23, 1477.
16. Flory, P.J. Principles of Polymer Chemistry; Cornell University Press: London, 1953.
17. Ullman, R.; Benoit, H.; King, J.S.; Macromolecules, 1986, 19, 183.
18. Earnest, T.R., Jr.; Higgins, J.S.; Handlin, D.L.; MacKnight, W.J.; Macromolecules, 1981, 14, 192.

RECEIVED January 11, 1989

Chapter 21

Plasticization Studies of Ionomers

A Review

C. G. Bazuin

Centre de Recherche en Sciences et Ingénierie des Macromolécules, Chemistry Department, Laval University, Québec G1K 7P4, Canada

Past and current research on ionomers plasticized by nonaqueous diluents are the focus of this paper, although hydrated ionomers are also considered. Studies involving viscoelastic properties, X-ray scattering and spectroscopic techniques are highlighted. The contrasting effects of polar and nonpolar plasticizers in ionomers are described, as well as the mixed effects of functionalized oligomeric plasticizers. The need for comprehensive and systematic investigations of ionomer-plasticizer systems in order to elucidate mechanisms of plasticization and structure-property relations in general in these systems is pointed out.

Ionomers have long been reputed for such properties as greatly increased melt viscosities, rubbery moduli and glass transition temperatures compared to equivalent nonionic polymers. However, these same properties, which are advantageous for a number of applications, render the processing of ionomers much more difficult.

The judicious use of plasticizers is an obvious way of resolving these difficulties. From this practical point of view, a thorough knowledge of plasticization of ionomers is desirable. Since it is possible for plasticizers to act upon the nonionic and ionic parts of the ionomer to differing extents, such knowledge should also be useful in tailoring the ultimate properties of ionomeric materials.

From a molecular point of view, the study of the effects of plasticizers on ionomers may lead to an increased understanding of the microstructure of ionomers, especially in terms of how this microstructure is modified by the presence of different plasticizers. To this end, certain well-defined parameters of the plasticizer or diluent can be varied, such as its concentration, its polarity and its size.

This paper summarizes various studies or parts of studies in which plasticizers or diluents were utilized, but without claiming

0097–6156/89/0395–0476$07.75/0

to be an exhaustive review. In particular, the many
investigations in which water has been employed as a diluent or
swelling agent, and more particularly the studies of hydrated
perfluorinated ionomer membranes, will not all be cited; nor will
water uptake studies be specifically addressed. Although studies
of ionomers in solution complement and can shed light on
plasticizer studies, these also will not be considered part of the
topic under discussion. It is recognized, however, that there is
no sharp demarcation between where "plasticization" ends and
"solution" begins as diluent concentration is increased; and thus
inevitably there will be a certain arbitrariness in deciding which
studies are relevant and which are not.

It is assumed that the reader has a basic understanding of the
field of ionomers - what they are, their characteristic properties,
microstructural aspects and models, etc. If not, he is referred to
review articles such as those given in references 1-5.

Historical Perspective

From the beginning of ionomer investigations, it was realized that
plasticizers notably water can have a profound effect on ionomer
properties and structure, even in small amounts. The presence of
water can reduce glass transition temperatures (6-8). It can
greatly affect the results of dynamic mechanical and dielectric
scans by reducing moduli, lowering the temperature of one or more
transitions, altering their intensities, and causing new
transitions to appear [see data on ionomers based on polyethylene
(1,3,9-10), polypentenamer (1,3,11), polysulfone (1,3,8,12) and
sulfonated EPDM (1,3,13), as well as perfluorinated (1,3,14-15) and
aromatic (16) ionomer membranes]. In some cases, the manner in
which water was observed to affect a transition helped in the
identification of that transition. In at least one study, a small
amount of water led to a breakdown in time-temperature
superposition (14). As will be discussed in greater detail in the
appropriate section of this paper, the presence of water also
drastically affects the small angle x-ray peak typical of ionomers,
to the point that the peak may disappear. [Because of the strong
influence of water on ionomers, all data on so-called dry ionomers
must be treated with some caution; and the question must always be
posed whether residual amounts of water remain in the material,
thus affecting what is being measured.]

Ionomer plasticization by diluents other than water have also
been studied, but to a more limited extent. Given the dual nature
of ionomers, that is ionic co-units in a nonpolar matrix,
plasticizers can be divided into two kinds: those which plasticize
primarily the matrix and those which are implicated in the ionic
interactions such that they plasticize the ionic domains. The
concept of "selective plasticization" was explicitly introduced by
Lundberg et al. (17). They described the contrasting effects of a
polar and nonpolar diluent on the melt viscosity and glass
transition temperature (Tg) of a sulfonated polystyrene ionomer:
the polar diluent (glycerol) is the more efficient in reducing the
ionomer melt viscosity (up to a certain concentration level),
whereas the nonpolar diluent (dioctyl phthalate or DOP), which

behaves more like usual plasticizers of homopolymers, is the more
effective by far in lowering the Tg of the ionomer. To reduce the
melt viscosity from 3.2 x 10^7 Pa·s for the unplasticized ionomer to
4 x 10^4 Pa·s for a plasticized ionomer, for example, more than ten
times more DOP than glycerol is required, with only 3.5 wt % of the
latter necessary. It was proposed that, whereas the nonpolar
diluent acts as a plasticizer of the polystyrene matrix ("backbone
plasticizer"), the polar diluent solvates the ionic interactions
which are responsible for the high melt viscosities of ionomers
("ionic domain plasticizer").

A specific case of the use of a plasticizer in facilitating
the processing of an ionomer without adversely affecting its
properties was illustrated by the effect of zinc stearate on the
melt rheology and the mechanical properties of EPDM ionomers (18-
21). Above its crystalline melting point, the zinc stearate
appears to solvate the ionic groups thereby allowing flow to occur
more easily. At ambient temperature, the plasticizer is
crystalline and acts as a filler with strong interactions with the
metal sulfonate groups; this leads to enhanced mechanical
properties. Plasticization by zinc stearate will be further
discussed in a subsequent section of this paper.

Effect of Plasticizers on the Viscoelastic Behaviour of Ionomers

Stress Relaxation. A stress relaxation study of plasticized
carboxylate styrene ionomers contrasts the effects of DOP and
diethyl phthalate (DEP) to those of dimethyl sulfoxide (DMSO) (22).
DOP and DEP, known to be good plasticizers for polystyrene (PS),
decrease the glass transition temperature and broaden the
transition zone of the ionomer just as they do that of PS. In
addition, it is reported that, at concentrations greater than ca.
15 wt %, DOP induces the breakdown of time-temperature
superposition for a 2.5 mol % ionomer, well below the critical ion
content of ca. 6 mol % above which superposition fails in the
unplasticized ionomer (23).

The polar plasticizer, DMSO, besides causing time-temperature
superposition to be inapplicable, results in a narrowing of the Tg
zone. This was attributed to the preferential association of DMSO
with the ionic groups in the ionomer, resulting in solvation of the
ionic regions (22-23). Thus, polar plasticizers essentially cancel
out the effect of the ionic co-units on ionomer properties, causing
the ionomer to tend back to the property characteristics of its
parent nonionic polymer. In a stress relaxation study of a rubber-
based ionomer, it was shown that swelling by glacial acetic acid
lowers the rubbery plateau modulus by a factor of three, while
modifying the glass transition temperature only slightly (24).

Dynamic Mechanical (DM) Analysis. One of the most straightforward
methods of investigating the effects of plasticizers on ionomer
properties is through measurements of dynamic mechanical
properties. Such studies allow observation of how plasticizers
influence the Tg of the ionomer as well as the higher temperature
transition (hereafter referred to as the ionic transition or Ti)
that is associated with the ionic aggregates in microphase-
separated ionomers. A number of such studies, as well as DSC or

calorimetric studies, involving the effect of water as a plasticizer have already been mentioned in the introductory part of this paper; these will not be further elaborated on here as these references are numerous, and many have been summarized in other review articles (e.g.1,3). [As a complement to DM studies, some dielectric studies of water in ionomers have also been described (1,3,10,14).]

In the first of more systematic studies reported of diluents other than water, diethylbenzene (DEB) and glycerol were used as plasticizers in carboxylate and some sulfonated styrene-based ionomers (25-26). It was found that in the range of diluent compositions investigated (up to ca. 45 wt % diluent), the nonpolar diluent DEB is equally effective in reducing Tg and Ti in the carboxylate ionomers, thus essentially displacing the curves to lower temperatures (Figure 1). Sub-Tg peaks also appear. The Ti in the sulfonated ionomers are reduced in temperature by DEB as well, but less than in the case of carboxylate ionomers, thus resulting in an extended rubbery zone for the plasticized sulfonated ionomers. The decrease in Tg due to DEB follows the Fox equation remarkably well for all the samples studied (Figure 2).

Incorporation of glycerol, on the other hand, as shown in Figure 3, causes the disappearance of the Ti loss peak, decreases the Tg much less effectively (see Figure 2), and results in a storage modulus curve resembling that of polystyrene. The intensity of the Tg loss peak increases with increasing glycerol content, indicating increasing participation of microphase-separated material in the glass transition relaxation. The dynamic mechanical curves comparing the effects of glycerol and DEB are shown in Figure 3. Weiss and coworkers (27-28), in a similar study contrasting DOP and glycerol as plasticizers, but concentrating on sulfonated polystyrene ionomers, report results that corroborate the observations just described.

Because evaporation of plasticizer at higher temperatures in DM studies is a problem when using diluents such as DEB, recourse was made to oligomers for a more thorough DM investigation of plasticizer effects on ionomer properties (29-30). Tg loss tangent data for a 5 mol % carboxylate styrene ionomer blended (through mutual dissolution in benzene/methanol, followed by freeze-drying) with a styrene oligomer of molar mass 800 (8-S) are shown in Figure 4. The variation in the transition temperatures as a function of blend composition are plotted in Figure 5.

Figure 4 shows the expected decrease in Tg with increasing oligomer concentration. It is of interest to note that, in addition, the width (as well as measured surface area) of the Tg loss tangent peak increases with increasing oligomer concentration, whereas the intensity of the peak initially decreases and then increases again with 8-S concentration, with the minimum appearing at ca. 40 wt % concentration. Furthermore, for the blend containing 80 wt % 8-S, an additional peak at ca. 10°C is present; as the ion content of the ionomer is increased, the intensity of this peak also increases. It is undoubtedly due to phase-separated oligomer, and confirms the visual observation that the higher the ion content of the ionomer, the greater the opacity of the blended samples for the 80/20 oligomer/ionomer (w/w) blends; all blends having a lower 8-S concentration were transparent (except the 60/40

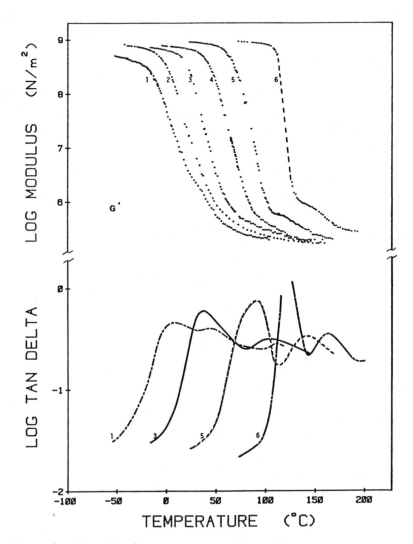

Figure 1. ·Dynamic shear storage modulus and loss tangent as a function of temperature for PS-0.02MAA-Na plasticized to varying degrees by diethylbenzene (DEB): curves 1, 75; 2, 80; 3, 84; 4, 88; 5, 92; 6, 100 wt % polymer (adapted from ref. 25).

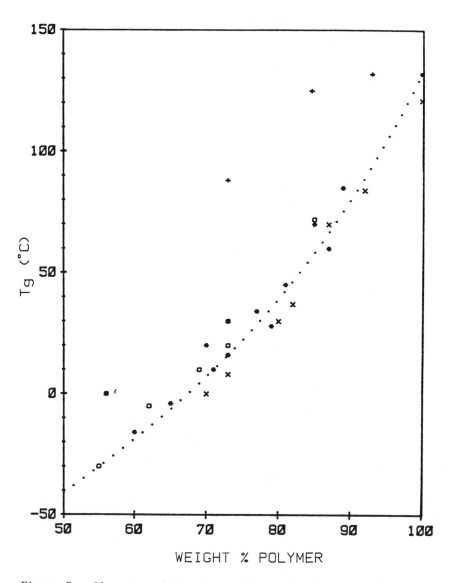

Figure 2. Glass transition temperature as a function of wt % ionomer: x, PS-0.02MAA-Na + DEB; *, PS-0.05MAA-Na + DEB; □, PS-0.05SSA-Na + DEB; ■, PS-0.09MAA-Na + DEB; +, PS-0.09MAA-Na + glycerol. The Fox equation for PS-0.05MAA-Na + DEB is represented by the dotted line. (Reproduced with permission from ref. 25. Copyright 1986 Wiley.)

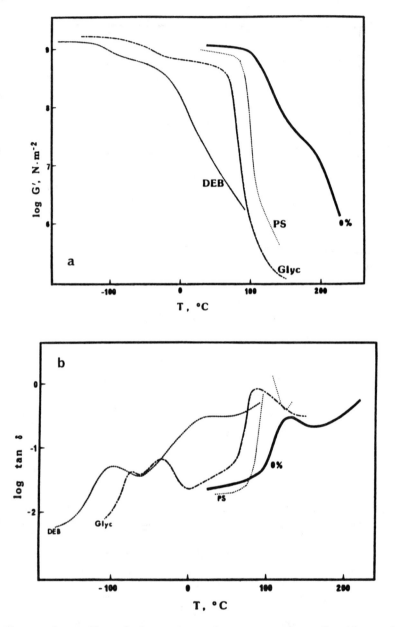

Figure 3.   G' and loss tangent curves as a function of temperature for polystyrene (i) and for PS-0.09MAA-Na without plasticizer (ii) and plasticized 25 wt % with DEB (iii) and glycerol (iv).   (Reproduced with permission from ref. 25. Copyright 1986 Wiley.)

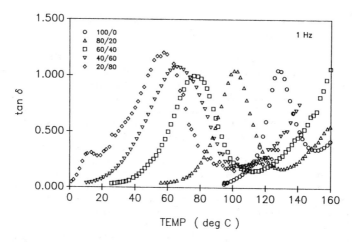

Figure 4. Loss tangents vs. temperature at 1 Hz for blends of PS-0.05MAA-Na and 8-S. Weight ratios (ionomer/oligomer) are 100/0, 80/20, 60/40, 40/60 and 20/80 for curves with Tg maxima in order of decreasing temperature.

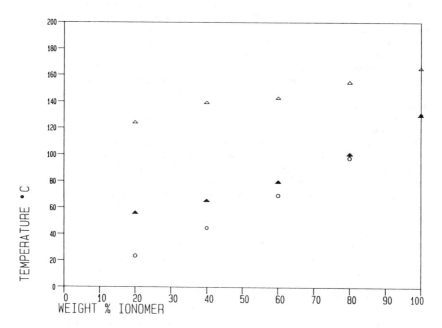

Figure 5. Transition temperatures from loss maxima (1 Hz) for the blends of Fig. 7 as a function of wt % ionomer. Open triangles: ionic transition temperatures from loss moduli maxima; closed triangles: glass transition temperatures from loss tangent maxima; open circles: Tg's calculated using the Fox equation.

blend of the 8 mol % ionomer which was translucent). This result indicates that there is a limit to the solubility of the oligomer in the ionomer (31).

From Figure 5, it is again evident that the Tg diminishes with increasing oligomer concentration. However, unlike the case for diethylbenzene (25), the Tg decrease is less than that predicted by the Fox equation. Moreover, the higher the oligomer concentration and also the higher the ion content, the greater the deviation of the Tg from the Fox prediction. This must reflect the diminishing solubility of the oligomer in the ionomer with increasing ion content of the ionomer and with increasing oligomer concentration. The difference in the effect of 8-S compared to DEB must be related, in turn, to the higher molecular weight of the former compared to the latter. Certainly similar studies on blends of polystyrene with the ionomer (32) indicate very limited compatibility between the two components (in the sense that the glass transitions of the two components are clearly distinct in DM analysis); this is naturally due to the difficulty of the large polystyrene molecules to associate completely with the styrene part of the ionomer molecules, the ionomer being a random copolymer.

It is also evident from Figure 5 that the ionic transition is decreased very little by the oligomer, to not more than ca. 20-25°C, and that the maximum extent of decrease is attained at relatively high oligomer concentrations. This, too, is in marked contrast to the effect of DEB on the same ionomer, where the Ti is diminished as much as the Tg in the concentration range studied. Presuming that the decrease in Ti is indeed due to incorporation of diluent molecules into the cluster regions, it is clear that the larger oligomer molecules are far less soluble in the microphase-separated regions of the ionomer than are the DEB molecules (31).

Preliminary data on the 8 mol % sodium ionomer blended with a monofunctionalized styrene oligomer (8-S with a sodium neutralized carboxylate group at one end, 8-S-COONa) show behaviour intermediate to a polar and a nonpolar plasticizer (33). The Tg is diminished, although (as for 8-S) to a lesser extent than that predicted by the Fox equation. However, the Tg loss tangent peak generally increases in intensity with increasing 8-S-COONa concentration (Figure 6), paralleling the effect seen with glycerol. The ionic regions are also clearly plasticized: the transition temperature is decreased and the rubber-like plateau is diminished in value with increasing plasticizer concentration. On the other hand, the intensity of the loss tangent peak for this transition is not reduced by addition of plasticizer. Not surprisingly, even at 80 wt % 8-S-COONa, compatibility appears to remain complete.

Zinc Stearate as a Special Case. The use of zinc stearate (ZnSt) as an ionic plasticizer for sulfonated EPDM (S-EPDM) is analogous to the use of the monofunctionalized styrene oligomer in that it is composed of an aliphatic chain attached to an ionic group of the same polarity as the ionic groups in the ionomer. However, a number of details distinguish this plasticizer from one like 8-S-COONa. In particular, the plasticizer is crystalline to above the Tg of the ionomer in question. In addition, two alkyl groups instead of one are attached to the ionic group, its ionic group

(carboxylate) differs from that of the ionomer (sulfonate), and its
aliphatic portion is not of identical nature to the nonionic parts
of the ionomer which is a terpolymer composed of three different
monomers. Thus, this plasticizer represents a special case of
sorts.

It has been the subject of several studies (18-21, 34-35), and
has been partially described earlier in the present paper. It is
notable that the blended samples are relatively transparent and
that the ZnSt does not migrate out of the blend over time. This is
in contrast to blends of ZnSt with nonionized EPDM, blends which
are opaque and in which the plasticizer does migrate out (18-19).
The effect of a variety of metal stearates on the melt flow and
mechanical properties of S-EPDM were investigated (18). Only the
zinc and lead stearates were found to improve both tensile
properties and melt flow. Ammonium stearate improves melt flow but
diminishes tensile strength. Other metal stearates were found to
be ineffective in improving melt flow. In addition, Zn stearate
substantially reduces water absorption in the ionomer. In
investigating the crystallization behaviour of ZnSt in S-EPDM,
using DSC, Weiss (34) found that, in cooling scans, crystallization
is evident only at higher concentrations of ZnSt, whereas, in
heating scans, melting of ZnSt was detected for all samples.
Furthermore, multiple transitions were evident, the intensities and
positions of which vary with ZnSt concentration and with aging
time. This is in contrast to the behaviour of ZnSt in ordinary
EPDM, where there is a single endotherm or exotherm whose position
remains unchanged and whose intensity varies normally with the
concentration of ZnSt. This anomalous crystallization behaviour
was considered as evidence of interactions between ZnSt and the
sulfonate groups in S-EPDM.

From X-ray and electron microscopy data (21), it was found
that the ZnSt in S-EPDM phase separates into crystalline domains of
less than 500 nm. It was also inferred from the data that the
interactions between the sulfonate groups and the ZnSt occur on
preferred crystal planes. In the context of stress relaxation
results, it was proposed that the long cell axis of ZnSt lies at
90° to the stress axis, and that relaxation of the stress occurs by
"interaction hopping". Some stress relaxation studies of ZnSt
plasticized S-EPDM were also reported by Granick (35).

Melt Rheology. Very few melt rheology studies of plasticized
ionomer systems have been reported to date. [For the purposes of
this review, viscoelastic studies of ionomer solutions such as
those reported by Agarwal and Lundberg (36-38) are not considered
relevant.] The effect of ZnSt on the melt rheology of S-EPDM has
already been mentioned. The main result is that melt flow is made
possible in the temperature range of 100-150° C through the presence
of ZnSt such that the plasticized ionomer behaves like a high
molecular weight amorphous polymer rather than as a cross-linked
polymer; without ZnSt the crosslink network persists to well beyond
150° C (20). This effect is shown in Figure 7. Bagrodia et al.
(39) relate similar melt rheological behaviour for triarm
polyisobutylene-based ionomers plasticized by ZnSt.

Preliminary results were described for a carboxylate styrene
ionomer plasticized by the nonpolar oligomer 8-S (29). Although

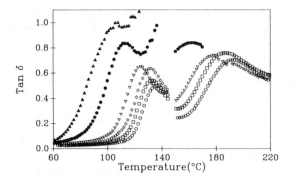

Figure 6. Loss tangents (1 Hz) vs. temperature for blends of PS-0.08MAA-Na and 8-S-COONa. Weight ratios (ionomer/oligomer) are: o, 100/0; □, 95/5; ◇, 90/10; ▽, 80/20; ●, 60/40; ▲, 40/60. The lower temperature maxima correspond to the Tg's (obtained in tensile mode), the upper to the Ti's (obtained in shear mode).

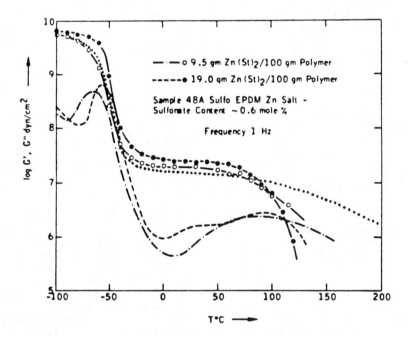

Figure 7. Dynamic moduli at 1 Hz as a function of temperature for zinc sulfonated EPDM unplasticized (dotted line, G' only) and plasticized with zinc stearate (adapted from ref. 20).

true melt flow is not achieved even at 75 wt % 8-S content, sufficient plasticization of the ionic regions of the ionomer allows the melt zone to at least be attained. It was noted that the relaxation behaviour of these systems is complex and that beyond a certain 8-S content there is increasing failure of time-temperature superposition with increasing 8-S content. A number of additional observations were made, but a much more thorough investigation of the rheology of plasticized ionomers is required before more definitive conclusions can be drawn.

X-Ray (and Neutron) Scattering Studies

In the numerous attempts to understand better ionomer morphology through x-ray scattering studies, diluents, primarily water, have played a (small) role. In particular, it was reported early in the history of ionomer research that the so-called small-angle x-ray "ionomer peak" (whose maximum generally corresponds to an apparent Bragg spacing that is between 1 and 10 nm) can be enhanced in intensity by low levels of humidity, but is "destroyed" by saturation with water (40-41). Marx et al. (42), in studies of ethylene- and butadiene-based ionomers, specified that addition of water causes the ionomer peak to shift to lower angles, such that it eventually disappears as the scattering angle becomes too low for detection. Other polar solvents, namely methanol, methacrylic acid, acetic acid and formic acid, decrease the intensity of the ionomer peak but do not affect its angular position (Figure 8).

In the case of methacrylic and acetic acids, as shown in Figure 8, a new peak appears at larger angles; this peak is stated to be characteristic of partially ionized molecular acids, which are formed through extraction of the metal ion from the ionomer by the diluent. It is hypothesized that methanol, whose hydroxyl group is compatible with the ionic scattering sites and whose methyl group is compatible with the surrounding hydrocarbon matrix, encircles the scattering sites and thus decreases the electron density in the system thereby causing the intensity of the ionomer peak to decrease. Water, on the other hand, is completely compatible with the ionic sites and thus apparently swells the ionic domains, thereby increasing the apparent Bragg spacing (42).

The decrease in scattering angle of the ionomer peak upon addition of water, and in some cases enhanced intensity, was also seen for carboxylated and sulfonated perfluorinated ionomer membranes (43-44), for magnesium-neutralized carboxy-telechelic butadienes and isoprenes (45), and for a Zn sulfonated polystyrene ionomer (46). In the last case, the sharpening and increased intensity of the ionomer peak with addition of water is particularly prominent. In general, it is possible to account for intensity variations by volume fraction changes of phase-separated material or by variations in the electron density differences between the two phases, or both (43,47). For example, in the case of Cs neutralized perfluorinated sulfonate ionomers, whose scattering curves for different water contents are shown in Figure 9, the electron density of the ionic aggregates is higher than that of the surrounding medium. Absorption of water by the ionic aggregates decreases their electron density so that the peak decreases in intensity, and disappears when the electron density of

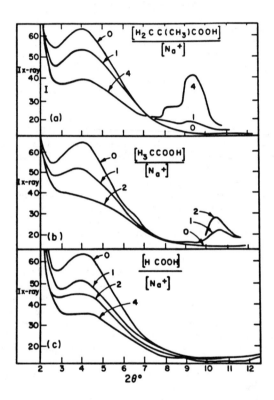

Figure 8. Effect of acid absorption by an ethylene-methacrylic
acid ionomer (16 wt % methacrylic acid, neutralized 37% with
Na) on the small angle x-ray scattering peak: absorption of
(a) methacrylic acid, (b) acetic acid, (c) formic acid.
(Reproduced from ref. 42. Copyright 1973 American Chemical Society.)

the ionic aggregates and the surrounding medium are comparable. As
still more water is added, the peak may reappear and increase in
intensity; this reflects an increasing difference in electron
density between the ionic aggregates and the surrounding matrix,
the electron density of the former now being less than that of the
latter (43).

The extent of swelling in polar solvents, as determined from
the position of the peak maximum converted to Bragg spacings, is
much greater than the increase in macroscopic linear dimensions
(43-44). This result was used by some (1,43) to support the
hypothesis that the origin of the peak is related to intraparticle
interference arising from the internal structure of the microphase-
separated, highly ionic aggregates. By contrast, if the ionic peak
is caused by interparticle interference, the Bragg spacing change
should correspond to the change in macroscopic linear dimensions,
assuming no change in the number of aggregates upon swelling.
However, others (46,48) argue that the swelling behaviour of the
ionomer peak is not inconsistent with interpretation of the SAXS
data as being due to interparticle interference: for it can be
shown that, if a change in the number of aggregates upon swelling
is considered, the difference between microscopic and macroscopic
swelling is not necessarily very large. In other words, without
additional details, the swelling behaviour of ionomers cannot be
used to resolve the debate concerning the origin of the SAXS
ionomer peak. Further discussion of this point can be found in
(49).

Besides the study (42) mentioned above, only Weiss and
coworkers have reported on the SAXS of polar diluents other than
water in ionomers. Fitzgerald et al. (27) found that methanol in a
3.85 mol % Zn sulfonated polystyrene ionomer shifts the scattering
maximum to slightly lower angles [besides decreasing its intensity,
as reported by Marx et al. (42)]. Glycerol, on the other hand,
increases the intensity of the peak while at the same time
sharpening it and shifting it to lower angles (28); this is shown
in Figure 10 for a 1.65 mol % Mn sulfonated polystyrene. It may be
concluded from the above studies that, in general, polar diluents
decrease the scattering angle of the ionomer peak (seemingly more
significantly for water than for the other polar diluents
investigated, although this comparison has not been made directly),
whereas the intensity of the peak may increase or decrease for the
reasons mentioned earlier.

In the case of nonpolar diluents, very few studies have
appeared in the literature. The Zn sulfonated polystyrene
plasticized by 3.4 wt % dodecane shows no change in either
position, intensity or shape of the ionomer peak compared to that
of the bulk polymer (27); this was taken to indicate that the ionic
microphase is unaffected by this diluent. The Mn sulfonated
polystyrene plasticized by up to 10 wt % DOP (Figure 11) shows a
decrease in intensity of the peak, but no change in its shape or
position (28). This was explained as a dilution effect of the
ionic phase due to swelling of the matrix by the plasticizer.

In contrast, for carboxy-telechelic butadiene swollen by
toluene, the small angle peak decreases both in intensity and in
scattering angle (45). Comparison of the microscopic with
macroscopic swelling indicated that the deformation in these

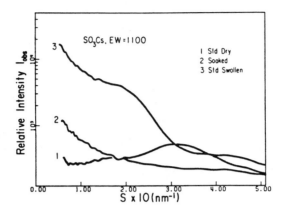

Figure 9. Effect on the SAXS curve of swelling by water of a Cs sulfonated perfluorinated ionomer membrane; water content is greater in the order 1, 2, 3. (Reproduced from ref. 43. Copyright 1981 American Chemical Society.)

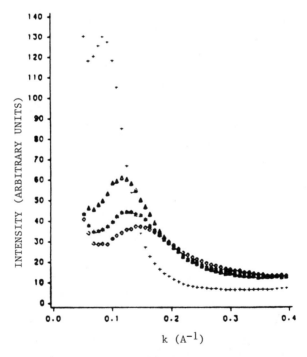

Figure 10. Effect on the SAXS curve of plasticization with glycerol of a 1.65 mol % Mn sulfonated polystyrene; glycerol content in order of increasing peak intensity is 0, 1, 5 and 10 wt %. (Reproduced with permission from ref. 28. Copyright 1987 Reidel.)

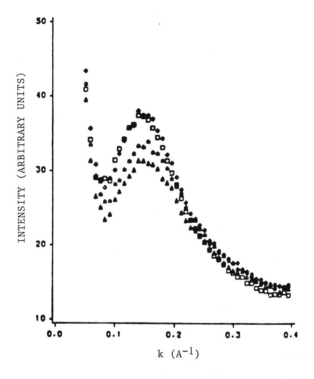

Figure 11. Effect on the SAXS curve of plasticization with dioctylphthalate (DOP) of a 1.65 mol % Mn sulfonated polystyrene; DOP content in order of decreasing peak intensity is 0, 1, 5 and 10 wt %. (Reproduced with permission from ref. 28. Copyright 1987 Reidel.)

samples is affine except at very high degrees of swelling (Figure
12). More detailed analysis of the scattering curves suggested
that, upon swelling, the size of the ionic domains remains
constant, but that there is a broader distribution of interdomain
distances as well as a network distortion, with some chain ends
forced into the hydrocarbon matrix. The reader is reminded that
these end-functionalized low molecular weight materials, of narrow
molecular weight distribution, allow for nearly complete microphase
separation between the ionic groups and the hydrocarbon chains.
Thus, certain results for these materials can be very different
from those for high molecular weight ionomers whose ionic co-units
are more or less randomly distributed along the polymer chains.

It is relevant at this point to mention small angle neutron
scattering (SANS) studies, SANS being a complementary technique to
SAXS. Again the studies involve water, this time deuterated,
specifically in a Cs carboxylated ethylene-based ionomer ($\underline{47}$), Cs
sulfonated polypentenamers ($\underline{47},\underline{50}$), and a perfluorinated ionomer
membrane ($\underline{51}$). In general, small amounts of heavy water enhance
the ionomer peak intensity without modifying its scattering angle,
whereas at water contents of greater than a water-to-ion pair ratio
of about 6, the peak position moves markedly to lower angles. The
ratio of 6 is interpreted as the number of water molecules that
form a primary hydration shell around each ion pair. Below this
number, the ionomer morphology is unaffected by the presence of the
water molecules; above it, rearrangement of the ionic microphase
probably takes place, possibly with randomly packed spheres of
water accounting for the scattering peak at very high water
contents. At low water contents, therefore, D_2O acts merely as a
contrast enhancer for the SANS experiment, and thus effectively
"tags" the location of the ionic groups ($\underline{50}$).

Spectroscopic Studies

It is advantageous to combine measurements of viscoelastic
properties and x-ray scattering studies with techniques that can
probe the microstructure of plasticized ionomers. This would allow
a better understanding of the interactions of the plasticizer with,
and its partition within, the ionomer. It is also indispensable
for elucidating the mechanisms of plasticization by different
diluents. A small beginning in this direction is being made with
recourse to spectroscopic techniques in particular, these being
powerful tools for deducing local structure on the molecular and
atomic levels. Once again, the main plasticizer to which these
studies have been applied is water. It is of course desirable to
be able to determine the presence, as well as the amount and
distribution, of water in the more or less hygroscopic ionomers;
for, it is clear from the foregoing that the presence of water can
profoundly affect the local structures and hence the spectroscopic
results obtained.

To this end, a number of infrared (IR) studies have addressed
the issue of water in ionomers either in passing or more
substantially ($\underline{28},\underline{52}-\underline{59}$). In perfluorosulfonated membranes, for
instance, at least two types of bonding environments for water have
been detected ($\underline{52},\underline{54}$). It is also clear that the extent and effect
of water absorption depends on the salt involved ($\underline{54},\underline{58}$).

Multistage association-dissociation equilibrium between bound and
unbound cations has been invoked to explain the observed variation
in certain vibration bands as a function of water content (55,60).
Fitzgerald and Weiss (59) have calculated that the cation and anion
in hydrated Zn sulfonated polystyrenes are separated by 3.6 water
molecules per cation.

In addition, Fitzgerald and Weiss reported some IR data for
7.6 mol % Na sulfonated ionomers that were gelled to about 50% in
tetrahydrofuran, dimethylformamide, DMSO and 95/5 (v/v)
toluene/methanol. The results indicate that all of the solvents
weaken the cation-anion bond, with toluene/methanol seeming to be
somewhat more effective than the other three.

Pineri and coworkers, and a few other groups, have used ESR
and Mossbauer spectroscopy as well as SANS, extended x-ray
absorption fine structure (EXAFS), and magnetization and
susceptibility data to analyze local structure in perfluorinated
ionomer membranes and the distribution of water within them [see,
for instance, (61-65)]. The application of the ENDOR (electron
nuclear double resonance) technique to deuteriated methanol-swollen
samples of these membranes has been reported recently (66).
Photophysical methods have also been applied in hydration studies
of these membranes (67-69). Finally, some NMR results on the same
hydrated perfluorinated ionomers (57,70), as well as on hydrated
styrene- and ethylene-based ionomers (71), have been reported.
Otherwise, except for some studies of unplasticized ionomers [e.g.
see (63)] and ionomers in solution [notably, ESR techniques by
Weiss and coworkers (27,59,72)], these techniques have been little
used to date on ionomer systems, and virtually unused on
plasticized ionomers other than hydrated ones as mentioned above.
[All of these studies are cited without further comment only to
demonstrate the applicability of a large number of techniques to
the study of plasticizer/ionomer systems. In the hydrated
perfluorinated ionomers, this multi-technique approach has yielded
much information regarding the types and location(s) of water
present in the membranes.]

A notable exception is the recent application of EXAFS to
plasticized Zn sulfonated polystyrene ionomers (73-74). This
technique was used to investigate the effect of various diluents on
the local structure of the Zn cation. Plasticizers termed "non-
coordinating" (because they do not complex metal cations) - in
particular, toluene, acetonitrile, and DOP - were shown to have a
minimal influence on the cation's local structure. At most, these
cause a slight increase in disorder, but they do not alter the
coordination structure around the cation. These diluents clearly
partition themselves into the nonpolar regions of the ionomer.

The coordinating plasticizers, on the other hand, were found
to interact strongly with the zinc cation. Specifically, it
appears that water molecules cause disorder and at least partially
displace sulfonate groups from the first coordination shell.
Glycerol induces a different local coordination structure. The
molecular model proposed for full solvation by glycerol is
illustrated in Figure 13, where the zinc cation is shown to be
coordinated by three glycerol molecules acting as bidentate
ligands. With this arrangement, the effective charge-to-radius
ratio of the cation is decreased by about two-thirds of its value

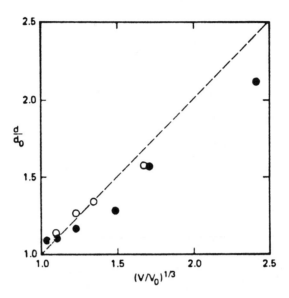

Figure 12. Equilibrium swelling by water of a Ti dicarboxylatopolybutadiene where the relative change in the separation distance of the ionic domains (d/d_0) is shown as a function of the macroscopic increase in volume. (Reproduced from ref. 45. Copyright 1986 American Chemical Society.)

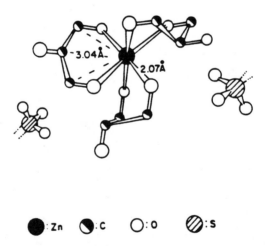

Figure 13. Postulated local coordination structure around the zinc atom for Zn sulfonated polystyrene ionomers fully solvated by glycerol. (Reproduced with permission from ref. 73. Copyright 1988 Wiley.)

in the unplasticized state, thereby reducing the strength of the coulombic interactions. This elucidates how glycerol plasticizes the ionic domains in the ionomer, and why its presence sharply reduces the melt viscosity while having much less effect than DOP on the Tg.

EXAFS analysis of an n-amyl alcohol swollen, Zn neutralized perfluorosulfonated ionomer (64) indicated that this solvent similarly interacts through its hydroxyl groups with the zinc cation. It was pointed out that this result does not exclude that there may also be formation of a hydrogen-bonded complex of the alcohol with the fluorocarbon ether linkage that is present in the side chains of the ionomer.

A Separate Class of Plasticizers

This section treats a class of plasticizers which simultaneously act as counterions for the ionic groups in ionomers. Representative of this class are oligomeric hydrocarbon molecules functionalized by one or more chemical groups that can interact specifically with the ionic or ionizable group in the ionomer. An example of such a specific interaction is the proton transfer that can occur between an amine group and a carboxylate or sulfonate group, one group being part of the ionomer, the other forming the functional group of the plasticizer. Long-chain substituted alkyl amines and anionic or cationic soaps are examples of the class of plasticizers described in this section. Mixtures of this type of plasticizer and an ionomer are analogous to blends of two different ionomers having groups that can interact (75), except that in the latter case both species are polymeric. A major advantage of this kind of plasticizer is that it suffers far less from "bleeding out" of the polymer, or from incompatibility, than do usual plasticizers. Thus, from both the processing point of view and that of molecular morphology, a thorough study of these plasticizers is in order. To date, the studies have been restricted primarily to measurements of certain physical and mechanical properties.

Some data on the modification of properties through the introduction of aliphatic diamines into ethylene-based carboxylate ionomers were reported by Rees (76). For example, the stiffness and yield point are increased progressively with increasing degree of neutralization by the diamine (in contrast, the maximum values of stiffness are achieved at low degrees of neutralization in the case of a metal salt as cation). The extent depends on aliphatic chain length and on ionomer acid content. Ultimate tensile strength is virtually unaffected (in contrast to metal salts), whereas ultimate elongation is decreased a little. The melt index is practically unaffected by the diamines at low ionomer acid content, and is decreased slightly at higher acid content (again in contrast to neutralization by metal salts for which the melt viscosity is markedly increased); in this respect the diamines appear to act similarly to the zinc stearates as described above. These results may reflect both the effects of (internal) plasticization by the aliphatic chains and/or the weaker ionic bonds formed with amine salts than with metal salts.

Brenner and Oswald (77) examined a variety of structural parameters in n-alkyl substituted quaternary phosphonium plasticizers in sulfonated EPDM. With increasing length of the n-alkyl substituents and with increasing number of substituents, the tensile properties generally decrease and melt flow increases. Divalent quaternary counterions can be either weaker or stronger than monovalent ones. The addition of water can either strengthen or weaken the ionomer depending on the structure of the counterion. Steric crowding was invoked to explain some of the observations.

Two studies involving sulfonated polystyrene ionomers (S-PS) have also been reported recently. Weiss et al. (78) examined S-PS stoichiometrically neutralized by mono-, di-, and tri-substituted alkyl amines with chain lengths varying from 0 to 20 carbon atoms. Thermal studies show that the Tg may increase with plasticizer content, remain constant, or decrease relative to the unneutralized ionomer, depending on the alkyl chain length and degree of substitution of the amine. As illustrated in Figures 14 and 15, the longer chain amines and a higher degree of substitution have greater plasticization effects. Melt index, viscosity, and thermal mechanical analysis results show parallel plasticization effects, leading to the reflection that there is much potential to easily tailor ionomers for specific ends with these types of plasticizers.

Dynamic mechanical studies by Smith and Eisenberg (79) of stoichiometric blends of S-PS and monosubstituted flexible amines corroborated the results of Weiss et al. regarding the Tg decrease with increasing alkyl chain length. Specifically, a linear decrease as a function of the number of carbon atoms in the chain was found. As far as an ionic transition is concerned, it seems that one may be present albeit greatly reduced in strength from the unplasticized ionomer. In the case of rigid amines, the Tg of the plasticized ionomer is increased by the plasticizer, with only a weak dispersion at higher temperatures. It was pointed out that, as far as the effects of the plasticizer on the Tg are concerned, there is a similarity between these plasticizer/ionomer systems and grafted polymers. IR evidence affirmed that proton transfer does in fact occur between the plasticizer and ionomer in these systems. Furthermore, although the plasticizer and ionomer appear miscible, scanning electron microscopy of a long-chain flexible amine indicated that additional microstructure due to small-scale phase separation is present.

Concluding Remarks

One of the striking impressions from the above is the variety of plasticizers that can be used effectively and for different purposes in ionomers; this, given the copolymeric nature of ionomers and the fact that the co-units are ionic. What is equally striking is that whereas a certain perspective of the effects of various plasticizers on ionomeric properties has been achieved, the studies have been limited to measurements of a restricted number of physical properties, to only a few types of ionomers, a few plasticizers (with the largest variety of techniques applied to water as a plasticizer), and to restricted concentration ranges (sometimes just one concentration). Many observations have been made, but many more questions remain. For example, how do the

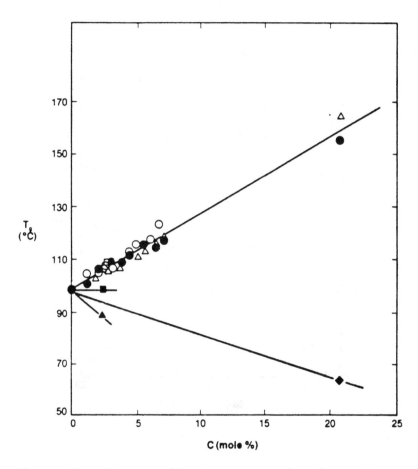

Figure 14.   Tg vs. sulfonate concentration for sulfonated polystyrene ionomers neutralized by alkyl ammonium salts of the form NRH_3^+ :   (●) SO_3H; (○) NH_3; (□) CH_3; (△) C_4H_9; (◇) C_8H_{17}; (■) $C_{12}H_{25}$; (▲) $C_{18}H_{37}$; (◆) C_{20-22}. (Reproduced with permission from ref. 78. Copyright 1984 Wiley.)

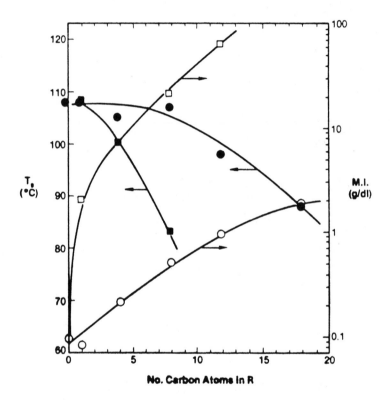

Figure 15. Tg and melt index vs. number of carbon atoms in the alkyl chain of primary and tertiary aliphatic ammonium salts of 2.7 mol % sulfonated polystyrene: circles, NH_3R^+; squares, NHR_3^+. (Reproduced with permission from ref. 78. Copyright 1984 Wiley.)

effects vary from one ionomer system to another (different backbones, different ionic co-units,...); how do they differ in random ionomers, block ionomers, and telechelics; to what extent if any, and how, do nonpolar plasticizers participate in the relaxation of the ionic domains (given that some of the above studies show that, at least in some cases, the ionic transition, Ti, in DM analysis, can be reduced in temperature by nonpolar plasticizers); how general is the observation in two or three systems that breakdown of time-temperature superposition can be induced by added plasticizer, and why does this occur?

Even less is known about ionomer/plasticizer interactions on a molecular level. A variety of scattering and spectroscopic techniques that can probe this level have been mentioned, but they have been applied primarily to the specific case of water in ionomers, and in particular to hydrated perfluorinated ionomers. At the least, these studies demonstrate the powerful potential of the techniques to contribute to a more complete understanding of structure-property relationships in plasticizer/ionomer systems. For example, to return to the question of the effect of nonpolar plasticizers on the ionic domains: how can the decrease in the ionic transition temperature be reconciled with the apparently minimal effect on the SAXS ionomer peaks and with the ESR studies that indicate (not surprisingly) that these plasticizers have essentially no influence on the local structure of the ions? Is it due to their association with the hydrocarbon component of the large aggregates or clusters? Or if these entities do not exist, as some researchers postulate, what is the interaction between the nonpolar plasticizer, the hydrocarbon component and the ionic domains? These questions are, of course, intimately related to the understanding of ionomer microstructure even in the absence of plasticizer. The interpretation of SAXS data in particular is subject to the choice of model used.

As for polar plasticizers, an enormous amount of work remains to be done to analyze the local structure introduced by different plasticizers around the cations and anions involved, to determine the coordination shells formed, and to generally elucidate the influence of these plasticizers on the ionic interactions in the ionomer. This knowledge, in turn, must contribute to the determination of correlations between structure and properties in the plasticized systems.

In conclusion, there is clearly a need for in-depth, systematic studies of a wide variety of plasticizer/ionomer systems. Such studies may additionally contribute to a better understanding of the ionomer microstructure in general.

Literature Cited

1. MacKnight, W.J.; Earnest, T.R., Jr. J. Polym. Sci. Macromol. Rev. 1981, 16, 41.
2. Bazuin, C.G.; Eisenberg, A. I&EC Prod. Res. Dev. 1981, 20, 271.
3. Longworth, R. In Developments in Ionic Polymers - 1; Wilson, A.D.; Prosser, H.J.; Eds. Applied Science: Essex, 1983; Ch. 3.

4. Pineri, M.; Eisenberg, A.; Eds. Structure and Properties of
    Ionomers (NATO ASI Series); Reidel: Dordrecht, 1987.
5. Fitzgerald, J.J.; Weiss, R.A. JMS - Rev. Macromol. Chem.
    Phys. 1988, C28, 99.
6. Williams, M.W. J. Polym. Sci. Symp. 1974, 45, 129.
7. Matsuura, H.; Eisenberg, A. J. Polym. Sci. Polym. Phys. 1976,
    14, 773.
8. Drzewinski, M.; MacKnight, W.J. J. Appl. Polym. Sci. 1985,
    30, 4753.
9. Longworth, R. In Ionic Polymers; Holliday, L., Ed.; Halstead
    Press, Wiley: New York, 1975; Ch. 2.
10. Read, B.E.; Carter, E.A.; Conner, T.M.; MacKnight, W.J. Brit.
    Polym. J. 1969, 1, 123.
11. Rahrig, D.; MacKnight, W.J. Adv. Chem. Ser. 1980, 187, 77 and
    91.
12. Noshay, A.; Robeson, L.M. J. Appl. Polym. Sci. 1976, 20,
    1885.
13. Neumann, R.M.; MacKnight, W.J.; Lundberg, R.D. Amer. Chem.
    Soc. Polym. Prepr. 1978, 19(2), 298.
14. Yeo, S.C.; Eisenberg, A. J. Appl. Polym. Sci. 1977, 21, 875.
15. Kyu, T.; Eisenberg, A. J. Polym. Sci. Polym. Symp. 1984, 71,
    203.
16. Besso, E.; Legras, R.; Eisenberg, A; Gupta, R.; Harris, F.W.;
    Steck, A.E.; Yeager, H.L. J. Appl. Polym. Sci. 1985, 30,
    2821.
17. Lundberg, R.D.; Makowski, H.S.; Westerman, L. Adv. Chem. Ser.
    1980, 187, 67.
18. Makowski, H.S.; Lundberg, R.D. Adv. Chem. Ser. 1980, 187, 37.
19. Makowski, H.S.; Agarwal, P.K.; Weiss, R.A.; Lundberg, R.D.
    Am. Chem. Soc. Polym. Prepr. 1979, 20(2), 281.
20. Agarwal, P.K.; Makowski, H.S.; Lundberg, R.D. Macromolecules
    1980, 13, 1679.
21. Duvdevani, I.; Lundberg, R.D., Wood-Cordova, C., Wilkes, G.L.
    ACS Symp. Ser. 1986, 302, 184.
22. Navratil, M.; Eisenberg, A. Macromolecules 1974, 7, 84.
23. Navratil, M. Ph.D. Thesis, McGill University, Montreal, 1972.
24. Tobolsky, A.V.; Lyons, P.F.; Hata, N. Macromolecules 1968, 1,
    515.
25. Bazuin, C.G.; Eisenberg, A. J. Polym. Sci. Polym. Phys. 1986,
    24, 1137.
26. Bazuin, C.G. Ph.D. Thesis, McGill University, Montreal, 1984.
27. Fitzgerald, J.J.; Kim, D.; Weiss, R.A. J. Polym. Sci. Polym.
    Lett. 1986, 24, 263.
28. Weiss, R.A.; Fitzgerald, J.J. In ref. 4, p. 361.
29. Bazuin, C.G.; Eisenberg, A.; Kamal, M. J. Polym. Sci. Polym.
    Phys. 1986, 24, 1155.
30. Bazuin, C.G.; Villeneuve, S. Am. Chem. Soc. Polym. Mater.
    Prepr. 1988, 58(1), 1069.
31. Villeneuve, S.; Bazuin, C.G. To be published.
32. Villeneuve, S.; Bazuin, C.G. IUPAC Macro 88 1988, to appear.
33. Caron, S.; Bazuin, C.G. To be published.
34. Weiss, R.A. J. Appl. Polym. Sci. 1983, 28, 3321.
35. Granick, S. J. Appl. Polym. Sci. 1983, 28, 1717.
36. Agarwal, P.K.; Lundberg, R.D. Macromolecules 1984, 17, 1918.
37. Agarwal, P.K.; Lundberg, R.D. Macromolecules 1984, 17, 1928.

38. Agarwal, P.K.; Garner, R.T.; Lundberg, R.D. Macromolecules 1984, 17, 2794.
39. Bagrodia, S.; Wilkes, G.L.; Kennedy, J.P. Polym. Eng. Sci. 1986, 26, 662.
40. Wilson,F.C.; Longworth, R.; Vaughan, D.J. Am. Chem. Soc. Polym. Prepr. 1968, 9, 505.
41. MacKnight, W.J.; Taggart, W.P.; Stein, R.S. J. Polym. Sci. Symp. 1974, 45, 113.
42. Marx, C.L.; Caulfield, D.F.; Cooper, S.L. Macromolecules 1973, 6, 344.
43. Fujimura, M.; Hashimoto, T.; Kawai, H. Macromolecules 1981, 14, 1309.
44. Fujimura, M.; Hashimoto, T.; Kawai, H. Macromolecules 1982, 15, 136.
45. Williams, C.E.; Russell, T.; Jérôme, R.; Horrion, J. Macromolecules 1986, 19, 2877.
46. Yarusso, D.J.; Cooper, S.L. Polymer 1985, 26, 371.
47. Roche, E.J.; Stein, R.S.; MacKnight, W.J. J. Polym. Sci. Polym. Phys. 1980, 18, 1035.
48. Yarusso, D.J.; Cooper, S.L. Macromolecules 1983, 16, 1871.
49. Kumar, S.; Pineri, M. J. Polym. Sci. Polym. Phys. 1986, 24, 1767.
50. Earnest, T.R., Jr.; Higgins, J.S.; MacKnight, W.J. Macromolecules 1982, 15, 1390.
51. Roche, E.J.; Pineri, M.; Duplessix, R.; Levelut, A.M. J. Polym. Sci. Polym. Phys. 1981, 19, 1.
52. Quezado, S.; Kwak, J.C.T.; Falk, M. Can. J. Chem. 1984, 62, 958.
53. Falk, M. In ref. 4, p. 141.
54. Barnes, D.J. In ref. 4, p. 501.
55. Lowry, S.R.; Mauritz, K.A. J. Am. Chem. Soc. 1980, 102, 4665.
56. Mattera, V.D., Jr.; Risen, W.M., Jr. J. Polym. Sci. Polym. Phys. 1984, 22, 67.
57. Komoroski, R.A.; Mauritz, K.A. J. Am. Chem. Soc. 1980, 100, 7487.
58. Brozoski, B.A.; Painter, P.C.; Coleman, M.M. Macromolecules 1984, 17, 1591.
59. Fitzgerald, J.J.; Weiss, R.A. ACS Symp. Ser. 1986, 302, 35.
60. Falk, M. ACS Symp. Ser. 1982, 180, 139.
61. Coey, J.M.D. In ref. 4, p. 117.
62. Pineri, M. ACS Symp. Ser. 1986, 302, 159.
63. Galland, D. In ref. 4, p. 107.
64. Pan, H.K.; Knapp, G.S.; Cooper, S.L. Coll. Polym. Sci. 1984, 262, 734.
65. Alonso-Amigo, M.G.; Schlick, S. J. Phys. Chem. 1986, 90, 6353.
66. Schlick, S.; Sjoqvist, L.; Lund, A. Macromolecules 1988, 21, 535.
67. Kelly, J.M. In ref. 4, p. 127.
68. Szentirmay, M.N.; Prieto, N.C.; Martin, C.R. J. Phys. Chem. 1985, 89, 3017.
69. Szentirmay, M.N.; Prieto, N.C.; Martin, C.R. Talanta 1985, 32, 745.
70. Starkweather, H.W., Jr.; Chang, J.J. Macromolecules 1982, 15, 752.

71. Dickinson, L.C.; MacKnight, W.J.; Connolly, J.M.; Chien, C.W. Polym. Bull. 1987, 17, 459.
72. Weiss, R.A.; Fitzgerald, J.J.; Frank, H.A.; Chadwick, B.W. Macromolecules 1986, 19, 2085.
73. Ding, Y.S.; Register, R.A.; Nagarajan, M.R.; Pan, H.K.; Cooper, S.L. J. Polym. Sci. Polym. Phys. 1988, 26, 289.
74. Cooper, S.L.; Ding, Y.L.; Yarusso, D.J.; Pan, H.K. Am. Chem. Soc. Polym. Prepr. 1984, 25(2), 307.
75. Eisenberg, A.; Hara, M. Polym. Eng. Sci. 1984, 24, 1306.
76. Rees, R.W. Am. Chem. Soc. Polym. Prepr. 1973, 14(2), 796.
77. Brenner, D.; Oswald, A.A. Adv. Chem. Ser. 1980, 187, 53.
78. Weiss, R.A.; Agarwal, P.K.; Lundberg, R.D. J. Appl. Polym. Sci. 1984, 29, 2719.
79. Smith, P.; Eisenberg, A. J. Polym. Sci. Polym. Phys. 1988, 26, 569.

RECEIVED April 27, 1989

INDEXES

Author Index

Affiliation Index

Subject Index

A

Absorbing properties and transition temperature, 264
Acid adsorption by ethylene–methacrylic acid ionomer, effect on SAXS, 487,488f
Acid–base interactions to improve miscibility, 21
Acid content, sulfonated star-block copolymers, 350
Acoustical properties
and frequency transition, 264
and transition temperature, 264
Aggregate size
small-angle neutron scattering, 466
sulfonated PS ionomers, 461
Aggregation behavior, ionomers in a low-polarity solvent, 453
Aliphatic ammonium salts, T_g and melt index vs. number of carbon atoms, 496,498f
Aliphatic diamines, effect on properties of carboxylate ionomers, 495
Alkyl amines as plasticizers, 496
Ammonium salts, aliphatic, T_g and melt index vs. number of carbon atoms, 496,498f
Angular frequency vs. complex viscosity, PE blends, 213,214f
Anionic block copolymerization, methacrylates, 54
Anionic polymerization techniques, to form living block copolymers, 336,338f
Anionomer, definition, 20
Anomalous SAXS, Ni sulfonated PS, 430
Apparent yield stress
polyethylene blends, 168
polypropylene blends, 195,197t
Applications, polypropylene blends, 158
Association phenomenon, sodium sulfonated polystyrene solutions, 472,474f
Atactic polystyrene blends, T_g behavior, 89

B

Bagley pressure correction term vs. shear stress, PP blends, 188,189f
Blend components, examples from industry, 69–71
Blend composition
vs. frequency shift factor, PE blends, 222,223f
vs. temperature shift factor, PE blends, 222,224f
vs. transition temperature, Zn sulfonated blend, 479,483f
Blend flow, PP blends, 193,194f

Blending
effect on boundary conditions, 101
function, 2
rationale, 16,18t
to develop new polymers, 39
Blends
compounding and processing, 16
formed by mechanical mixing, 3
formed by reactive mixing, 4
formed from ternary systems, 3
general characteristics, 67
major technological problem, 3
miscible, formation, 84
mixing by various methods, 3
phase transition, 145
PSVP–CTB, properties, 356
rheology, 12
See also Incompatible polymer blends
Block copolymers
interfacial activity, 40–50
phase transition, 145
Boundary conditions, effect of blending, 101
Breakup, polymer drops in a matrix, 7

C

^{13}C NMR spectra, PS–PB star-block copolymer, 346,348f
Cadmium and calcium ionomers, core radius values, 426
Capillary flow, PP blends, 188
Capillary flow entrance, PE blends, 166,167f
Carbonyl peak position vs. temperature, 7,8f
Carboxy telechelics, X-ray analysis, 420–436
Carboxyl-terminated polybutadiene, reaction to form Cu salt, 361
Carboxylate ionomers, preparation, 20
Cation, effect on SAXS of sulfonated polystyrene, 22,23f
Cationomer, definition, 20
Characterization, SINs, one-shot vs. prepolymer procedure, 314
Cloud temperatures and exothermic interaction, 97
Cluster size, effect on temperature shift factor, 226
Clusters, definition, 21
Cocontinuous morphology, polymer blends, 4
Cole–Cole plot
homogeneous compared to heterogeneous systems, 216
polyethylene blends, 174,177f,216
polypropylene blends, 195,200f
Compatible polymer blends, definition, 2

Production: Rebecca Hunsicker
Indexing: Janet S. Dodd
Acquisition: Cheryl Shanks

Elements typeset by Hot Type Ltd., Washington, DC
Printed and bound by Maple Press, York, PA